STATISTICS IN THE *environmental* & EARTH SCIENCES

edited by

Andrew T Walden
Department of Mathematics,
Imperial College of Science, Technology and Medicine,
London

and

Peter Guttorp
Department of Statistics, University of Washington,
Seattle

Edward Arnold
A division of Hodder & Stoughton
LONDON MELBOURNE AUCKLAND

Copublished in the Americas by Halsted Press,
an imprint of John Wiley & Sons, Inc.
New York – Toronto

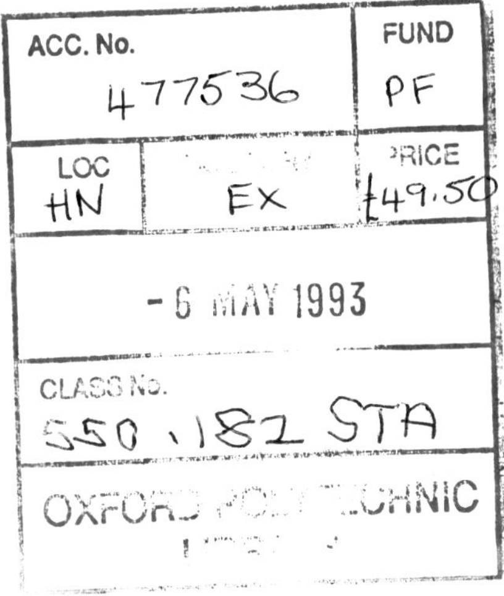

First published in Great Britain 1992
Copublished in the Americas by Halsted Press, an imprint of John Wiley & Sons Inc., 605 Third Avenue, New York, NY 10158

British Library Cataloguing in Publication Data

Walden, A.T.
Statistical methods in environmental and earth sciences: New developments.
I. Title II. Guttorp, P.
550.1

ISBN 0-340-54530-5

Library of Congress Cataloging-in-Publication Data

Available upon request

ISBN 0 470 21946 7

Typeset in Times New Roman 569 by H Charlesworth & Co Ltd, Huddersfield
Printed in Great Britain for Edward Arnold, a division of Hodder and Stoughton Limited, Mill Road, Dunton Green, Sevenoaks, Kent TN13 2YA by St Edmundsbury Press Ltd, Bury St Edmunds, Suffolk and bound by Hartnolls Ltd, Bodmin, Cornwall.

Contents

Preface

There have been many exhortations in recent years – on both sides of the Atlantic – directed towards increasing the role of statistical methodology in substantive scientific investigations. The intended consequence for science is an improvement in terms of precision in both design and analysis of experiments. One hoped for result from the point of view of statistics is a better appreciation of the power of statistical reasoning and an improved status for statistics as a discipline. We believe this volume captures the spirit of such endeavor. The uniting factor of all the papers in this book is that they try to tackle important practical problems, but in doing so create, develop or review many intriguing strands of statistical thinking and methodology.

The intended audience of this book consists of graduate students and statistical researchers, looking for interesting new areas of research, as well as experts in the fields of environmental and earth sciences applications, who will find extensions, innovations, and reviews of recent work in these areas of statistical applications.

The large number of different data sets discussed in these papers is refreshing, and should surely help steer statisticians away from the Nth analysis of an old data set with $m < N$ points, towards some of the challenging new data readily available now. With the arrival of data-collection devices with extremely high acquisition rates, such as satellite imagers and seismic networks, we have currently, in contradistinction to earlier statistical problems, the frequent need to make careful and often large-scale data reduction. Stochastic models are finding increasing use as tools to summarize complex data sets.

The idea for this volume originated in a conference co-sponsored by the International Statistical Institute, the Institute for Mathematical Statistics, and the Bernoulli Society, in Belgium during the summer of 1989. The topic of the conference was *Statistics: Space and Earth Sciences*. It was a most successful gathering, with discussions of a variety of statistical applications in areas ranging from astronomy to seismology. It is quite common that this type of conference results in proceedings volumes. This, however, is not such a volume. Although many of the authors were speakers at the conference, they were specifically invited by the editors to submit papers in the general area covered by environmental and earth sciences. The submitted papers have all been refereed.

The volume is divided into two parts, one on environmental applications and one on earth sciences applications. (This division is perhaps somewhat artificial, and reflects more the interests of each of the editors. Most of the techniques discussed could find additional applications outside either of these subject areas.) Each part has a separate introduction with summaries of the papers, giving a flavor of the wide variety of subjects and techniques.

The editors would particularly like to thank our reviewers (listed on page ix) for their hard work and diligence. Indeed, many of the authors commented on the high quality of the reviews. The patience of James Griffin, our publisher, has been outstanding, in the face of innumerable delays and communication problems.

Andrew Walden,
London,
England

Peter Guttorp,
Seattle,
USA

Reviewers

Dr E. Foufoula-Georgiou
St Anthony Falls Hydraulic Lab.
University of Minnesota
USA

Dr J. Haslett
Dept of Statistics
Trinity College Dublin
IRELAND

Prof. D. Oldenburg
Dept of Geophys. & Astron.
University of British Columbia
CANADA

Dr D. B. Percival
Applied Physics Lab.
University of Washington
USA

Dr E. Riskin
Dept of Electrical Eng.
University of Washington
USA

Prof. R. L. Smith
Dept of Statistics
University of North Carolina
Chapel Hill
USA

Prof. R. E. White
Dept of Geology
Birkbeck College
University of London
UK

Dr T. Haas
Dept of Mathematics
University of Wisconsin-Milwaukee
USA

Dr Y. Ogata
Institute of Statistical Mathematics
Tokyo
JAPAN

Prof. J. Petkau
Dept of Statistics
University of British Columbia
CANADA

Prof. A. E. Raftery
Dept of Statistics
University of Washington
USA

Prof. N. J. Shackleton
Godwin Lab. for Quaternary Res.
Cambridge University
UK

Prof. D. M. Titterington
Dept of Mathematics
University of Glasgow
Glasgow
UK

Prof. W. Zucchini
Dept of Statistics
University of Cape Town
SOUTH AFRICA

Contributors List

W. F. Caselton
Department of Civil Engineering, University of British Columbia, 2324 Main Mall, Vancouver V6T 1W5, Canada

K. Conradsen
Institute of Mathematical Statistics and Operations Research, Technical University of Denmark, DK 2800 Lingby, Denmark

Z. A. Der
ENSCO Inc., 5400 Port Royal Rd., Springfield, Virginia 22151, USA

P. Guttorp
Department of Statistics, GN-22, University of Washington, Seattle, Washington 98195, USA

L. Kan
Department of Statistics, University of British Columbia, 2021 West Mall, Vancouver V6T 1W5, Canada

A. C. Lees
175 Glenwood St., Manchester, Connecticut 06040, USA

C. Loader
AT & T Bell Laboratories, 600 Mountain Ave (Rm 2C-279), Murray Hill, New Jersey 07974, USA

K. L. McLaughlin
S-Cubed, PO Box 1620, LaJolla, California 92038, USA

K. Newman
Department of Statistics, GN-22, University of Washington, Seattle, Washington 98195, USA

A. Nielsen
Institute of Mathematical Statistics and Operations Research, Technical University of Denmark, DK 2800 Lingby, Denmark

J. Park
Dcpartment of Geology and Geophysics, PO Box 6666, Yale University, New Haven, Connecticut 06511, USA

M. J. Phelan
Department of Statistics, The Wharton School, University of Pennsylvania, Philadelphia, PA 19104, USA

B. D. Ripley
Department of Statistics, University of Oxford, 1 South Parks Road, Oxford OX1 3TG, UK

P. D. Sampson
Department of Statistics, GN-22, University of Washington, Seattle, Washington 98195, USA

J. D. Scargle
MS 245-3, Theoretical Studies Branch, Space Science Division, National Aeronautics and Space Administration, Ames Research Center, Moffett Field, California 94035, USA

R. H. Shumway
Division of Statistics, University of California, Davis, California 95616, USA

P. Switzer
Department of Statistics, Stanford University, Stanford, California 94305-4065, USA

A. D. Thrall
Electric Power Research Institute, Palo Alto, California, USA

D. Vere-Jones
Institute of Statistics and Operations Research, Victoria University of Wellington, Wellington, New Zealand

A. T. Walden
Department of Mathematics, Imperial College of Science, Technology and Medicine, London SW7 2BZ, UK

K. Windfeld
Institute of Mathematical Statistics and Operations Research, Technical University of Denmark, DK 2800 Lingby, Denmark

D. A. Woolhiser
US Department of Agriculture, Agricultural Research Service, 2000 E. Allen Rd., Tucson, Arizona 85719, USA

J. V. Zidek
Department of Statistics, University of British Columbia, 2021 West Mall, Vancouver V6T 1W5, Canada

Introduction to statistics in the environmental sciences

During the last fifteen years there has been much attention focused on environmental problems, such as tree and lake death from acidic precipitation, and global warming due to increased carbon dioxide concentration and a possible reduction of the ozone layer in the stratosphere. In order to monitor the acid precipitation problem, several models of long-range transport of atmospheric pollutants have been developed. For example, the United Nations' Economic Commission for Europe, together with the World Meteorological Organization, is maintaining a routine program for calculation of long-range transport of sulfur in Europe, called EMEP (Co-operative Programme for Monitoring and Evaluation of the Long Range Transmission of Air Pollutants in Europe). The goal is to provide the participating governments with information about transboundary fluxes of pollutants. In the USA, the ten-year intensive assessment program called NAPAP (National Acid Precipitation Assessment Program) resulted in a major model of acid precipitation called RADM (Regional Acidic Deposition Model). This model, much more complex than EMEP, is intended for the evaluation of effects of changes in environmental policy.

In Chapter 1, Thrall discusses a very fundamental question, namely precisely what role the statistician may play in the analysis and evaluation of these kinds of deterministic data-driven forecasting models. In particular, the question of what confidence can be put in model predictions following 'what if' scenarios is one of considerable difficulty. Even if a model of, say, long-range transport of atmospheric pollutants has been found adequate in describing observed data, one has to extrapolate the validity of the model to a different range of model parameters. Naturally, there are no data available to assess the model performance over such different ranges.

Chapter 2, by Caselton, Kan, and Zidek, contains new material regarding the optimal design of a monitoring network. A criterion of optimizing the information (in the sense of Shannon) in a network, originally proposed by the two senior authors, has been modified and revised. In particular, they discuss in some detail strategies for removing stations from an existing network, and replacing them with new stations. The methods are applied to a large monitoring network for wet deposition of sulfate and other pollutants in the continental USA.

One important aspect of the design methodology in Chapter 2 is the dependence on spatial covariance. Chapters 3 and 4 deal with the statistical problem of estimating such covariances. While Guttorp, Sampson, and Newman implement a nonparametric method for estimating nonstationary spatial covariances, Loader and Switzer develop an empirical Bayes method where a prior parametric covariance model is modified by shrinking it towards the data. Guttorp *et al.* apply their procedure to a design problem similar to that of Caselton *et al.* Loader and Switzer apply their methodology to produce interpolated covariance maps around the same stations as those considered by Guttorp *et al.*, although for a different pollutant.

While Chapters 1–4 have been concerned with general statistical issues, albeit all applied to environmental problems, Chapters 5 and 6 are more directly addressing concrete scientific problems, both involved with stochastic models for rainfall. Woolhiser studies the effect of major climatic factors, such as the Southern Oscillation, on precipitation in the western USA. His approach is to use a Markov chain with periodic time-dependent transition probabilities, which are modified (linearly on the logit scale) by a lagged version of the Southern Oscillation Index. Phelan, finally, applies sophisticated martingale estimation techniques to account for the aging of convective cells in precipitation. He uses data from the GATE study of tropical rainfall to illustrate the need for the new techniques.

Chapter 1

Establishing a statistical context for the evaluation of air quality models

A. D. Thrall

Several types of air quality models are described, along with past and current efforts to evaluate them. Although statistical measures have been used to compare model output to field observations, all too often it is not clear what questions are really being asked, and whether current methods of statistical interference can help to answer them. We need to distinguish more clearly the respective goals of scientific understanding and environmental policy. Among the technical challenges perhaps the most important is to ascertain whether and how statistics can help to determine the appropriate degree of model complexity.

1.1 Introduction

I embark on this chapter with some trepidation. Statistical presentations usually proceed from the assumption that the topic at hand is suited to statistical analysis, whereas I propose to discuss the assumption itself. At one time or another most statistical consultants have needed to clarify the question being asked as a prerequisite to productive analysis. My aim in this chapter is to encourage such discussion.

One might argue that the process of clarifying the question, although necessary at times, is outside the discipline of statistics, i.e. of applying or developing appropriate methods to address well-posed statistical problems. Moreover such clarifications, even if acknowledged to be a proper part of statistical practice, may not be suited to discussion at a technical conference.

Nevertheless, there are grounds for using this forum to review the statistical context of air quality model evaluation. First, the problem is in the public domain – the computer simulation models to which I will refer provide technical guidance for environmental policy. Second, clarification is needed; in particular, scientific goals need to be distinguished from policy goals. Third, model evaluation, no matter how construed, poses significant challenges to statistical methods. Finally, statisticians have a responsibility to determine whether and how the evaluation of air quality (or other environmental) models is a statistical problem, for only then can we ensure the relevance and credibility of the statistical methods applied to this increasingly important policy area.

To make the statistical issues more concrete I will describe some broad categories of air quality models, along with past and current efforts to evaluate them.

1.2 Development and evaluation of air quality models

The term 'air quality' typically denotes the chemical composition of the air at ground level, in particular, the concentration of certain particles, aerosols, or gases that can affect human health or the environment. This covers a lot of territory, which I will not attempt to summarize. Instead, I will briefly describe some of the larger modeling efforts so that we can begin to distinguish the statistical issues.

1.2.1 Smokestack plumes

The factory chimney, which once signified progress, has come to symbolize air pollution. To help to estimate the public health effects of this pollution, atmospheric scientists have developed numerical models of the dispersion of a plume of gases, radionuclides, or other particles emanating from a 'point source' such as an industrial smokestack. Turner (1979) describes the early modeling work, and the continuing development of more realistic representations of both the plume (e.g. the degree of plume buoyancy or chemical reactivity) and the environmental media (e.g. plume interactions with the terrain or with upper layers of the atmosphere).

Plume models provide technical guidance for a variety of environmental decisions, whose scope ranges from individual facilities to industrial sectors. Consequently, great effort has been expended to diagnose and remedy model weaknesses, and to quantify the reliability of model output somehow.

The initial protocols for evaluating plume models – cooperatively developed by the American Meteorological Society (AMS), the US Environmental Protection Agency (EPA), and the Electric Power Research Institute (EPRI) – called for a wide range of statistical measures to compare model calculations to field measurements (Fox, 1981; Londergan, 1981). Unfortunately, this multitude of measures tends to blur rather than distinguish the performance of a model compared to its competitors (Smith, 1984).

Over the years the EPA has sought to focus these measures on practical criteria for accepting a model to guide a specified, pending, regulatory decision. One suggestion (EPA, 1984) is to form a weighted average of the various model performance scores. A simpler alternative, suggested by Cox and Tikvart (1986), is to restrict attention to the extreme quantiles or other statistics most relevant to the regulatory decision.

Let me elaborate on the latter suggestions. Health-based regulations for certain air pollutants address the highest concentration of the pollutant that occur in a specified period of time, i.e. the upper tail of the data. The suggested procedure for comparing the respective tails of the measured and model-calculated concentrations consists of the following steps:

- compute the average and standard deviation of the tails (e.g. highest 25 values) of the measured and model-calculated concentrations, respectively;
- compute the fractional bias, i.e. the difference divided by the average of the measured and model-calculated tail-average;
- similarly compute the fractional difference of the respective standard deviations, i.e. the difference divided by the average of the measured and model-calculated tail-standard deviation;
- bootstrap the above three steps, i.e. draw at random and without replacement pairs of measurements and model calculations from the original data set to construct new samples and compute the fractional statistics as above for each sample;
- judge that model best whose bootstrap fractional statistics (fractional bias, fractional difference of tail-SDs) are most tightly distributed about the origin.

The procedure has the merit of concentrating on the relevant regions of the data while avoiding the large variability of extreme data points. It may therefore help one to decide which model among several competitors is best suited for a certain application. Thus the procedure may work well within the context of administrative decisions for which it is designed.

But granted that we have selected the most reliable model, just how reliable is it? How reliable are the measurements to which we are comparing model calculations? Is our rendition of the bootstrap an adequate surrogate for the historical data we don't have? Could we do better? I think such questions deserve more attention than they have received so far.

1.2.2 Urban ozone

During each year in certain parts of Los Angeles, and other US metropolitan areas, the concentration of ozone exceeds the level specified by the EPA's health-based national ambient air quality standard.

Simulation models are used to determine the effectiveness of various proposed measures to reduce urban ozone. Because of the complexity of emissions, chemical reactions, and meteorology in urban airsheds, the models are quite demanding of both input data and computing resources.

This complexity has two noteworthy consequences: (1) the models sometimes produce results that run counter to lay intuition, e.g. increased ozone concentration resulting from the proposed reduction of certain ozone precursors; and (2) for most practical purposes only a few carefully selected scenarios (days) can be modeled.

Thus model reliability is quite important, but has been evaluated primarily by expert opinion, since the cost of model runs prohibits extensive comparisons of model calculations to ozone measurements. The data analysis is limited to checking whether the model differs by no more than, say, 30 per cent (10–30 per cent is about as close as the most sophisticated ozone models get, in general) from measurements of high ozone episodes in the urban area.

Because of the paucity of data for comparing model calculations to measurements, Seinfeld (1988) suggests that such statistical comparisons be de-emphasized in favor of testing model components.

Perhaps at the time of their introduction complex technologies, like the atom bomb or the space shuttle, tend to elude statistical evaluation for practical reasons. Is this the case for current ozone models? If so, is there any merit in simpler alternative models whose reliability can be estimated statistically? Such questions have yet to be addressed by any body of statisticians.

1.2.3 Acid rain

For more than a decade scientists have investigated whether and to what extent sulfur and nitrogen compounds emitted by fossil-fuel combustion increase the acidity of precipitation and thereby damage the environment. The US government is now completing its ten-year National Acid Precipitation Assessment Program (NAPAP), whose mission has been to provide scientific information to help to identify effective strategies to control acidic deposition and mitigate its adverse effects.

Computer simulation models have been designed to calculate the reduction in acidic deposition that would result from a specified reduction in man-made emissions. Foremost among these models are the Acid Deposition and Oxidant Model (ADOM) developed by the Canadians, and the Regional Acid Deposition Model (RADM) developed by the Americans.

ADOM and RADM are very complex (cf. Chang *et al.*, 1987), and thus obtaining the data to run or evaluate them requires an enormous effort. Hansen and Mueller (1990) note that on completion of the modeling program approximately $20m will have been spent on the development of these models and another $35m to collect the data necessary to evaluate them.

Like the development of ADOM and RADM, the work to design protocols for evaluating the models has been intense in order to meet deadlines set by policy considerations. The principal goal of the evaluation has been to determine whether the models can reliably predict the annual acidic deposition that would result from specified emission scenarios (Seilkop, 1988).

But the models are so complex that simple comparisons of measurements versus calculations of acidic depositions are believed to be insufficient for evaluating the models' predictive abilities. Even the substantial measurement program previously mentioned appears to be inadequate for this purpose. To make the most of the available data, NAPAP will augment simple comparisons of measured and calculated pH, for example, with multivariate comparisons of relationships among analytes over space and time.

Thus the protocol is to compute a host of statistics that compare model calculations to measurements; the synthesis of these statistics into a qualitative notion of the reliability of the models is to be accomplished by atmospheric scientists. Admittedly, this entails more subjective judgment, albeit expert judgment, than many would like; but this is deemed to be the best we can do under the circumstances.

It would be premature to comment on the evaluation of ADOM and RADM. Much hard work has been done and much remains; recently completed field measurements are still being compiled into a data base. But it seems to me that we statisticians have been too quick to throw in the towel – for example, little effort has been expended so far on whether or how one should construct prediction intervals. Even if it turned out that such intervals required too many questionable assumptions, the effort might well be revealing.

1.2.4 Climate change

Many scientists are convinced that the continuing build-up of carbon dioxide and other so-called greenhouse gases (e.g. methane, nitrous oxide, and chlorofluorocarbons) in the atmosphere will cause significant global warming by the middle of the next century. And some analysts believe that even moderate changes in climate could disrupt national economies and infrastructures. Consequently, there is a mounting pressure to reduce greenhouse gas emissions somehow, as will be discussed by the United Nations in 1992.

The prediction of global warming is based to some extent on analyses of data, e.g. temperature records and estimated changes in the chemical composition of the atmosphere. But the primary basis for projections of global warming are the calculations of general circulation models (GCMs).

The various GCMs are in agreement that the global temperature would increase if the atmospheric concentrations of CO_2 were to double, for example. But the projected increases range from less than 1·5 °C to more than 4·5 °C depending on which model is used. The models also differ in the changes that would occur in specific regions of the globe, e.g. whether the North American mid-west would become wetter or drier.

Ongoing efforts to determine the sources of model differences (cf. Grotch, 1988) are designed to uncover model weaknesses, promote scientific exchanges among climate scientists, and ultimately make the models more reliable. Efforts to evaluate just how reliable these models are have begun only recently. The GCMs are even more complex than the models of acidic deposition, thus the cost and difficulty of such efforts are much greater.

To cite one example of these first steps toward GCM evaluation, Santer and Wigley (Santer and Wigley, 1990; Wigley and Santer, 1990), compute several test statistics that summarize how well four GCMs agree with field measurements of monthly mean sea level pressure. In their concluding remarks, the authors state that the purpose of their analysis is to guide subsequent investigations of model sensitivities, and to demonstrate the general usefulness of 'rigorous significance testing'. This seems to me to be a promising beginning, provided that someone eventually raises the question of how we might use such statistical findings to deduce the reliability of model predictions.

1.3 Purposes and protocols of model evaluation

It seems to me that air quality modeling serves three broad purposes:

- *Setting environmental policy.* The primary reason given for the development and improvement of air quality models seems to be to guide environmental decisions. Models of one sort or another (implicit or explicit) are the only technical means of estimating the environmental consequences of maintaining or changing our management of the environment.
- *Scientific understanding.* We may regard computer simulations as elaborate thought experiments, i.e. surrogates for actual experiments that are impossible or impractical to conduct. Depending on the adequacy of this substitution, the careful planning, analysis, and interpretation of the computer simulations may yield scientific insights as valuable as those obtained from actual experiments. And even if the simulations are not fully adequate surrogates, they may yet help to identify gaps in information useful to the design of future experiments.
- *Planning future research.* In the short term, the analysis of model sensitivities may help to design the next field or laboratory experiment; in the longer term such sensitivity studies may help to identify major scientific investigations that are worth funding.

Arguably, all three purposes are served by the statistical comparison of model output to field observations. But I would like to focus discussion on the first of these purposes, for I believe that establishing a factual foundation for government policy – in this case, determining the degree to which model projections should be trusted to guide environmental decisions – is a central function of statistical practice.

A rough consensus has been established about how models should be evaluated. Much thought and effort has gone into the development of protocols for evaluating simulation models of smokestack plumes, acidic deposition, and urban ozone. For example, Dennis *et al.* (1989) distinguish various aspects and stages of examination to determine whether a model is acceptable for a specified purpose.

The general idea is to ascertain whether a model meets some criteria, rather than estimate how far off model predictions might be. (By predictions I mean model calculations of air quality under conditions for which little or no air quality measurements exist.) Moreover, the statistical testing – consisting of test statistics and perhaps their p-values – is only one part of this assessment.

Such statistical summaries describe model–data agreement, but they fail to distinguish the possible inadequacies of model formulation from other possible sources of error that may not pertain to the reliability of model predictions, e.g. measurement error in the model's input data or in the field measurements to which the model is compared. The rejection or acceptance of a model is therefore not determined solely by the detection or failure to detect 'statistically significant' discrepancies between model output and field measurements, and wisely so.

1.4 How can statistical inference become relevant?

It is true that 'statistics' in some sense are used to evaluate air quality models. For example, model–data discrepancies are sometimes summarized in statistical terms to rank competing models. But it is generally difficult to discern the use of statistical interference to interpret model calculations that portray the environmental consequences of alternative policies. It seems that the question, 'How reliable are these model calculations?' is generally deemed to be either irrelevant – perhaps because the analyst reasons that some model must be used and presumably the best model has already been selected – or impossible to answer quantitatively.

Indeed, the question of model reliability poses some technical difficulties that need to be overcome. As just mentioned, inferential methods need to be developed to distinguish the errors relevant to model predictions from the other sources of model–data discrepancies. Another important statistical issue is the determination of whether the geographic sites and periods of observation comprising an evaluation data base are an adequate basis from which to infer the reliability of specified model calculations (and if not, whether some statistical technique, e.g. some form of resampling or data-smoothing, might help).

Certainly the failure to surmount such technical obstacles has limited the relevance of statistical evaluations. But there are nontechnical obstacles that are equally important to mention:

- First, the question of model reliability must be recognized at the administrative level as deserving an answer, and statisticians should help to promote the importance of this question.
- Second, the various purposes of model evaluation must be distinguished more clearly. The primary use of model–data comparisons has been to diagnose and thereby improve model performance, although such comparisons have often been conducted with the declared intent of somehow helping someone at some time to interpret model output for policy purposes.
- Third, the technical staff concerned with the reliability of model predictions must be willing to take responsibility, if necessary, for summarizing the model output into a single number or a few numbers pertinent to environmental decisions. To estimate model reliability quantitatively it is essential to define 'model prediction' narrowly, and to focus evaluation efforts on this aspect of the model.
- Fourth, an attempt must be made to quantitatively describe the reliability of the model prediction, e.g. in the form of a standard error or prediction interval. If such quantitative description is deemed to be impossible, it

is important to state the explanation, and to search for other models or methods whose reliability can be estimated quantitatively.

1.5 Model complexity

The evaluation of model reliability challenges many aspects of statistical inference, but perhaps the most profound challenge concerns the issue of model complexity. All the types of models previously mentioned (even the plume models in their more complex form) are highly complex, and consequently expensive and difficult to operate or evaluate.

An argument can be made for the use and development of complex models on the grounds that air quality models must faithfully represent important atmospheric processes to be scientifically plausible. This is a utilitarian as well as aesthetic issue. In the arena of climate forecasting, for example, plans to improve models entail large-scale scientific data collection as well as major improvements in computing capabilities, thus representing a major national investment.

We have yet to establish, however, whether and to what degree complex models provide more reliable policy guidance than simpler models. It would seem that at some point the benefits of increased realism in the model would be overwhelmed by the cost and quality of massive amounts of input data.

Current statistical methods may not be up to the tasks of defining and indentifying an optimal or near-optimal degree of model complexity. This is a fairly new statistical issue that has been addressed primarily in the limited context of nested models, such as polynomial regression. But we cannot afford to dismiss real-world decisions about complex modeling or other research investments as being forever beyond our ken. To do so would further jeopardize the relevance of what we do claim to know.

References

Chang, J.S., Brost, R.E., Isaksen, I.S.A., Madronich, S., Middleton, P., Stockwell, W.R., and Walcek, C.J. (1987). A three-dimensional Eulerian deposition model: physical concepts and formulation. *J. Geophys. Res.*, **92**, 14 681–700.

Cox, W.M. and Tikvart, J.A. (1986). Assessing the performance level of air quality models. In *Air Pollution Modeling and Its Applications V*, DeWispelaere, C., Schiermeier, F.A., and Gillani, N.V. (eds), Plenum, New York, pp. 425–40.

Dennis, R.L., Barchet, W.R., Clark, T.L., Seilkop, S.K., and Roth, P.M. (1989). *State of Science and Technology Report Number 5: Evaluation of Regional Acidic Deposition Models*, National Acid Precipitation Assessment Program.

EPA (1984). *Interim Procedures for Judging Air Quality Models (Revised)*. EPA-450/4-84-023.

Fox, D.G. (1981). Judging air quality model performance – review of the Woods Hole workshop. *Bull. Amer. Met. Soc.*, **62**, 599–602.

Grotch, S.L. (1988). *Regional Intercomparisons of General Circulation Model Predictions and Historical Climate Data.* Technical report TR041 prepared under contract number W-7405-ENG-48 for the Carbon Dioxide Research Division of the US Department of Energy.

Hansen, D.A. and Mueller, P.K. (1990). Acid deposition model evaluation. *EPRI Journal*, October/November, 42–44. Electric Power Research Institute, Palo Alto, California.

Londergan, R.J. (1981). *Validation of Plume Models: Statistical Methods and Criteria*. Report EA-1673-SY. Electric Power Research Institute, Palo Alto, California.

Santer, B.D. and Wigley, T.M.L. (1990). Regional validation of means, variances, and spatial patterns in general circulation model control runs. *J. Geophys. Res.*, **95**, 829–50.

Seilkop, S.K. (1988). *Measures and Interpretation for Regional Model Evaluation*. Report of a workshop held 19–20 August 1988 by the US EPA's Atmospheric Research and Exposure Assessment Laboratory. EPA-68D80063.

Seinfeld, J.H. (1988). Ozone air quality models: a critical review. *J. Air Poll. Control Assoc.*, **38**, 616–45.

Smith, M.E. (1984). Review of the attributes and performance of 10 rural diffusion models. *Bull. Amer. Met. Soc.*, **65**, 554–8.

Turner, D.B. (1979). Atmospheric modeling: a critical review. *J. Air Poll. Control Assoc.*, **29**, 502–19.

Wigley, T.M.L. and Santer, B.D. (1990). Statistical comparison of spatial fields in model validation, perturbation, and predictability experiments. *J. Geophys. Res.*, **95**, 851–65.

Chapter 2

Quality data networks that minimize entropy

W. F. Caselton, L. Kan and J. V. Zidek

2.1 Introduction

Designers of environmental monitoring networks face a number of difficulties. There are usually many objectives rather than just one. Moreover, many of the important future uses of the data may not be foreseen so that the network is often perceived as being of only limited value and therefore of low financial priority. At the same time the cost of establishing and operating a network is substantial. The most effective design of a cost-constrained network, which involves specifying how many stations should be installed and where they should be located, is not intuitively obvious. If, in any objective sense, the network designer seeks to derive the best value from a limited assignment of funds, he must resort to some kind of formal network design optimization.

The general problem described in the last paragraph is of great concern to national data-gathering agencies, like the Environmental Protection Agency and Environment Canada. They have the responsibility of monitoring areas which are continental in scale and must reckon with the demand for increasingly sophisticated environmental quality measurments, such as those related to the acid rain phenomenon. Furthermore, the crisis nature of environmental issues is often accompanied by strident demands for high data quality. But in this context, 'quality' means not only accuracy in a given measurement but also fitness for intended use, and this has implications concerning the choice of monitoring locations, the addition of new locations to existing networks, and the discontinuance of others to permit reallocation of funds to better sites. Consumers of environmental data include both enforcement agencies as well as private concerns and other agencies who are attempting to operate profitably under legislated emission or impact standards. They will all have very different requirements for environmental data. The designers of data-collection networks therefore must try to satisfy these individual requirements on the one hand but inescapably must compromise between them on the other.

A way round this problem, proposed by Caselton and Husain (1984) and further developed by Caselton and Zidek (1984), begins with the recognition that all data have the fundamental purpose of reducing uncertainty about some aspect of the world, regardless of how that need may be expressed in any particular situation. Both Bayesian and entropy ideas are involved. Bayesian theory argues that all individuals who aspire to be rational in a certain sense will seek to behave as if their uncertainties had been quantified by probability distributions. The use of entropy is dictated by certain basic requirements as the only way of representing the overall uncertainty about quantities with uncertain values which have probability distributions attached to them. This reasoning then leads inevitably to network designs which minimize certain entropies as the optimal solution to the designer's problem. While these designs may well fail to be optimal with respect to any single given objective, they seek to best meet the common features of the multiplicity of uses to which the resulting data may be put.

There is an alternative justification for adopting entropy in this situation, based on utility arguments, which comes out of the work of Bernardo (1979). The justification is also pursuasive in the monitoring network context and is discussed, together with other aspects of the normative theory of network designs, in Caselton and Zidek (1984).

A different approach to designing monitoring networks is taken by Fedorov and Mueller (1989) who cite related work. Their approach forces the network design problem into that of the classical theory of optimum design by assuming a regression model and by assuming, as an objective function, any one of a number of functionals of the covariance matrix of the estimates of the regression coefficients. Computational issues are addressed in Fedorov *et al.* (1987) where a design for monitoring sulfate concentrations is constructed. The regression model regresses the response, y_i, on (x_{1i}, x_{2i}), the geographical coordinates of station i through a polynomial of second degree (in the x's).

This approach does not seem appropriate in the context addressed here or, more generally, in Caselton and Zidek (1984). It is difficult to conceive of a model which would accurately relate concentrations of sulfate over the whole USA to design variables of stations (like their geographical coordinates). In the Gaussian case addressed in Section 2.3, we do inherit from our theory a linear prediction model, but this is a dynamic model which relates unobserved concentrations to observed concentrations in the spirit of kriging. And this linear predictor is not central to our basic theory which is really intended to address more fundamentally the need to select future sites so as to maximize the reduction of uncertainty. This issue does not appear to be addressed by the Fedorov–Mueller theory.

Of course, there will be situations where the classical theory of optimal design will apply and then it should be used in preference to the theory described in this paper.

2.2 General theory

2.2.1 Entropy

To conclude our introduction we will review the elements of entropy theory. For a more thorough and recent discussion, the reader is referred to Theil and Fiebig (1984).

Let $-\log(p)$ represent the decrease in uncertainty that results from learning that an uncertain event, say E, with probability p has in fact occurred. This measure of the reduction in uncertainty is dictated by certain weak axioms such as the requirement that this measure be increasing in p and that it be additive when applied to the intersection of two independent events. Before it is known whether the event or its complement has occurred the expected reduction in uncertainty is the average, $p[-\log p]+(1-p)[-\log(1-p)]$ which is the entropy of the simple two-point distribution with probabilities p and $(1-p)$ on the two points. This definition has an obvious extension to the case of a many-point distribution. The entropy is always nonnegative for distributions on a set of discrete points and is maximized by the uniform distribution over these points when the number of such points is finite.

Unfortunately, it is not possible to find the corresponding quantity in the case of points in a continuum of values. Simply taking the limit through a series of progressively finer discrete approximations, representing as it were a sequence of successively more accurate scales of measurement, results in a limit which is not finite. So the entropy for the continuous case is defined by analogy and uses integrals in place of sums and probability density functions in place of probabilities. The obvious analogy, $E[-\log f(Y)]$, where f is the density of random variable Y, is however, unsatisfactory because it is not invariant under transformations of Y. This is because f, unlike p, is not in units of probability but rather probability per unit of Y. Jaynes (1968) proposes the alternative adopted here:

$$H(Y)=E[-\log f(Y)/m(Y)],$$

where, in the terminology of Jaynes (1968), m is a *measure* representing *complete ignorance*. He recognizes the difficulty this entails, since m is not unambiguously defined, and this issue has not been fully resolved.

In this paper, we adopt for m an invariant measure, the one which, if used in a conventional Bayesian analysis, yields the conventional statistical estimation procedures based on frequency theory, at least up to constant multiples, depending how they are defined. This distinguishes m among the candidates for this role. However, the ambiguity about the choice of m means the so-called *optimal* designs derived in this report must be viewed as tentative.

The entropy in this report has some of the properties of its discrete counterpart, like, for example, additivity under independence (assuming m is chosen appropriately), but it loses others like nonnegativity. Indeed, the entropy for a continuous distribution approaches $-\infty$ as the distribution approaches the distribution concentrated on a single point and a state of complete certainty is reached. It is, however, widely used in the continuous case and seems, modulo the uncertainty about the choice of m, to be a natural index of uncertainty.

2.2.2 Network entropy

It is assumed that interest focuses on a random field generated by some natural processes. To simplify our presentation, we will view this field as dispersed over a two-dimensional space as it is in the application given in Section 2.4, but in the theory described in Sections 2.3 and 2.4 this is not required. In this section some basic theory will be reviewed and developed.

It greatly simplifies the problem under consideration to suppose that there is a fairly dense grid of sites spread out over space. Events occurring at all sites in this dense grid will be considered to be representative of events throughout the region. Not all of the sites will necessarily be candidates for permanent monitoring stations. These sites could be determined by latitude and longitude, as in the case considered by Zidek (1984). The density of this grid is limited by practical considerations discussed in Caselton and Zidek (1984) and in Zidek (1984). Or these sites can be in an existing network being considered for possible reductions in size, as in this chapter.

Let X denote the random vector whose coordinates represent the values of the spatial field at the various grid points ordered in any reasonable way. Monitoring stations are to be established at some of these grid points and the designer's problem reduces to deciding which of the grid points should be selected and 'gauged'. For any proposed design, the partitioning of X can be represented as $X = (U, G)$, where U stands for the values of interest at the unmonitored sites, and G the value at the monitored sites. For simplicity we will write $f(U, G)$ for the joint probability density function of $X = (U, G)$, and $m(U, G)$ for the corresponding measure of complete ignorance described in Section 2.1. Assume $m(G)$ has also been specified and that $m(U \mid G) = m(U, G)/m(G)$. Initially G will be considered as being measured without error.

If the underlying stochastic model is certain, the overall *a priori* uncertainty in the grid is fixed and is given by the entropy of X,

$$H(X) = H(U, G)$$

which is easily shown to decompose as

$$H(U, G) = H(U \mid G) + H(G)$$

where $H(U \mid G) = E[-\log f(U \mid G)/m(U \mid G)]$, and in general, $f(y \mid z)$ denotes the conditional density function of Y given that Z is known to be z. To interpret this decomposition, suppose G has been obtained. Then the remaining uncertainty, that about U, is $E[-\log f(U \mid G)/m(U \mid G) \mid G]$. However, *a priori* G is not certain, so the anticipated *a posteriori* uncertainty about U is

$$E\{E[-\log f(U \mid G)/m(U \mid G) \mid G]\} = H(U \mid G),$$

which is the first term in the above decomposition of the entropy of X.

One design strategy arising from the decomposition is that X should be partitioned so that $H(U \mid G)$ is minimized. We will refer to the minimization of $H(U \mid G)$ as the modeling objective. But, since the total uncertainty about X is fixed *a priori*, minimizing term 1 in the decomposition is equivalent to maximizing the second term. This second term represents the *a priori* uncertainty about G and is, in fact, eliminated by monitoring G. So it is intuitively reasonable that this term should be made as large as possible by suitably partitioning X and thereby maximizing the benefit of measurement as distinct from modeling, the objective addressed by term 1.

It is easily shown that

$$H(U \mid G) = E[-\log f(U \mid G)/m(U \mid G)] < E[-\log f(U)/m(U)] = H(U),$$

provided that $m(U \mid G) \equiv m(U)$ so that the same baseline is used to index the uncertainty in U whether G is given or not. Under these conditions, obtaining G by measurement is never expected to increase the amount of uncertainty about U. Now, assuming $m(U \mid G) \equiv m(U)$,

$$H(U \mid G) = H(U) - I(U; G)$$

where

$$I(U; G) = E\left[\log \frac{f(U, G)}{f(U) \cdot f(G)}\right]$$

is called the *mutual information* in U and G; the observation in the last paragraph implies $I > 0$ and obviously it is symmetric.

The objective addressed by the paper of Caselton and Zidek (1984) is in the context of installing a new network or extending an existing one. In either case an extensive, but relatively small number of potential monitoring stations are to be selected from a vast number of locations spread out over the region of interest. Under these circumstances, and excepting pathological circumstances, the reduction in uncertainty attributable directly to measurement alone would seem to be small. At the same time the potential to infer U from G becomes of critical importance as the monitoring network will be used repeatedly over time for this purpose. Thus the criterion postulated by Caselton and Zidek (1984) is the maximal degree to which the uncertainty about U may be reduced by obtaining G, i.e. the maximization of

$$H(U) - H(U \mid G) = I(U; G) \equiv I,$$

say. This approach has the deficiency that it ignores the consequences to $H(U)$ and to $H(G)$ in partitioning X to minimize I. The possibility of including in U highly uncertain values is thereby admitted on the basis that they are relatively easy to infer from G. We would observe that this approach has close links with Shannon's theory of information transmission which adopts precisely the same criterion. The role of the input message is played by U, and that of the receiving signal by G.

The situation confronted in the sequel is different from that described in the last paragraph, however. Here there is an existing network which will be described more fully in the next section, and interest lies in reducing the number of monitoring stations with the least loss of monitoring benefits in the light of the data accumulated to date. Because these networks consist of a relatively small number of stations, between 9 and 81, the uncertainty represented in $H(G)$ is a significant fraction of the overall uncertainty. As noted earlier, maximizing $H(G)$ is equivalent to minimizing $H(U \mid G)$.

This also coincides with the criterion used by Caselton and Husain (1984) in a hydrology application, although it was arrived at by a somewhat different, communications engineering inspired and distinctly non-Bayesian approach.

In practice the stochastic models required for the computation of the various expectations given above will themselves be uncertain. Much of the theory of statistics is devoted to evaluating and describing this uncertainty when expressed as uncertainty about a vector of parameters, say θ, which indexes the requisite stochastic models. We may incorporate this source of uncertainty in a straightforward fashion. The total *a priori* uncertainty is now given by $H(U, G, \theta)$, which may be decomposed as

$$H(U, G, \theta) = H(U, G \mid \theta) + H(\theta)$$

where the second term represents the uncertainty about the model parameters while the first term is the total stochastic uncertainty in X analyzed earlier.

In the application considered in the next section, and often in practice, there is some data available from candidate monitoring sites. This is easily incorporated via Bayes' theorem; all the expectations and probability densities given above now become conditional on these data. Even if no data were available the theory described here still applies but the validity of the results would then obviously depend solely on the quality of the prior information available.

Instrument and other forms of measurement error are also issues which a designer must recognize in practice, although surprisingly there are networks without provision for replicate measurements. Of course, uncertainty about $X = (U, G)$ is the only uncertainty of interest to policy makers, planners etc., but these quantities may often be conceptual; and G, which is supposed to be given by the network, is given operational meaning only by defining the process to be used for its measurement. Measurement introduces error and replicate measurements are desirable as a way of assessing the extent of this error. Suppose data, D, represents the vector of all the measurements including replicates to be made on all the monitoring sites represented in G. Then, the *a priori* reduction in expected uncertainty attributable to D is

$$H(U, G, \theta) - H(U, G, \theta \mid D) = H(G) - H(G \mid D)$$

provided that $m(U, G, \theta \mid D) = m(U, \theta \mid G)m(G \mid D)$. Clearly this should be maximized by partitioning X. Since $H(U, G, \theta)$ is fixed and independent of the choice of the process of measuring the values in G, and of the way in which $X = (U, G)$ is partitioned, we deduce that $H(U, G, \theta \mid D)$ must be mimimized by a combination of the choice of the process of measurement and the way of partitioning X.

A natural assumption about measurement error is that $f(U, \theta \mid G, D) = f(U, \theta \mid G)$, and so, if $m(U, G, \theta \mid D) = m(U, \theta \mid G)m(G \mid D)$,

$$H(U, G, \theta \mid D) = H(U, \theta \mid G) + H(G \mid D)$$

Thus, the choice of the process of measurement affects only the second term whereas the choice of network affects both potentially. Ideally, the choice of the network and the process which generates D will be made together to optimize the design. And this seems reasonable from a practical point of view. The measurement process might need to include, for example, the designation of an analytical laboratory. The quality of the data may depend not only on the laboratory but also on how long it takes to transport the material for analysis to the lab. In other words, the size of the measurement errors could in some situations depend on the distance between a monitoring site and the laboratory site.

Finally, observe that

$$H(G \mid D) = H(G) - [H(D) - H(D \mid G)]$$

if $m(D \mid G) = m(G \mid D)m(D)/m(G)$; this may be useful for computational purposes since it is reasonable to expect that, conditional on $G = G_i$, the associated D-vectors, D_i, are independent and hence that

$$\log f(D \mid G) = \Sigma \log f(D_i \mid G_i).$$

The first equation given in the last sentence shows that once the design has been specified, a good measurement is one for which $H(D) - H(D \mid G)$ is maximized, i.e. one for which a knowledge of G would be expected to substantially reduce the expected amount of uncertainty; this seems natural; it says the process is sensitive to variation in G, as it should be.

2.3 Multivariate normal theory

Suitably transformed, many enviromental data series can have distributions which are approximately Gaussian in character (cf. Wu and Zidek, 1989). In this section, the approach to network design outlined in Section 2.2 is implemented for the situation where all uncertainty may be represented by the multivariate Gaussian distribution (referred to in the sequel as the multinormal distribution). With the application of Section 2.4 in mind, an existing network is assumed. At each of a sequence of times, a vector of (possibly transformed) measurements, X, is obtained, the coordinates of which correspond to the sites in the network. The columns of the data matrix, $D = (X_1, \ldots, X_n)$, are assumed *a priori* to be (conditionally) independent and identically distributed, a realistic assumption for many environmental data series (Wu and Zidek, 1989). Another reasonable assumption is that the $\{X_i\}$ have a multivariable normal distribution with mean vector, μ, and covariance matrix, Σ, or more simply,

$$X_i \mid \mu, \Sigma \sim \text{ind } N_p(\mu, \Sigma),$$

for $i = 1, \ldots, n$, where p denotes the dimension of the $\{X_i\}$, i.e. the total number of sites in the existing monitoring network.

An uncertain future X and the uncertain model parameters are taken to be the designer's concern, where X is conditionally independent of, and distributed like, the other X_is. The problem to be addressed in this section is that of partitioning X as (U, G) effectively so that obtaining G by monitoring will maximally reduce the designer's uncertainty about the *state of the world*. The accomplishment of this goal entails the specification of the designer's uncertainty about μ and Σ. This will depend heavily on the context of the problem and the designer's background knowledge. Without a prescription of the designer's uncertainty our analyses could not proceed further than the very general theory outlined in Section 2.2.

We will proceed here with an illustrative analysis based on the supposition that the designer's uncertainty may be approximately described by a member of the class of conjugate prior distributions for μ and Σ. In effect, this is equivalent to assuming the designer's prior knowledge is equivalent to his or her having observed a sequence of prior Xs before the monitoring network began operation. The validity of this assumption would have to be assessed in particular contexts for particular designers. The assumption may well be found to hold approximately in a variety of situations, however, since the conjugate class of priors is rather large.

Thus, let the designer's uncertainty about μ and Σ be representable by the prior distributions given by

$$\mu \mid \nu, \Sigma, k \sim N_p\left(\nu, \frac{\Sigma}{k}\right) \tag{2.1}$$

and

$$\Sigma \mid \Psi, m \sim W_p^{-1}(\Psi, m) \tag{2.2}$$

where $\sim$ means 'distributed as' and in general $R \mid S$ means the conditional distribution of the random vector R given S. Also $W_p^{-1}(\Psi, m)$ denotes the inverted Wishart distribution with dimension p and parameters Ψ and m (see Anderson, 1984, p. 268 for a definition). Equivalently, $\Sigma^{-1} \mid \Psi \sim W_p(\Psi^{-1}, m)$, and unless $m > p - 1$, this distribution will be *improper*. That is, the densities of Σ or equivalently, Σ^{-1}, will have infinite integrals. Nevertheless, the density may, even when $m \leqslant p - 1$, be combined with the likelihood function to obtain a posterior density for Σ which is integrable, provided that $n + m > p - 1$, as will now be assumed in the ensuing analysis.

It will be convenient to reparametrize Σ as $(\Sigma_{GG}, \Sigma_{U|G}, \tau)$ where in general, for any two random vectors R and S,

$$\Sigma_{RS} \equiv \text{cov}\,(R, S^{\mathrm{T}}) \equiv E[R - E(R)][S - E(S)]^{\mathrm{T}},$$

and

$$\Sigma_{R|S} \equiv \Sigma_{RR} - \Sigma_{RS}\Sigma_{SS}^{-1}\Sigma_{SR},$$

i.e. the residual covariance of R conditional on S. The matrix $\tau \equiv \Sigma_{UG}\Sigma_{GG}^{-1}$ denotes the *slope* of the optimal linear predictor of U on G, i.e. $E(U) + \tau[G - E(G)]$. This transformation is achieved through the Bartlett (1933) decomposition, $\Sigma = T\Delta T^{\mathrm{T}}$, where

$$\Delta \equiv \begin{bmatrix} \Sigma_{U|G} & 0 \\ 0 & \Sigma_{GG} \end{bmatrix} \quad \text{and} \quad T \equiv \begin{bmatrix} I & \tau \\ 0 & I \end{bmatrix}.$$

Thus, $\Sigma^{-1} = (T^{\mathrm{T}})^{-1}\Delta^{-1}T^{-1}$, where

$$\Delta^{-1} \equiv \begin{bmatrix} \Sigma_{U|G}^{-1} & 0 \\ 0 & \Sigma_{GG}^{-1} \end{bmatrix} \quad \text{and} \quad T^{-1} \equiv \begin{bmatrix} I & -\tau \\ 0 & I \end{bmatrix}.$$

The analogous factorization of the prior parameter matrix is $\Psi = N\zeta N^{\mathrm{T}}$, where

$$\zeta \equiv \begin{bmatrix} \Psi_{U|G} & 0 \\ 0 & \Psi_{GG} \end{bmatrix} \quad \text{and} \quad N \equiv \begin{bmatrix} I & \eta \\ 0 & I \end{bmatrix}.$$

with $\eta = \Psi_{UG}\Psi_{GG}^{-1}$.

Now, the differential element of the (possibly improper) prior density for Σ is proportional to

$$|\Sigma|^{-(m+p+1)/2} \exp\left[-\tfrac{1}{2} \operatorname{tr} \Psi\Sigma^{-1}\right] \mathrm{d}\Sigma,$$

where Σ is positive definite, i.e. $\Sigma > 0$. But $|\Sigma| = |\Sigma_{GG}||\Sigma_{U|G}|$. At the same time, $\operatorname{tr} \Psi\Sigma^{-1} = \operatorname{tr} N\zeta N^{\mathrm{T}}(T^{\mathrm{T}})^{-1}\Delta^{-1}T^{-1} = \operatorname{tr} \zeta[T^{-1}N]^{\mathrm{T}}\Delta^{-1}[T^{-1}N]$, where

$$T^{-1}N = \begin{bmatrix} I & -\tau \\ 0 & I \end{bmatrix}\begin{bmatrix} I & \eta \\ 0 & I \end{bmatrix} = \begin{bmatrix} I & \eta - \tau \\ 0 & I \end{bmatrix}.$$

Thus, in terms of the new parameters,

$$\operatorname{tr} \Psi\Sigma^{-1} = \operatorname{tr} \Psi_{U|G}\Sigma_{U|G}^{-1} + \operatorname{tr} \Psi_{GG}\Sigma_{GG}^{-1} + \operatorname{tr} \Psi_{GG}(\eta - \tau)^{\mathrm{T}}\Sigma_{U|G}^{-1}(\eta - \tau).$$

Finally it is easy to show that with $u = \dim(U)$,

$$\mathrm{d}\Sigma = |\Sigma_{GG}|^{u}\, \mathrm{d}\Sigma_{GG}\, \mathrm{d}\Sigma_{U|G}\, \mathrm{d}\tau.$$

Thus, the differential density element for the new parameters factors is

$$f(\Sigma_{GG} \mid \Psi_{GG})\, \mathrm{d}\Sigma_{GG} \cdot f(\Sigma_{U|G}, \tau \mid \Psi)\, \mathrm{d}\Sigma_{U|G}\, \mathrm{d}\tau \tag{2.3}$$

where the density of Σ_{GG} is that of $W_g^{-1}(\Psi_{GG}, m - u)$.

This last result is well known (cf. Dempster, 1969) and proves incidentally another well-known result that Σ_{GG} is independent of $(\Sigma_{U|G}, \tau)$ when the prior distribution is proper. This latter result is obtained, since the decomposition, $\Sigma = T\Delta T^{\mathrm{T}}$, implies that for any vector $x = (x_U, x_G)^{\mathrm{T}}$, $x\Sigma x^{\mathrm{T}} = y_U\Sigma_{U|G}y_U^{\mathrm{T}} + y_G\Sigma_{GG}y_G^{\mathrm{T}}$ with $y_U \equiv x_U$ and $y_G \equiv \tau x_U + x_G$, which proves that $\Sigma > 0$ if and only if $\Sigma_{U|G} > 0$ and $\Sigma_{GG} > 0$. Thus, the factorization in (2.3) is over the space $\{\Sigma_{GG} > 0, \Sigma_{U|G} > 0, \tau \text{ arbitrary}\}$, and so in particular, the range of Σ_{GG} is unrestricted by $\Sigma_{U|G}$ or τ.

Anderson (1984, p. 270) shows that under the prior distribution described in (1) and (2), and the realized data $d = (x_1, \ldots, x_n)$,

$$\mu \mid \Sigma, \nu, k, D \sim \mu \mid \Sigma, \hat{\mu} \sim N_p\left(\hat{\mu}, \frac{\Sigma}{n + k}\right) \tag{2.4}$$

$$\Sigma \mid \Psi, m, D \sim \Sigma \mid \hat{\Psi} \sim W_p^{-1}(\hat{\Psi}, n + m) \tag{2.5}$$

where

$$\hat{\mu} \equiv \frac{n\bar{x} + k\nu}{n + k} \quad \text{and} \quad \hat{\Psi} \equiv \Psi + (n - 1)s + \frac{nk}{n + k}(\bar{x} - \nu)(\bar{x} - \nu)^{\mathrm{T}},$$

and $\bar{x}$ and s denote, respectively, the usual unbiased estimators of μ and Σ. This result holds even if the prior density is improper as long as it has the appropriate form. The posterior density will be proper under the fundamental assumption stated earlier that $n + m > p - 1$. Since (3) obtains whatever be Ψ, it follows that *a posteriori* $\Sigma_{GG} \sim W_g^{-1}(\hat{\Psi}_{GG}, n + m - u)$ is also independent of $(\Sigma_{U|G}, \tau)$.

The factorization in (2.3), which as just noted above, applies to the posterior distribution of μ and Σ, also shows the well-known results that

$$\tau \mid \Sigma_{U|G}, \hat{\eta}, \hat{\Psi}_{GG} \sim N_{ug}(\hat{\eta}, \Sigma_{U|G} \otimes \hat{\Psi}_{GG}^{-1}) \tag{2.6}$$

and

$$\Sigma_{U|G} \mid \hat{\Psi}_{U|G} \sim W_u^{-1}(\hat{\Psi}_{U|G}, n + m).$$

We can now decompose the designer's uncertainty given the realized data $d = (x_1, \ldots, x_n)$. To effect this decomposition entails the specification of the reference measure $m(U, G, \mu, \Sigma \mid d)$. This will be taken to be independent of d, thereby keeping to the goal of providing a baseline representing a state of complete ignorance. In as much as U, G and μ are elements of an affine invariant space, letting each of them have a uniform reference measure over their respective spaces seems natural. But it must be remembered that these uniform measures would be transformed under transformations of their associated vector variables. Finally for the reason given in Section 2.1, the reference measure for Σ is taken to be

$$|\Sigma|^{-(p+1)/2} = |\Sigma_{GG}|^{-(p+1)/2} |\Sigma_{U|G}|^{-(p+1)/2}.$$

Under the transformation which carries Σ into $(\Sigma_{GG}, \Sigma_{U|G}, \tau)$, this reference measure transforms into $|\Sigma_{GG}|^{-(g-u+1)/2} |\Sigma_{U|G}|^{-(p+1)/2}$, where $g = \dim(G)$, $u = \dim(U)$, and $g + u = p$. Thus, in summary, after transformation of Σ,

$$m(U, G, \mu, \Sigma \mid d) = m(U)m(G)m(\mu, \Sigma) = |\Sigma_{GG}|^{-(g-u+1)/2} |\Sigma_{U|G}|^{-(p+1)/2}.$$

The general decomposition of uncertainty given in Section 2.2 is

$$E[-\log f(X, \mu, \Sigma \mid d)/m(X, \mu, \Sigma \mid d) \mid d] = PRED + MODEL + MEAS \tag{2.7}$$

where

$$PRED = E[-\log f(U \mid G, \mu, \Sigma) \mid d]$$

$$MODEL = E[-\log f(\mu, \Sigma \mid G, d)/m(\mu, \Sigma) \mid d]$$

and

$$MEAS = E[-\log f(G \mid d) \mid d].$$

In obtaining this decomposition, the conditional independence of U and d given μ and Σ has been employed in evaluating $PRED$. This term represents the *a priori* residual uncertainty about U after the (*a priori*) uncertain G is employed to predict U through the optimal linear model $E(U \mid \mu, \Sigma) + \tau[G - E(G \mid \mu, \Sigma)]$. It should be noted that $MEAS$ is the uncertainty (in G) which is to be eliminated by measurement and is to be maximized by optimally partitioning X. Equivalently, this partition must be chosen to minimize $PRED + MODEL$, the *a priori* expected uncertainty remaining in U and (μ, Σ) given G, the latter being described by $MODEL$.

A further decomposition is

$$MODEL = UMEAN + GMEAN + COV \tag{2.8}$$

where

$$UMEAN = E[-\log f(\mu_U \mid \mu_G, G, \Sigma, d) \mid d]$$

$$GMEAN = E[-\log f(\mu_G \mid G, \Sigma, d) \mid d]$$

and

$$COV = E[-\log f(\Sigma \mid G, \Psi, m, d)/m(\Sigma) \mid d],$$

with $\mu = (\mu_U, \mu_G)$ in conformity with the earlier (U, G) decomposition of X. It is easily shown that, in general, the multinormal distribution, $N_p(\mu, \Sigma)$, has entropy $\frac{1}{2}\log|\Sigma| + K_p$, where $K_p = (p/2)(\log 2\pi + 1)$. Thus, in (2.7), $PRED = E[\frac{1}{2}\log|\Sigma_{U|G}| + K_u \mid d]$, where $u = \dim(U)$. Hence,

$$PRED = \tfrac{1}{2}\beta_{U|G} + K_u \tag{2.9}$$

where $\beta_{U|G} \equiv E[\log|\Sigma_{U|G}|\,|d]$.

Display (2.4) entails

$$UMEAN = E\left[\frac{1}{2}\log\left|\frac{1}{n+k}\Sigma_{U|G}\right| + K_u \mid d\right]$$

or

$$UMEAN = \frac{1}{2}\beta_{U|G} - \frac{u}{2}\log(n+k) + K_u. \tag{2.10}$$

Now the posterior distribution (given Σ) in (2.4) can be taken as a prior as G is being obtained, and the marginal 'prior' of μ_G may be combined with the likelihood given by G by reapplying (2.4) with the 'new' $n = n + 1$. The result gives

$$GMEAN = E\left[\frac{1}{2}\log\left|\frac{1}{n+k+1}\Sigma_{GG}\right| + K_g \mid d\right],$$

or

$$GMEAN = \frac{1}{2}\beta_G - \frac{g}{2}\log(n+k+1) + K_g, \tag{2.11}$$

where $\beta_G \equiv E[\log|\Sigma_{GG}|\,|\,d]$.

Finally,

$$f(G, \mu, \Sigma \mid \nu, \Psi, d) = f(G \mid \mu_G, \Sigma_{GG}) \cdot f(\mu, \Sigma \mid \nu, \Psi, d).$$

And by (2.4) and (2.5),

$$f(\mu, \Sigma \mid \nu\nu, \Psi, d) = f(\mu_U \mid \mu_G, \hat{\mu}, \Sigma_{U|G}, \tau) \cdot f(\mu_G \mid \hat{\mu}_G, \Sigma_{GG}) \cdot f(\Sigma \mid \hat{\Psi}).$$

Thus by (2.3) and these last two equations,

$$f(G, \Sigma \mid \nu, \Psi, d) = f(G \mid \hat{\mu}_G, \Sigma_{GG}) \cdot f(\Sigma_{GG} \mid \hat{\Psi}_{GG}) \cdot f(\Sigma_{U|G}, \tau \mid \hat{\Psi})$$
$$= f(G, \Sigma_{GG} \mid \hat{\mu}_G, \hat{\Psi}_{GG}) \cdot f(\Sigma_{U|G}, \tau \mid \hat{\Psi}).$$

So,

$$f(G \mid \nu, \Psi, d) = f(G \mid \hat{\mu}_G, \hat{\Psi}_{GG}).$$

This last density may be found explicitly, since

$$G \mid \mu_G, \Sigma_{GG} \sim N_g(\mu_G, \Sigma_{GG})$$

and

$$\mu_G \mid \hat{\mu}_G, \frac{1}{n+k}\Sigma_{GG}\ (a\ posteriori) \sim N_g\left(\hat{\mu}_G, \frac{1}{n+k}\Sigma_{GG}\right)$$

yield

$$\mu_G \mid \hat{\mu}_G, \Sigma_{GG} \sim N_g\left(\hat{\mu}_G, \frac{n+k+1}{n+k}\Sigma_{GG}\right).$$

This combined with

$$\Sigma_{GG} \mid \hat{\Psi}_{GG} \sim W_g^{-1}(\hat{\Psi}_{GG}, n+m-u)$$

implies (see Anderson 1984, p. 272) that given d, ν, k, Ψ, and m, G has the multivariate Student's t-distribution with density proportional to

$$\left[1 + \frac{1}{n+m+1-p}(G-\hat{\mu}_G)^{\mathrm{T}} \times \left[\frac{n+k+1}{n+k}\hat{\Psi}_{GG}\right]^{-1}(G-\hat{\mu}_G)\right]^{-(n+m+1-u)/2}$$

Thus,

$$Y \equiv \left[\frac{n+k}{n+k+1}\right]^{1/2}\hat{\Psi}_{GG}^{-1/2}(G-\hat{\mu}_G)$$

has the standardized multivariate student's t-distribution with $n+m+1-p$ degrees of freedom, and density proportional to

$$\left[1 + \frac{1}{n+m+1-p}Y^{\mathrm{T}}Y\right]^{-(n+m+1-u)/2} \tag{2.12}$$

regarded as a function of the g-dimensional vector Y. Finally,

$$f(\Sigma \mid G, \nu, \Psi, d) = f(\Sigma_{GG} \mid G, \hat{\mu}_G, \hat{\Psi}_{GG}) \cdot f(\Sigma_{U|G}, \tau \mid \hat{\Psi}).$$

Thus,

$$COV = GCOV + UCOV$$

where

$$GCOV = E[-\log f(\Sigma_{GG} \mid G, \hat{\mu}_G, \hat{\Psi}_{GG})/m(\Sigma_{GG}) \mid \hat{\mu}_G, \hat{\Psi}_{GG}]$$

and

$$UCOV = E[-\log f(\Sigma_{U|G}, \tau)/m(\Sigma_{U|G}, \tau) \mid \hat{\Psi}], \tag{2.13}$$

where we have taken $m(\Sigma_{GG}) = |\Sigma_{GG}|^{-(g-u+1)/2}$ and $m(\Sigma_{U|G}, \tau) = \Sigma_{U|G}|^{-(p+1)/2}$.

Now

$$UCOV = UCOV1 + UCOV2, \tag{2.14}$$

where

$$UCOV1 = E[-\log f(\tau \mid \Sigma_{U|G}, \hat{\Psi})/m(\tau) \mid \hat{\Psi}],$$
$$UCOV2 = E[-\log f(\Sigma_{U|G} \mid \hat{\Psi})/m(\Sigma_{U|G}) \mid \hat{\Psi}],$$

and we have without loss of generality factored $m(\Sigma_{U|G}, \tau)$ as $m(\Sigma_{U|G})m(\tau)$ with $m(\Sigma_{U|G}) = |\Sigma_{U|G}|^{-(p+1)/2}$ and $m(\tau) \equiv 1$ for simplicity. But display (2.6) implies, using the notation in eq. (2.9),

$$\begin{aligned} UCOV1 &= \tfrac{1}{2}E[\log |\Sigma_{U|G} \otimes \hat{\Psi}_{GG}^{-1}| \mid \hat{\Psi}] + K_{ug} \\ &= \tfrac{1}{2}E[\log |\Sigma_{U|G}|^{g} |\hat{\Psi}_{GG}|^{-u} \mid \hat{\Psi}] + K_{ug} \\ &= \frac{g}{2}E[\log |\Sigma_{U|G}| \hat{\Psi}] - \frac{u}{2}\log |\hat{\Psi}_{GG}| + K_{ug}. \end{aligned}$$

That is,

$$UCOV1 = \frac{g}{2}\beta_{U|G} - \frac{u}{2}\log |\hat{\Psi}_{GG}| + K_{ug}. \tag{2.15}$$

At the same time, using the form of the inverted Wishart density function (cf. Anderson 1984, p. 268),

$$\begin{aligned} UCOV2 = -\frac{n+m}{2}\log |\hat{\Psi}_{U|G}| &+ \frac{n+m-g}{2}E[\log |\Sigma_{U|G}| \mid \hat{\Psi}] \\ &+ \tfrac{1}{2}E[\mathrm{tr}\, \hat{\Psi}_{U|G}\Sigma_{U|G}^{-1} \mid D] + C_u(n+m), \end{aligned}$$

where for any integers $l, r = 1, 2, \ldots, l \leqslant r$,

$$C_l(r) = l\left(\frac{r}{2}\right)\log 2 + \frac{1}{4}l(l-1)\log \pi + \sum_{i=1}^{l} \log \Gamma\left(\frac{r-i+1}{2}\right). \tag{2.16}$$

But given d, $\Psi_{U|G}\Sigma_{U|G}^{-1} \sim W_u(I_u, n+m)$ according to the definition of the inverted Wishart distribution. Thus, $E[\mathrm{tr}\, \Psi_{U|G}\Sigma_{U|G}^{-1} \mid d] = (n+m)u$ and

$$\begin{aligned} UCOV2 = -\frac{n+m}{2}\log |\hat{\Psi}_{U|G}| &+ \frac{n+m-g}{2}\beta_{U|G} \\ &+ \frac{(n+m)u}{2} + C_u(n+m). \end{aligned}$$

$GCOV$ can be found by treating $f(\Sigma_{GG} \mid \hat{\Psi}_{GG})$ as an (updated) prior which combines with $G \mid \mu_G, \Sigma_{GG}$ to give

$$\Sigma_{GG} \mid G, \nu, \Psi, d \sim W_g^{-1}(\hat{\Psi}_{GG,1}, n+m+1-u),$$

where

$$\hat{\Psi}_{GG,1} \equiv \hat{\Psi}_{GG} + \frac{n+k}{n+k+1}(G-\hat{\mu}_G)(G-\hat{\mu}_G)^{\mathrm{T}}.$$

Thus (cf. Anderson 1984, p. 268),

$$\begin{aligned}
&\log f(\Sigma_{GG} \mid G, \hat{\mu}_G, \hat{\Psi}_{GG})/m(\Sigma_{GG}) \\
&= \frac{n+m+1-u}{2}\log|\Psi_{GG,1}| - \frac{n+m+1}{2}\log|\Sigma_{GG}| \\
&\qquad\qquad - \tfrac{1}{2}\operatorname{tr}\hat{\Psi}_{GG,1}\Sigma_{GG}^{-1} - C_g(n+m+1-u),
\end{aligned}$$

where $C_g(n+m+1)$ is as defined in eq. (2.16). But given G,

$$\hat{\Psi}_{GG,1}\Sigma_{GG}^{-1} \sim W_g(I_g, n+m+1-u).$$

Thus,

$$E[\hat{\Psi}_{GG,1}\Sigma_{GG}^{-1} \mid \hat{\mu}_G, \hat{\Psi}_{GG}] = (n+m+1-u)I_g$$

and in terms of the notation given in (2.11) and (2.16),

$$\begin{aligned}
GCOV &= E[-\log f(\Sigma_{GG} \mid G, \hat{\mu}_G, \hat{\Psi}_{GG})/m(\Sigma_{GG}) \mid \hat{\mu}_G, \hat{\Psi}_{GG}] \\
&= -\frac{n+m+1-u}{2}\gamma_G + \frac{n+m+1}{2}\beta_G + \frac{(n+m+1-u)g}{2} \\
&\quad + C_g(n+m+1-u), \qquad (2.17)
\end{aligned}$$

where $\gamma_G \equiv E[\log|\hat{\Psi}_{GG,1}| \mid \hat{\mu}_G, \hat{\Psi}_{GG}]$. But

$$\begin{aligned}
|\hat{\Psi}_{GG,1}| &= \left|\hat{\Psi}_{GG} + \frac{n+k}{n+k+1}(G-\hat{\mu}_G)(G-\hat{\mu}_G)^{\mathrm{T}}\right| \\
&= |\hat{\Psi}_{GG}|\left[1 + \frac{n+k}{n+k+1}(G-\mu_G)^{\mathrm{T}}\Psi_{GG}^{-1}(G-\mu_G)\right].
\end{aligned}$$

In other words,

$$\gamma_G = \log|\hat{\Psi}_{GG}| + E[\log(1+Y^{\mathrm{T}}Y)], \qquad (2.18)$$

where Y has the multivariate t density given in expression (2.12). But, $Z \equiv (1/g)Y^{\mathrm{T}}Y$ has F distribution with degrees of freedom, g and $n+m+1-p$ (cf. Muirhead 1984, p. 48). Thus,

$$E[\log(1+Y^{\mathrm{T}}Y)] = E[\log(1+gZ)]. \qquad (2.19)$$

Let us summarize the results which have been obtained. The quantity to be minimized by optimally partitioning X is

$$PRED + MODEL \qquad \text{(eq. (2.7))}$$

where

$$MODEL = UMEAN + GMEAN + COV \qquad \text{(eq. (2.8))}$$

$$COV = GCOV + UCOV \qquad \text{(eq. (2.13))}$$

and

$$UCOV = UCOV1 + UCOV2. \qquad \text{(eq. (2.14))}$$

Furthermore,

$$PRED = \frac{1}{2}\beta_{U|G} + K_u \qquad \text{(eq. (2.9))}$$

$$UMEAN = \frac{1}{2}\beta_{U|G} - \frac{u}{2}\log(n+k) + K_u, \qquad \text{(eq. (2.10))}$$

$$GMEAN = \frac{1}{2}\beta_G - \frac{g}{2}\log(n+k+1) + K_g, \qquad \text{(eq. (2.11))}$$

$$GCOV = -\frac{n+m+1-u}{2}\gamma_G + \frac{n+m+1}{2}\beta_G + \frac{(n+m+1-u)g}{2} + C_g(n+m+1-u) \qquad \text{(eq. (2.17))}$$

$$UCOV1 = \frac{g}{2}\beta_{U|G} - \frac{u}{2}\log|\hat{\Psi}_{GG}| + K_{ug} \qquad \text{(eq. (2.15))}$$

and

$$UCOV2 = -\frac{n+m}{2}\log|\hat{\Psi}_{U|G}| + \frac{n+m-g}{2}\beta_{U|G} + \frac{(n+m)u}{2} + C_u(n+m), \qquad \text{(eq. (2.16))}$$

where γ_G, K_l, and $C_l(r)$, for integer l and r, are given respectively in expressions (2.18), (2.19) and (2.16). Combining these results gives

$$PRED + MODEL = \left[\frac{n+m+2}{2}\right]\beta_{U|G} + \left[\frac{n+m+2}{2}\right]\beta_G - \left[\frac{n+m+1-u}{2}\right]\gamma_G - \left[\frac{u}{2}\right]\log|\hat{\Psi}_{GG}| - \left[\frac{n+m}{2}\right]\log|\hat{\Psi}_{U|G}| + R(u, g, k, m, n)$$

where

$$R(u, g, k, m, n) = -\left[\frac{u}{2}\log(n+k) + \frac{g}{2}\log(n+k+1)\right] + \frac{(n+m)p-(u-1)g}{2} + [2K_u + K_g + K_{ug}] + [C_u(n+m) + C_g(n+m+1-u)].$$

This expression can be reduced somewhat using eqs. (2.18), (2.7), (2.16) and (2.19) to get

$$
\begin{aligned}
PRED + MODEL = \log|\hat{\Psi}| - \tfrac{1}{2}\log|\hat{\Psi}_{GG}| \\
&- \frac{n+m+2}{2}\sum_{i=1}^{p}\xi_{n+m-i+1} \\
&- \frac{n+m+1-u}{2}E[\log(1+gZ)] \\
&+ R(u, g, k, m, n), \qquad (2.20)
\end{aligned}
$$

where $\xi_i = E[\log\chi_i^2]$ and χ_i^2 denotes a chi-squared random variable with i degrees of freedom. Equation (2.20) is obtained from

$$
\begin{aligned}
\beta_{U|G} + \beta_G = [\log|\Sigma|\,|\,\hat{\Psi}] &= \log|\hat{\Psi}| + E[\log|\Sigma\hat{\Psi}^{-1}|] \\
&= \log|\hat{\Psi}| - \sum_{i=1}^{p}\xi_{n+m-i+1}
\end{aligned}
$$

since given $\hat{\Psi}, \Sigma\hat{\Psi}^{-1} \sim W_p(I_p, n+m)$ and hence $|\Sigma^{-1}\hat{\Psi}| = \chi_{n+m}^2 \cdots \chi_{n+m-p+1}^2$ for independent chi-squared random variables.

Equation (2.20) yields the remarkably simple conclusion that for each fixed g, the optimum partition of X_{n+1} is found by maximizing $|\hat{\Psi}_{GG}|$.

To complete the section we turn to the evaluation of the overall uncertainty. It is

$$
\begin{aligned}
&E[-\log f(X, \mu, \Sigma\,|\,d)/m(X, \mu, \Sigma\,|\,d)\,|\,d] \\
&= E[-\log f(X\,|\,\mu, \Sigma)\,|\,d] + E[-\log f(\mu\,|\,\Sigma, d)\,|\,d] \\
&\qquad + E[-\log f(\Sigma\,|\,d)/m(\Sigma)\,|\,d].
\end{aligned}
$$

But

$$
E[-\log f(X\,|\,\mu, \Sigma)\,|\,d] = \tfrac{1}{2}E[\log|\Sigma|\,|\,d] + K_p, \qquad (2.21)
$$

and

$$
\begin{aligned}
E[-\log f(\mu\,|\,\Sigma, d)\,|\,d] = \tfrac{1}{2}E[\log|\Sigma|\,|d] + K_p, \\
&- \frac{p}{2}\log(n+k). \qquad (2.22)
\end{aligned}
$$

Finally,

$$
\begin{aligned}
E[-\log f(\Sigma\,|\,d)/m(\Sigma)\,|\,d] = &- \frac{n+m}{2}\log|\hat{\Psi}| \\
&+ \frac{n+m}{2}E[\log|\Sigma|\,|d] \\
&+ \frac{(n+m)p}{2} + C_p(n+m) \qquad (2.23)
\end{aligned}
$$

Combining eqs. (2.21)–(2.23) yields

$$E[-\log f(X, \mu, \Sigma \mid d)/m(X, \mu, \Sigma \mid d) \mid d]$$

$$= \log |\Psi| - \frac{n+m+2}{2} \sum_{i=1}^{p} \xi_{n+m-i+1} - \frac{p}{2} \log (n+k)$$

$$+ \frac{(n+m)p}{2} + 2K_p + C_p(n+m),$$

on applying the result stated as eq. (2.23).

2.4 An illustrative example

In this section we apply the theory developed above and analyse a wet deposition monitoring network. The network was established in the United States in 1978 by the Association of Agricultural Experimental Stations and called the National Acidic Deposition Program (NADP) monitoring network. Over time this network has grown to include more than 200 stations and it is the largest of all such networks. As an outgrowth of the National Acid Precipitation Assessment Program (NAPAP) the National Trends Network (NTN) has been established and incorporates many of the former NADP sites. Indeed the Network is now often referred to as the NADP/NTN network. It collects weekly samples, and the concentrations of various chemical components in wet deposition are routinely measured. In this section we investigate the network from the point of view of its capacity to reduce uncertainty about the levels of sulfate, one of the major components of acid rain.

Our study uses data from the Acid Deposition System (ADS), a very large data base created and maintained by Batelle's Pacific Northwest Laboratory, in particular, the data collected there from the NADP/NTN network. Monthly volume weighted average levels of sulfate for the period 1983–1986 were very kindly provided for this study by Dr Tony Olsen and his colleagues at the Pacific Northwest Laboratory. In the end only $p = 81$ stations were investigated, those with no more than 5 missing monthly values. We thereby avoided the need to deal with extensive amounts of missing data. (See Table 2.1 and Figure 2.1.).

An analysis of these data by Wu and Zidek (1989) show that the values obtained by taking logarithms of the sulfate concentration data values have an approximately normal form even when aggregated over time and clusters of sites. Eynon and Switzer (1983) found an apparent absence of autocorrelation of daily pH values over time after correcting for the (small) effects of seasonality and rainfall volume. Egbert and Lettenmaier (1986) determined that for weekly measurements from the NADP network, temporal correlation is small or absent, for both pH and sulfate measurements. It seems plausible that the monthly data values upon which our data are based would also have little serial correlation, and we assume they are independent as an approximation.

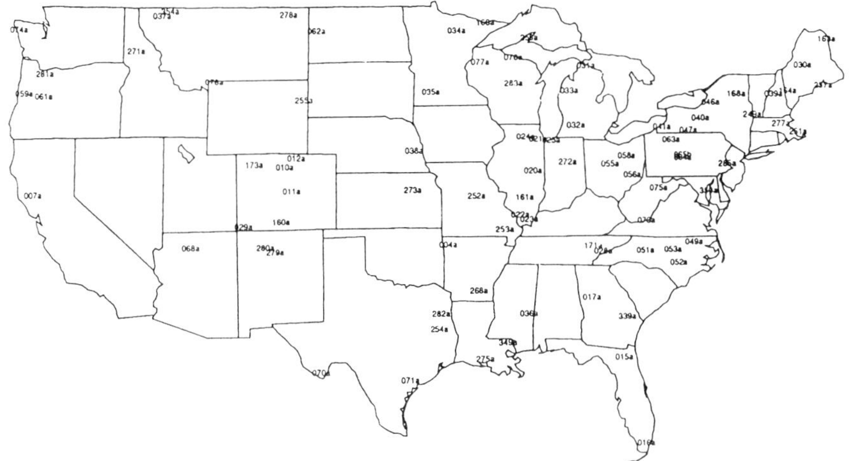

Figure 2.1

A referee pointed to the possibility of long-term memory in these data series. Their short length precludes an adequate assessment of this hypothetical possibility. In any case, our approach is based upon the relative performance of stations in their spatial context as measured by entropy; the data are effectively used only to estimate the normal model parameters rather than to predict future data values. Intuitively, it seems unlikely that long-term memory would significantly affect the relative quality ranking of the stations.

Central to the method of the previous section is the spatial covariance matrix for these 81 stations. This cannot be reliably estimated from these data using just the $n = 48$ monthly observations, since the Wishart distribution of the sample covariance has a negative number of $n - p + 1 = -32$ degrees of freedom. Indeed as the calculations described below will show, our lack of knowledge about this covariance matrix is a principal part of our overall uncertainty. Our theory, however, incorporates this uncertainty and allows us to proceed by specifying the *a priori* covariance matrix 'estimate' and the number of months, m, of hypothetical data on which this prior estimate is based. The total number of months of data including these hypothetical data becomes $n + m - p + 1 = m - 32$ and as long as $m - 32 > 0$, the resulting 'likelihood', which becomes in fact the posterior density function of the unknown Σ, is proper, i.e. integrable. Thus we look below at the implications of using $m = 33$, 36, 42 and 48.

Even the minimally acceptable value of $m = 33$ implies an improbably high degree of prior certainty on the part of the authors, although it may well be realistic for someone with greater substantive knowledge. Therefore, the results of this section must be treated as a mere indication of the quality of data provided by the various stations to be taken in conjunction with other evidence in any applications which might be made. An alternative approach, which decomposes these stations into homogeneous subgroups of less than 48 stations wherein more realistic choices of m will be acceptable, will be used in subsequent work.

At the same time our lack of substantive knowledge led us to choose a prior covariance of the intra-class correlation form, that is, we assumed that the distribution of concentration levels had the same variance from station to station, and that the correlation between any pair of stations was the same as that between any other pair. This prior covariance matrix is conservative in that it does not single out, *a priori*, pairs of stations which have an unusually high correlation relative to other pairs, so that one of its members would be potentially redundant, and a candidate for elimination. There are alternative approaches like that suggested in a different context by Professor Paul Switzer in his presentation at the Statistics Earth and Space Sciences Conference held in Leuven, Belgium, August 1989, that of adopting and fitting a parametric model for the covariance which declines with increasing spatial separation. We would rule out such approaches because of the need to be conservative, and more importantly, because we do not have any *a priori* basis for making such a strong assumption.

The *a priori* covariance matrix involves only two parameters, the common variance and inter-station correlation. These were estimated from the data in the obvious way. As indicated in the last section, the prior mean is estimated by the sample mean.

For each choice of m indicated above, we computed for each successive $g = 80, \ldots, 1$, the optimal station to 'remove' and so create a new network, G, with g stations. The single station so identified at each stage is shown in Table 2.2 in terms of its ADS identification number (see Table 2.1 for the identification code, and Figure 2.1 for a map showing the location of the station) and it is the one which, in conjunction with those added to U before it, minimizes $PRED + MODEL$, in our earlier notation. Dropping one station at a time like this leads to a suboptimal design, not the G which among all possible G's of size g minimizes this objective function, our index of residual uncertainty, but the best attainable with the computing power and time available to the authors for this task. The discrepancy between this suboptimal design and the optimal design deserves further study in future work. But in low-dimensional examples, the two designs are in close agreement. At any rate, with $m = 33$ and $g = 80$, there is only one unobserved value corresponding to any given choice of U. It turns out to be (see Table 2.2) Station 065b or Penn State, Pennsylvania (see Table 2.1). This means that of all the 81 stations, Station 065b is the best station to drop from the monitoring network if it is to be reduced by 1 to just 80 stations.

Continuing with this example, we anticipate that the expected residual uncertainty about the U-values, given the 80 observed G-values and the model parameters, must be small. It is just 1, to the nearest integer, according to Table 2.3(a), where some of the components of uncertainty are displayed. This is smaller than the successively increasing entries under 'prediction' ($PRED$) in that table for $g = 60$, 40, 20 and 1, which grow because the amount of residual uncertainty about the U-values must increase as the dimension of U increases. Observe in Table 2.3(a) that the amount of model uncertainty deriving from the unknown 'mean' ($MEAN$) is essentially

Table 2.1 Inventory of wet deposition sites, ordered by ADS site identification

ADS site ID	Site name	ADA site ID	Site name
004a	Fayetteville, Arkansas	065b	Penn State, Pennsylvania
007a	Hopland (Ukiah), California	068a	Grand Canyon, Arizona
010a	Rocky Mt. Net Park, Colorado	070a	K-bar, Texas
011a	Manitou, Colorado	071a	Victoria, Texas
012a	Pawnee, Colorado	073a	Horton's Station, Virginia
015a	Bradford Forest, Florida	074a	Olympic Nat. Park, Washington
016a	Everglades Nat. Pa, Florida	075a	Parsons, West Virginia
017a	Georgia Station, Georgia	076a	Trout Lake, Wisconsin
020a	Bondville, Illinois	077a	Spooner, Wisconsin
021a	Argonne, Illinois	078a	Yellowstone, Wyoming
022a	Southern Ill U, Illinois	160a	Alamosa, Colorado
023a	Dixon Springs, Illinois	161a	Salem, Illinois
024a	NIARC, Illinois	163a	Caribou (a), Maine
025a	Indiana Dunes, Indiana	164a	Bridgton, Maine
028a	Elkmont, Tennessee	166a	Fernberg, Minnesota
029a	Mesa Verde, Colorado	168a	Huntington, New York
030a	Greenville Station, Maine	171a	Walker Branch, Tennessee
031a	Douglas Lake, Michigan	172a	American Samoa, American Samoa
032a	Kellogg, Michigan		
033a	Wellston, Michigan	173a	Sand Spring, Colorado
034a	Marcell, Minnesota	249a	Bennington, Vermont
035a	Lamberton, Minnesota	251a	NACL, Massachusetts
036a	Meridian, Mississippi	252a	Ashland, Missouri
037a	Glacier Nat Park, Montana	253a	University Forest, Missouri
038a	Mead, Nebraska	254a	Forest Seed Ctr, Texas
039a	Hubbard Brook, New Hampshire	255a	Newcastle, Wyoming
		257a	Acadia > 11/81, Maine
040a	Aurora, New York	258a	Chassell, Michigan
041a	Chautauqua, New York	268a	Warren ZWSW, Arkansas
046a	Bennett Bridge, New York	271a	Headquarters, Idaho
047a	Jasper, New York	272a	Purdue U Ag Farm, Indiana
049a	Lewiston, North Carolina	273a	Konza Prairie, Kansas
051a	Piedmont Station, North Carolina	275a	Iberia, Louisiana
		277a	East, Massachusetts
052a	Clinton Station, North Carolina	278a	Give Out Morgan, Montana
053a	Finley (A), North Carolina	279a	Bandelier, New Mexico
055a	Delaware, Ohio	280a	CUBA, New Mexico
056a	Caldwell, Ohio	281a	Bull Run, Oregon
058a	Wooster, Ohio	282a	Longview, Texas
059a	Alsea, Oregon	283a	Lake Bubay, Wisconsin
061a	H.J. Andrews, Oregon	285a	Washington Xing, New Jersey
062a	Teddy Roosevelt NP, North Dakota	339a	Bellville, Georgia
		349a	Southeast, Louisiana
063a	Kane, Pennsylvania	350a	Wye, Maryland
064a	Leading Ridge, Pennsylvania	354a	St Mary Ranger St, Montana

Table 2.2 Successive stations to be dropped to reduce the network one at a time to *g* stations

	m					*m*			
g	33	36	42	48	*g*	33	36	42	48
80	065b	065b	065b	078a	40	172a	172a	172a	022a
79	078a	078a	078a	037a	39	046a	022a	022a	172a
78	037a	037a	037a	271a	38	022a	046a	046a	046a
77	271a	271a	271a	065b	37	055a	055a	055a	055a
76	061a	061a	061a	061a	36	272a	272a	272a	272a
75	059a	059a	059a	059a	35	251a	251a	251a	251a
74	173a	173a	173a	173a	34	282a	282a	282a	024a
73	166a	166a	010a	010a	33	051a	029a	029a	029a
72	010a	010a	166a	166a	32	029a	051a	024a	282a
71	011a	011a	011a	011a	31	024a	024a	051a	051a
70	349a	349a	349a	349a	30	028a	028a	028a	028a
69	354a	354a	354a	354a	29	163a	163a	163a	163a
68	004a	004a	004a	004a	28	038a	038a	038a	038a
67	030a	030a	030a	030a	27	075a	075a	063a	063a
66	257a	257a	257a	257a	26	063a	063a	075a	075a
65	258a	258a	275a	275a	25	350a	350a	350a	254a
64	275a	275a	258a	036a	24	074a	074a	254a	350a
63	036a	036a	036a	258a	23	254a	254a	074a	074a
62	252a	252a	252a	252a	22	249a	249a	277a	277a
61	164a	068a	068a	068a	21	277a	277a	035a	035a
60	068a	020a	020a	020a	20	035a	035a	031a	031a
59	020a	039a	039a	034a	19	031a	031a	249a	171a
58	034a	034a	034a	039a	18	171a	171a	171a	249a
57	077a	077a	077a	077a	17	040a	040a	040a	040a
56	039a	164a	164a	253a	16	058a	058a	058a	058a
55	253a	253a	253a	164a	15	168a	168a	168a	168a
54	047a	052a	052a	052a	14	056a	056a	056a	056a
53	052a	047a	047a	047a	13	161a	161a	161a	161a
52	025a	025a	280a	280a	12	255a	255a	255a	255a
51	073a	273a	025a	025a	11	070a	070a	070a	070a
50	273a	280a	273a	273a	10	053a	053a	053a	053a
49	280a	073a	076a	076a	9	017a	017a	017a	017a
48	076a	076a	073a	023a	8	032a	032a	032a	032a
47	023a	023a	023a	073a	7	041a	041a	041a	041a
46	033a	033a	033a	033a	6	160a	160a	160a	160a
45	283a	283a	278a	278a	5	021a	021a	021a	021a
44	049a	278a	283a	283a	4	268a	268a	268a	268a
43	278a	049a	049a	049a	3	285a	285a	285a	285a
42	281a	281a	281a	281a	2	012a	012a	012a	012a
41	279a	279a	279a	279a	1	064a	064a	064a	064a

Table 2.3 The components of uncertainty for various levels, m, of prior precision and optimal network sizes, g

(a) $m = 33$ and total uncertainty is 127·03

g	*PRED*	*MEAN*	*COV*	*MEAS*
80	0·93	−15·01	−89·39	230·50
60	23·75	−14·81	−69·47	187·57
40	51·38	−14·61	−47·17	137·43
20	86·72	−14·41	−23·16	77·88
1	138·49	−14·22	−2·44	5·20

(b) $m = 36$ and total uncertainty is −321·52

g	*PRED*	*MEAN*	*COV*	*MEAS*
80	0·99	−19·18	−598·35	295·03
60	24·20	−18·98	−551·81	225·06
40	52·01	−18·78	−508·97	154·22
20	86·98	−18·58	−471·11	81·19
1	135·50	−18·39	−443·52	4·88

(c) $m = 42$ and total uncertainty is −815·19

g	*PRED*	*MEAN*	*COV*	*MEAS*
80	1·07	−22·84	−1089·59	296·16
60	24·94	−22·63	−1045·10	227·60
40	53·02	−22·43	−1003·15	157·36
20	87·40	−22·23	−964·99	84·63
1	132·38	−22·04	−931·46	5·93

(d) $m = 48$ and total uncertainty is −1160·13

g	*PRED*	*MEAN*	*COV*	*MEAS*
80	1·12	−24·95	−1442·71	306·41
60	25·53	−24·75	−1397·02	236·12
40	53·81	−24·55	−1354·23	164·84
20	87·72	−24·34	−1314·37	90·87
1	130·51	−24·15	−1273·77	7·29

independent of g. This is somewhat curious. Intuitively, uncertainty should grow as the amount of information conveyed by G is reduced, as in the case of the 'covariance' (*COV*) displayed in Table 2.3(a). This increase in uncertainty is especially notable in Table 2.3(a), where m has its minimally acceptable value of 33. In this case, the data to be collected through G will add substantially to the minimal amount of knowledge about the model already available through the 33 hypothetical prior months of collection and the 48 months of real data already in hand. However, in Tables 2.3(b)–(d), the choice of g has comparatively little effect on the index of uncertainty about the 'covariance' as might be anticipated on intuitive grounds.

The column headed *MEAS* for 'measurement' in Tables 2.3(a)–(d) represents the amount of uncertainty that will, prospectively, be eliminated by observing G. It is found by subtracting from the total uncertainty the sum of the columns headed 'prediction', 'mean', and 'covariance'. It decreases because as g decreases, less uncertainty is being 'measured out'. What is reassuring to the authors is the fact that the uncertainty ascribed to measurement is qualitatively the same for all of the various values of m considered. To put this differently, it would seem the last column in Table 2.3 is fairly robust against the choice of m. A surprising result can be seen in the 'measurement' column of Table 2.3, namely that the decline in uncertainty is approximately linear, the reduction per station in Table 2.3(a) being 2·15, 2·51, 2·98, and 3·82 respectively, for $60 < g \leqslant 80$, $40 < g \leqslant 60$, $20 < g \leqslant 40$, and $1 < g \leqslant 20$. As might be expected, the 'marginal' stations, the first to be eliminated, contributed less to the network than those eliminated last. However, there is no transition point in $1 < g \leqslant 80$ at which the quality of information provided by this network about the concentrations of sulfate declined abruptly.

As we have already argued in this section, our conservatively chosen prior covariance matrix will have favored marginal stations in the network. Indeed our choice would have helped to induce this observed decline, the bigger the m, the more nearly linear the decline. In the extreme case where m is very large, a linear down trend must emerge because, with our prior covariance now heavily weighted over that of the data and the network essentially homogeneous, entropy will be approximately symmetric in the various stations; they will contribute equally to the overall uncertainty, in other words. With more prior knowledge we could have made a more decisive choice of a prior covariance matrix and a more accurate evaluation of the relative quality of the data emerging from the various stations in the network. An advantage of the approach we have taken is that our model learns from additional data and in time, our chosen prior covariance matrix would be 'corrected' for any misinformation or lack of information which went into its formulation.

2.5 Network design policy and implementation

The adoption of a network design criterion involving the reduction of uncertainties represented by entropies has practical implications both in network design policy and in implementation. In this section, we will further contrast some of these implications for the two alternative entropy-based criteria descibed in Section 2.2.

2.5.1 Design policy

The criterion previously proposed in Caselton and Zidek (1984) is based on maximizing the reduction in uncertainty in U based on the measurements of G; this is equivalent to maximizing the mutual information $I(U; G)$. In effect, this criterion designs for a single purpose, the prediction of U from G, using the network data.

The criterion used in this paper embraces two network purposes: the prediction of events at unmonitored locations, and the estimation of the parameters of the multivariate distribution of the random vector of measurements for all locations. This network design objective addresses both purposes by minimizing the sum of the uncertainties confronting each of them, as measured by their entropies. These are, for any given choice of G, the residual uncertainty in U after its prediction based on G, and the uncertainty in the parameters of the distribution of X after their estimation based on G.

Minimizing the sum of these two uncertainties also reflects a third and equally important purpose of the network. It is equivalent to maximizing the uncertainty which is resolved only at the locations included in G by monitoring events at those locations. That is, it is equivalent to maximizing the uncertainty component we have labelled *MEAS*, which depends only on the characteristics of events at locations in G. While not without considerable appeal, it would also be reasonable to conclude that maximizing *MEAS* was too narrow to be the only objective as it appears to ignore all consideration of U. It was partly this reaction that led us from an original criterion based on maximizing an entropy solely determined by the choice of G (Caselton and Husain 1980) to the maximum $I(U; G)$ criterion (Caselton and Zidek, 1984). Our recognition that maximizing *MEAS* is logically equivalent to minimizing *PRED* + *MODEL* has substantially revised our view of this criterion.

Optimizing design for more than one purpose raises the issue of the relative weights which might be given to the individual purposes. It is not clear what the relative weights of the *PRED* and *MODEL* uncertainties should be other than the unit weights which arise naturally from the decomposition of the total uncertainty, as presented in Section 2.3. This question is resolved, we feel, by the equivalence of minimization of *PRED* + *MODEL* to the maximization of *MEAS*, the third component of uncertainty. If other than equal weights were adopted for *PRED* and *MODEL*, then *MEAS* would no longer be maximized, and the network would therefore be deficient in that respect. In the face of the many unspecified future uses of the network data, an issue which was mentioned in the Introduction, there is considerable appeal to a criterion which optimizes design for as many sensible purposes as can be accommodated.

While seeking the best prediction performance of U from G through the maximum $I(U; G)$ criterion still has some relevance, it is clearly incomplete in the context of general-purpose network design. It fails to recognize the direct benefits inherent in the measurements of G, and judges them only in the performance of a prediction process dependent upon them. It would, however, be an appropriate criterion for selecting those stations of an existing network which should be used in the estimation of events at an individual location in U. And it avoids the somewhat controversial issue of choosing a reference measure. Moreover, as noted earlier, it has a foundation in utility theory.

Implementation of the *PRED* + *MODEL* based criterion ultimately focuses exclusively on events at the network stations, either in the multivariate normal case of minimizing $|\hat{\Psi}_{GG}|$ or, more generally, in equivalently maximizing *MEAS*. It would therefore also be the easiest criterion to defend when only the network measurements themselves are considered to have any worth. This could arise, for example, where environmental measurements have the primary purpose of ensuring that industrial activities are conforming to legislated standards and estimates of environmental consequences would not be entertained in any litigation to determine otherwise.

2.5.2 Implementation

In the following we will consider the ramifications of the two criteria discussed above, the maximization of $I(U; G)$ and the minimization of $PRED + MODEL$, solely from the implementation standpoint. Three commonly occurring network design situations will be discussed: the initiation of a new network, the expansion of an existing network, and the termination of stations in an existing network.

(a) Initiation of a new network

The initiation of a monitoring network often occurs under conditions where no prior data have been collected in any substantial and systematic fashion in the subject region. Network design must then be based entirely on an *a priori* knowledge of the region. Often this knowledge is limited, and largely subjective. This effectively precludes any rational design plan. Under these circumstances, the adoption of monitoring locations on a uniform grid would often appear to be the most appropriate choice. In practice, this approach lacks authority, and is easily over-ridden by more pragmatic considerations of accessibility, security, and the desire to minimize operating costs or maximize the number of monitoring stations under a fixed budget. In the past, this has frequently resulted in environmental networks with monitoring locations which correlate more with population density than any environmental factors.

When, as would typically be the case, prior knowledge about a spatial field is vague, an objective and rational basis for the design of a new monitoring network clearly requires that some data be collected prior to committing to a permanent network design. Unless remotely sensed data are feasible, such as from satellites, these data could only be obtained through the use of a network of temporary monitoring stations. This is not a traditional precursor of permanent network design and would incur significant costs. It is therefore one of the issues considered in the following discussion of implementation. Furthermore it will be useful to draw a distinction between locations which are potential sites for permanent monitoring stations and locations which would only be monitored temporarily to obtain better statistical information on events throughout the region in support of the design process. For example, a situation might arise where a five-station permanent network is being designed. An adequate assessment of the character of events in the region might be judged to require temporary monitoring at a total of say 40 locations dispersed throughout the region. For reasons of access, ownership, winter conditions, etc. only 12 of these locations could be considered to be potential sites for permanent monitoring stations.

The factors of cost and limited lead times would inevitably limit the time of operation for the temporary network. The merging of both prior knowledge and the more objective temporary network data, through the Bayesian scheme described in Section 2.3, makes an important contribution to achieving an adequate statistical description of regional environmental phenomena with the shortest possible duration for the temporary network.

The two alternative entropy criteria under consideration are:

(1) *Maximize* $I(U; G)$. Implementation of this criterion places heavy demands on the temporary network. It requires data on events at all locations both in U and in G in order to establish $f(U)$ and $f(U \mid G)$. The temporary monitoring must therefore be extensive to provide a full assessment of U, even if this includes many locations which would not be considered potential permanent monitoring sites.

The value of maximum $I(U; G)$ will initially increase with the number of network stations but, after reaching a peak value, will subsequently decline as the number of locations in G begins to dominate the number in U. The criterion is perhaps meaningful for design purposes only when the desired size of network g is substantially smaller than the size of temporary network p. If no constraint is imposed on the choice of locations at any network size, then the curve shown in Figure 2.2 is symmetrical. No indication as to the optimal number of stations in the network is provided by this curve.

(2) *Minimize PRED + MODEL.* This criterion places the least demand on the temporary network when it is used to design a permanent network with a prescribed number of stations. As we have demonstrated for the multivariate normal case, the equivalent criterion under these conditions is to maximize $|\hat{\Psi}_{GG}|$. The criterion need only be evaluated for subsets of locations drawn from the set of candidate locations. Temporary monitoring at other than candidate locations would contribute nothing to the network design process, although the supporting theory explicitly considers the prediction of events at ungauged locations.

If the *PRED + MODEL* criterion is used to make comparisons between optimal networks of different sizes then certain additional terms, detailed in Section 2.3, must be computed. The estimates for the values of these additional terms would be improved with extensive monitoring

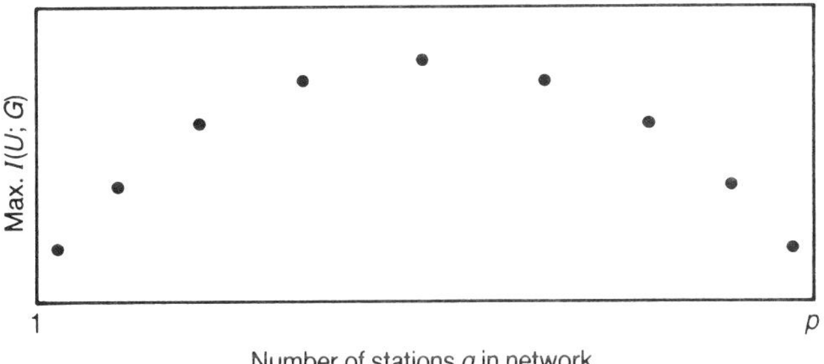

Figure 2.2

at other than candidate locations. The computational advantage is not entirely lost since these additional terms are invariant over all choices of network for a given number of stations.

As the number of monitored stations is increased, the minimum $PRED + MODEL$ component of uncertainty would decline while the maximum $MEAS$ component would grow monotonically as indicated in Figure 2.3.

The issue of how many monitoring stations are 'necessary' would be resolved if, for example, a minimum reduction in the $PRED + MODEL$ component of uncertainty for the last location added could be specified. The practical significance of a specified minimum reduction in uncertainty is not clear. It might, however, provide a reasonable basis for 'balancing' the effectiveness of monitoring expenditures across a number of regional networks operated for the same purpose.

As we have stated, an equivalent approach to the minimizing of $PRED + MODEL$ approach is to maximize $MEAS$ directly. But, as is indicated in the concluding portion of Section 2.2, this involves the estimation of distributions like $f(G \mid D)$. In practice, this would necessitate obtaining replicate measurements from the temporary network to quantify the measurement error statistics – a requirement which is rarely met even in well-funded permanent networks.

(b) Expansion of an existing network

This situation is very similar to the initiation of a network except that the first group of locations is predetermined by the existing stations. The minimum $PRED + MODEL$ criterion is again favored over maximum $I(U; G)$ for the same reasons given under network initiation.

There would be a substantial imbalance in the quality of information at existing and potential new locations in this case. The existing network locations will inevitably have a longer history than that provided by temporary monitoring at potential new locations.

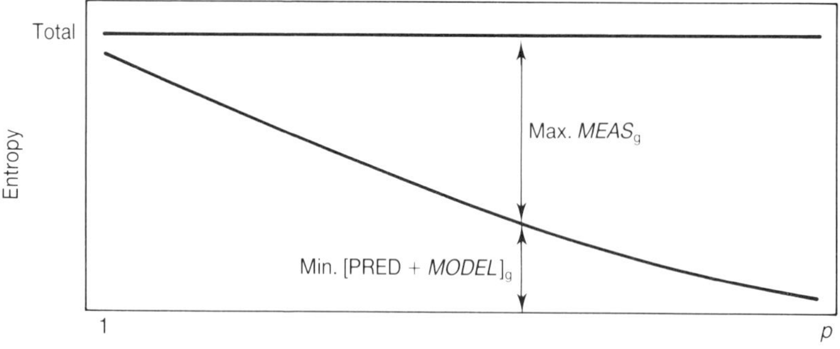

Figure 2.3

(c) Termination of some stations in an existing network

This is the problem presented in our case study. It can be approached in the same way as the network initiation problem except that the existing network stations already provide history for all locations which need to be considered. This should, of course, result in less dependence on subjective prior information. But, as we have seen in the case of the acid rain network, with a large number of existing stations, the historic record can still be inadequate to form proper posterior distributions for the parameters.

(1) *Maximize $I(U; G)$.* It is unlikely in a situation which has led to reducing the number of permanent network stations that temporary monitoring of any other locations would be entertained. Thus, through practical necessity, X would include only the existing network locations. The application of the maximum $I(U; G)$ criterion then leads to questionable results. For example, the station which would be the optimal choice if only a single station were retained would also be the optimal choice if just one station were to be discontinued.

(2) *Mimimize $PRED + MODEL$.* The minimum $PRED + MODEL$ criterion can be applied without difficulty as all previously monitored locations in the permanent network are now considered as candidate locations for the new reduced network.

Termination of stations does not necessarily occur only after the network has matured in a statistical sense. The *MODEL* uncertainty will be far from fully resolved as we have found in the acid rain monitoring experience. And the *PRED* component is not diminished in any way, but will presumably increase in importance with less stations. Stations which upon termination would result in the smallest increase in residual $PRED + MODEL$ uncertainties are favored. This would appear to be a sensible approach to revising the network design.

2.6 Conclusion

The entropy-based approach to network design presented here gives an operational meaning to network purpose, as well as to data quality, when the network will serve diverse and often unforeseen needs in the future.

The illustrative example results show promise when applied to a very large continental acid rain network with limited data. The theory is now being implemented with smaller groups of stations, and for a variety of acid rain ions.

The methodology relies on the spatial structure of the acid rain field as the main determinant of relative quality of the stations in the network. Our analysis should be extended to include cases, unlike that considered here, in which a significant temporal structure is also present, to determine its impact on the ranking of the stations.

References

Anderson, T.W. (1984). *An Introduction to Multivariate Statistical Analysis*, Wiley, New York.

Bartlett, M.S. (1933). On the theory of statistical regression. *Proc. Roy. Soc. Edinburgh*, **53**, 260–83.

Bernardo, J.M. (1979). Expected information as expected utility. *Ann. Statist.*, **7**(3), 686–90.

Caselton, W.F. and Husain, T. (1980). Hydrologic networks: information transmission. *J. Water Resour. Planning and Management Division, ASCE*, **106**(WR2), 503–20.

Caselton, W.F. and Zidek J.V. (1984). Optimal monitoring network designs. *Statist. & Prob. Lett.*, **2**, 223–7.

Dempster, A.P. (1969). *Elements of Continuous Multivariate Analysis*, Addison-Wesley, London, p. 296.

Egbert, G.D. and Lettenmaier, D.P. (1986). Stochastic modelling of the space–time structure of atmospheric chemical deposition. *Water Resour. Res.*, **22**, 165–79.

Eynon, B.P. and Switzer, P. (1983). The variability of rainfall acidity. *Canad. J. Statist.*, **11**, 11–24.

Fedorov, V. and Mueller, W. (1989). Comparison of two approaches in the optimal design of an observation network, *Statistics*, **20**, 339–51.

Fedorov, V., Leonov, S. and Pitovranov, S. (1987). Experimental design technique in the optimization of a monitoring network. In *Model-oriented Data Analysis*, Proceedings, Eisenach, GDR, 1987, Fedorov, V. and Lauter, H. (eds), Springer Verlag, New York.

Jaynes, E.T. (1963). Information theory and statistical mechanics, *Statistical Physics*, Vol. 3, Ford, K.W. (ed), Benjamin, New York, pp. 102–218.

Jaynes, E.T. (1968). Prior probabilities, *IEEE Trans. Syst. Sci. and Cyber.*, SSC-4, 227–41.

Muirhead, R.J. (1982). *Aspects of Multivariate Statistical Theory*, Wiley, New York.

Theil, H. and Fiebig, D.G. (1984). Exploiting continuity: maximum entropy estimation of continuous distributions, Series on *Economics and Management Science*, Vol. 1, Cooper, W.W. and Theil, H. (eds), Ballinger, Cambridge.

Wu, S. and Zidek, J.V. (1989). *Recent Trends in the Chemistry of Precipitation in the United States. Part 2: Concentrations of Sulfate.* SIMS Technical Report, No. 130, University of British Columbia, Department of Statistics.

Zidek, J.V. (1984). Detailed statistical approach to sediment chemistry monitoring. Appendix B of *Beaufort Sea Monitoring Program Workshop Synthesis and Sampling Design Recommendations.* Report to National Oceanic and Atmospheric Administration, National Ocean Systems, Juneau, Alaska, prepared by Dames and Moore, SEAMOcean, Inc., and University of Washington, Department of Statistics.

Chapter 3

Nonparametric estimation of spatial covariance with application to monitoring network evaluation

P. Guttorp, P. D. Sampson and K. Newman

We discuss the evaluation of monitoring networks for spatially dependent data. Candidate sites to remove from an existing network may be defined as those that are best predicted by the rest of the network, or as those which provide the least (additional) information about the spatial process of interest at a dense network of potential additional monitoring sites. We also consider which sites to replace or add from this potential dense network by maximizing the expected information provided by the additional sites. This maximization problem, and approximations to it, depend explicitly on the spatial covariance. Given measurements at preliminary stations we estimate the covariance structure for the area of interest using a nonparametric estimator developed specifically with monitoring data in mind. This estimator does not require assumptions of stationarity or isotropy for networks monitored over time. An application to evaluation of a precipitation chemistry network is discussed.

3.1 Introduction

Long-term monitoring of lakes, streams, and forests is a crucial tool in assessing the effect of atmospheric pollutants and other environmental insults. Monitoring networks must be well designed, so as to provide adequate spatial coverage, reasonable temporal resolution, and measurements on a variety of biologically and chemically significant variables. Over the last decade there have been several implementations of monitoring networks aimed at assessing acidity of precipitation in North America. Because very few stations have been operating longer than ten years, it is very difficult to assess trends in the data; yet trends are what policymakers are most interested in. Sound statistical design and analysis techniques can reduce the variability in estimates based on monitoring data by including meteorological covariates, co-located samplers, and well-documented laboratory techniques. However, the remaining variability is often still too large to find trends (Stensland *et al.*, 1986, Pollack *et al.*, 1989).

A key feature of a long-term environmental monitoring network is that it will likely be used to assess a variety of environmental problems. Therefore it is important to include components that assess ecological balance (such as measuring changes in species composition among phytoplankton in lakes; Guttorp, 1990) in addition to chemical analyses of precipitation, soil, and water. The former component need not be temporally intensive relative to the latter from the point of view of precision. However, the location of measuring stations, particularly for the temporally intensive measurements, must be carefully considered.

Many design criteria address the problem of detecting sudden changes in some measured variable (Green, 1979, Millard *et al.*, 1985). These are usually obtained by maximizing the power of a test of the hypothesis of no change. The most common design criterion associated with the kriging technique in geostatistics (Cressie *et al.*, 1990) aims to minimize prediction variability at a fixed, unobserved location, or to minimize the maximum prediction variability among a group of locations. Another design perspective seeks to optimize the estimate of the variogram (Warrick and Myers, 1989) for kriging purposes. Haas (1990) combines these objectives in a composite objective function. The most likely (or most important) effects for which long-term monitoring networks are generally designed are not of the sudden change type, and they are likely to influence a large area subtly, rather than be distinctly observable at a single location. Therefore, a compromise solution between different design criteria is useful. Caselton and Zidek (1984) proposed designing networks that maximize the information (in the sense of Shannon, 1948) expected over the network. To implement this, one must specify a number of potential sites, and determine the spatial covariance among these sites for the measures of interest. The optimal design of size n is the set of n stations that maximize the expected information about the unmonitored potential sites.

The estimation of spatial covariance from monitoring data has lagged far behind the development of interpolation methods, which usually depend on knowing the covariance structure for their implementation (Sampson and Guttorp, 1990; see also Chapter 4). Most techniques necessarily assume stationarity and isotropy in order to make efficient use of limited amounts of data. In the monitoring situation, as opposed to the geological applications for which kriging was developed, there are usually time series of observations at each site. This allows relaxation of the severe (and often unwarranted) assumptions of stationarity and isotropy.

The existence of monitoring data for a subset of the potential sites is sufficient to compute estimates of spatial covariance among unmonitored sites using the technique of Sampson and Guttorp (1990). The method of Chapter 4 could also be used, although it appears less general (and, to us, less straightforward). The focus of our method is actually the spatial covariance structure as manifest in var $(Z(\mathbf{x}_i, t) - Z(\mathbf{x}_j, t))$ for observations at geographic locations $\mathbf{x}_i$ and $\mathbf{x}_j$ at time t. We call this the *spatial dispersion* function. The basic idea is to deform the geographic region into a two- or three-dimensional image wherein spatial dispersion is a simple monotonically increasing function of distance. That is, in the deformed image of the geographic region, spatial dispersions are *stationary* and *isotropic*, and therefore relatively straightforward to model and estimate (as in common variogram estimation). A smooth function maps geographic coordinates to transformed coordinates.

In this paper we apply the Caselton–Zidek design approach, using the Sampson–Guttorp covariance estimation technique, to the problem of deciding which of the stations in a given network can most usefully be deleted or replaced – in the sense of providing the least information about a potentially dense network of additional sites – and finding the sites among a set of potential sites which would most enhance the network in terms of information. We first discuss the network design and evaluation method in Section 3.2, then the covariance estimation technique in Section 3.3, and apply these to the UAPSP network of monitoring stations in Section 3.4. The final section contains discussion and extensions.

3.2 Information transmission network design

The future benefits that may be derived from a network cannot be specified in advance. Even when a network is designed with a particular objective in mind, it is quite common that answers to very different questions must be elicited from the data once the network is operational. Caselton and Zidek (1984) suggest circumventing these difficulties by an approach which may be suboptimal in specific cases, but has overall merits for these types of networks. For a more detailed description, and some critical discussion, see Chapter 2.

We let **Z** denote a random field of measurable quantities indexed by potential site labels i. We decompose **Z** into the gauged sites $G = \{Z_i, i \in D\}$ and the ungauged sites U. The choice of design D will be made to maximize the amount of information in G about U. Here the information measure is taken to be Shannon's index of information transmission $I(U, G) = \mathbf{E}(\log (f(U \mid G)/f(U)))$, where $f(U \mid G)$ is the conditional density of U given G, and $f(U)$ the *a priori* density of U. Bernardo (1979) shows that this criterion is unique under some appropriate regularity conditions (cf. Lindley, 1956). A simple special case is when the random field is multivariate normal; then $I(U, G) = -\frac{1}{2}\log|I - R|$, where I is the identity matrix, and R the diagonal matrix with elements the squared canonical correlation coefficients between U and G. These can be obtained from estimates of the spatial covariances by diagonalizing the matrix

$$\Sigma_{UU}^{-1/2}\Sigma_{UG}\Sigma_{GG}^{-1}\Sigma_{GU}\Sigma_{UU}^{-1/2}$$

where Σ_{xy} denotes the covariance matrix between x and y. In the case of a known mean field, and when U consists of a single site, this is a simple function of the kriging variance for prediction at the ungauged station U (cf. Cressie, 1990). However, the Caselton–Zidek design criterion is more general than this simplest of kriging design criteria since it allows for optimization with respect to a large set of ungauged sites. In fact, for multiple ungauged sites Caselton and Zidek note that $I(U, G)$ also represents the generalized variance of the normalized prediction residuals at the ungauged sites.

For particular patterns of the covariance matrix of **Z** derived from random effects models of acidic deposition (such as that used by Vong *et al.*, 1988), it is possible to develop workable approximations to the canonical correlations and solve the design problem in terms of signal-to-noise ratios at gauged and ungauged sites, respectively. The analysis suggests the importance of replicate measurements at gauged sites (Caselton and Zidek, 1984; Guttorp *et al.*, 1987, Section 3.3).

In order to eliminate a site from an existing network there are two possible scenarios. If no other data exist – for example if the network was a pilot network in an area not previously covered by this type of measurement – one might simply remove the site with the highest information about the remaining stations (i.e. the station best predicted by those remaining). On the other hand, if inference about a larger (dense) network of sites is of interest, one can instead maximize the information from $n-1$ sites in the current network, viewed as the gauged stations G, about all the relevant sites (typically those in the same geographic region) in the larger network, viewed as the ungauged stations, U.

Adding a site to an existing network will be subject to similar considerations. If there is no alternative network, a suitable additional site should be based on a contour plot of estimated information values – which will be equivalent to a contour plot of standard errors of prediction. The site should be located in an area of low information, computed from the current network viewed as gauged sites, and any point of interest viewed as a single ungauged site. It is usually convenient to compute these information numbers using a regular grid of points, unless one is only interested in a small number of alternative sites.

We also consider the case where sites from an alternative network are available, and the goal is to replace a site in the current network with a site from the alternative network. A reasonable strategy is to compute the information from the current network, with one station replaced by a station from the other network, about the remainder of the other network. These computations are done for all possible replacements, and that with the highest information is chosen.

3.3 Nonparametric estimation of spatial covariance

Monitoring data have several characteristic features that are important to consider in defining and estimating spatial covariance. There are repeated measurements at each site. Although these may be temporally correlated, they still contain more information about the underlying random field than the single observation per site that is characteristic of the geological applications of kriging (Matheron, 1971; Journel and Huijbregts, 1978). The assumptions of stationarity and isotropy are common in the geostatistical literature, where they are necessary in order to fit a parametric model of spatial covariance. However, these assumptions are highly suspect in environmental monitoring of mesoscale (10–100 km in extent) regions, where the spatial covariance is usually induced by meteorological and orographic features. On the other hand, the covariance can often be assumed constant over time (within seasons in some applications). In other words, the space–time covariance factors into a product of spatial and temporal components.

Let $\mathbf{x}_1, \ldots, \mathbf{x}_n$ be the locations of monitoring stations, and $(\mathbf{z}_1, \ldots, \mathbf{z}_n)$ the time series of observations at each station. We assume that the $\mathbf{z}_i$ are observations of a spatial random field $Z(\mathbf{x})$ which can be written

$$Z(\mathbf{x}) = Z^*(\mathbf{x}) + E(\mathbf{x})$$

where $Z^*(\mathbf{x})$ is a random field with covariance function $c^*(\mathbf{x}, \mathbf{y}) = \text{cov}\,(Z^*(\mathbf{x}), Z^*(\mathbf{y}))$, and $E(\mathbf{x})$ is an error process, spatially uncorrelated, and assumed independent of Z^*. $E(\mathbf{x})$ usually represents measurement error and small-scale spatial variation. The covariance function of the observed process is then

$$c(\mathbf{x}, \mathbf{y}) = \text{cov}\,(Z(\mathbf{x}), Z(\mathbf{y})) = c^*(\mathbf{x}, \mathbf{y}) + \delta_{\mathbf{xy}}\sigma^2_{\mathbf{x}}$$

where $\delta_{\mathbf{xy}}$ is the Kronecker delta. In what follows we will consider spatial *dispersion*

$$D^2(\mathbf{x}, \mathbf{y}) = \text{var}\,(Z(\mathbf{x}) - Z(\mathbf{y})) = c(\mathbf{x}, \mathbf{x}) + c(\mathbf{y}, \mathbf{y}) - 2c(\mathbf{x}, \mathbf{y}).$$

In the geostatistical literature this is called the *variogram*. We are avoiding that term because of its connotation of stationarity and isotropy assumptions. The quantities $D^2_{ij} = D^2(\mathbf{x}_i, \mathbf{x}_j)$ can be estimated by the empirical dispersions

$$d^2_{ij} = \frac{1}{T}\sum_{1}^{T}(z_{it} - z_{jt})^2$$

(after first centering the z_{it}, i.e. subtracting out spatial and temporal trends from the observed Z process). Because of missing data, we chose to compute the d^2_{ij} from empirical covariances c_{ij} computed from all pairwise complete data.

The first step in estimating $D^2(\mathbf{x}, \mathbf{y})$ is to transform the geographic map of the monitoring stations – the 'G-plane' – to a map (possibly in higher dimension than two) where dispersion increases (approximately) monotonically with distance – the 'D-image'. This is accomplished using nonmetric multidimensional scaling (MDS). The D-image configuration of stations is chosen so that interpoint distances h_{ij} minimize the stress measure

$$\min_{\delta} \frac{\sum_{i<j}(\delta(d^2_{ij}) - h_{ij})^2}{\sum_{i<j} h^2_{ij}}$$

where the minimum is taken over all monotone functions δ, so that $\delta(d^2_{ij})$ represents a least-squares monotone regression of the h_{ij} on the d^2_{ij} (Mardia *et al.*, 1979). In this application we have introduced weights inversely proportional to geographic distance in the summations of the stress formula in accord with the prior assumption that the spatial dispersions d^2_{ij} and their standard errors increase with spatial separation, suggesting that dispersions for distant stations should be down-weighted.

A plot of MDS distances h_{ij} versus dispersions d_{ij}^2 indicates the quality of this spatial model. (See Figure 3.4 below.) To this plot we fit a general model for stationary and isotropic dispersions (a variogram model), since by construction the measurements, when referred to the D-image rather than the G-plane, correspond to a stationary and isotropic random field. We use a mixture of Gaussian variograms, namely

$$\gamma(h) = \int_0^\infty (1 - \exp(-th^2))\, \mathrm{d}\mu(t)$$

where μ is a finite measure on the positive reals having an atom at the origin to account for measurement error (the nugget effect). We chose these mixtures of Gaussians because they define the class of all legitimate (i.e. conditionally non-positive definite) variograms valid in any dimension; see Sampson and Guttorp, 1990. Results of Böhning (1982) generalize to produce an algorithm for the fitting of this model (Sampson and Guttorp, 1990, Appendix).

The final step is to compute a smooth function f that maps the G-plane into the D-image. For this we use a pair of thin-plate smoothing splines, one for each D-image coordinate, and write $f(x) = x^*$. In the limit, as the spline smoothing parameter gets arbitrarily large, the mapping f is linear and the resulting spatial dispersion model becomes stationary. By random subsampling, or by cross-validation to an independent network, it is possible to estimate the appropriate amount of smoothing.

The composition of the computed smoothing spline f and the fitted variogram γ provides our model for the spatial dispersion structure. In order to estimate the dispersion between two points x and y (monitored or unmonitored), one first computes their locations in the D-image, $\mathbf{x}^* = f(\mathbf{x})$ and $\mathbf{y}^* = f(\mathbf{y})$, and then estimates their dispersion by $\gamma(|\mathbf{x}^* - \mathbf{y}^*|)$. Cross-validation or resampling methods can also be used to estimate the accuracy of the disperion estimates.

3.4 An application

Since 1979 a network funded by electric utilities has been collecting and analyzing daily precipitation samples. The Electric Power Research Institute (EPRI) operated the Sulfate Regional Experiment (SURE) during 1979 and 1980, and the Utility Acid Precipitation Study Program (UAPSP) network continued on from the beginning of 1981 until the end of 1987. Several stations have been in operation continuously since 1979. At the beginning of 1988 the EPRI Operational Evaluation Network (OEN) replaced the UAPSP network. The location of the UAPSP stations we used is shown in Figure 3.1.

We use only those stations with data for the entire seven-year operational period of UAPSP, and further exclude the Brookings, SD, station, since there is very little deposition there. We analyze four-weekly deposition of hydrogen ions at each of the seventeen resulting stations. The data were

Figure 3.1 Locations of 18 UAPSP sites in the eastern United States with four-weekly acid deposition data available from 1981 through 1987. Crosses (+) mark the locations of 80 sites from NADP monitoring network. Three of these are identified by name.

obtained through the courtesy of Tony Olsen at Batelle's Pacific Northwest Laboratory, which maintains the Acid Deposition System data base.

Quality control limits suggested by EPRI for this network (UAPSP, 1987) requires assessment of four data quality measures: percentage of time the sampler is in operation (>90), per centage of total precipitation attributed to valid events (>70), per centage of valid events (>70), and a measure of collection efficiency (>60 per cent). These criteria result in rejection of 10 per cent of the data (52 observations out of 43 four-weekly observations on each of 17 sites) due to low quality. Pollack *et al.* (1989) present a discussion of sources of variability in the UAPSP data. Cressie *et al.* (1990) and Loader and Switzer (this volume, Chapter 4) have analyzed different subsets of these data.

Preliminary data analysis showed that a logarithmic transformation suitably reduced the skewness and heterogeneity of variance of four-weekly hydrogen ion deposition. We thus analyze $y_{it} = \log(1 + d_{it})$, where d_{it} represents deposition at site i in time interval t. We observed seasonality both in the mean hydrogen ion deposition patterns *and* in the spatial covariance structure of deposition. Figure 3.2 presents a sample of time series plots of log hydrogen ion deposition along with fitted additive combinations of trigonometric seasonal components and long-term smooth trends. These models were fitted, site-by-site, using an iterative backfitting procedure (Hastie and Tibshirani, 1986, Section 7) to fit a two-parameter trigonometric component to residuals from a long-term nonparametric smooth trend, and

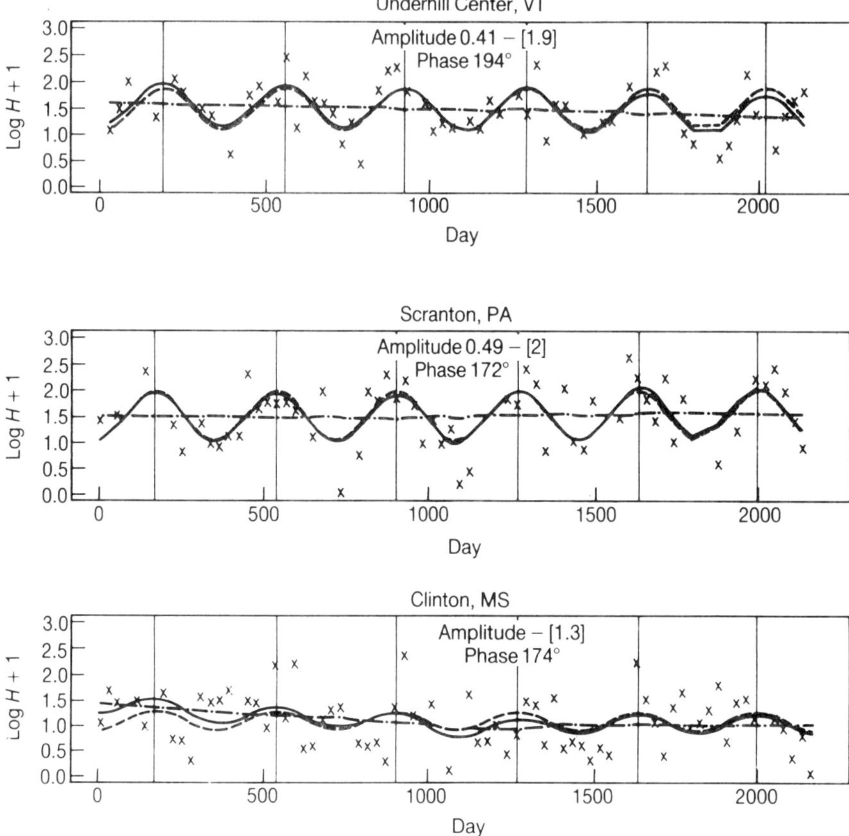

Figure 3.2 Examples of time series plots of log hydrogen ion deposition for three sites from the UAPSP network. Dash-dotted lines represent estimated long-term smooth trends; dotted lines represent a stationary seasonal component; solid lines represent the sum of these seasonal and long-term trends.

to fit the long-term trend with a non-parametric scatterplot smoother applied to residuals from the trigonometric component. Although the trigonometric components are quite significant in most cases, these models clearly leave substantial unexplained variation in hydrogen ion deposition. Generally, the long-term trend (where it is noticeable) is decreasing.

Residuals from these fitted trends, $z_{it} = y_{it} - s_i(t)$, had uniformly greater variances and spatial dispersions during the summer months (April–September) than during the winter months. The following analysis is based only on the spatial covariance structure for the winter months, computed from the deseasonalized, detrended data. Because of missing data, sample sizes for the computation of covariances for station pairs varied from 13 to 42 cases with an average of 30 cases per station pair.

Figure 3.3(a) presents a square grid in the geographic coordinates of the

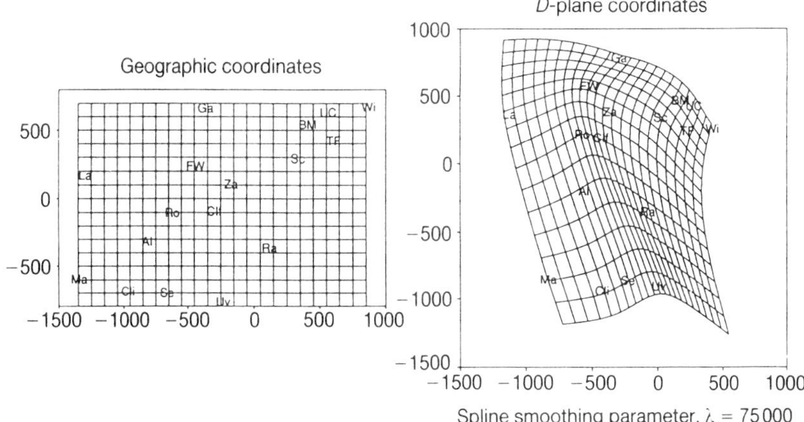

Figure 3.3 Depiction of a thin-plate smoothing spline mapping the geographic representation of the 17 UAPSP sites into their D-plane representation encoding the spatial dispersion structure.

UAPSP stations, while Figure 3.3(b) presents the D-image of the grid. The computed coordinates $\mathbf{x}_i^*$, obtained by applying MDS to the matrix of sample dispersions and then computing smoothing splines to relate the original geographic configuration to the two-dimensional configuration produced by MDS, are also shown in Figure 3.3(b).

The distorted image illustrates the nature of the estimated nonstationary, anisotropic covariance structure. Note that the D-image represents primarily a compression of the east–west axis relative to the north–south axis with some shear of the geographic map compressing the southwest–northeast axis relative to the northwest–southeast axis. Thus, there is generally weaker spatial dispersion (stronger covariance) along the southwest–northeast direction. The deviation of this grid from a regular parallelogram expresses nonstationarity. (Such a regular grid corresponds to a perfectly stationary but anisotropic model. For further discussion, see Sampson and Guttorp, 1990.) The smoothing parameter for this fitted model was chosen by eye to provide a smooth spatial model and refrain from overfitting the observed spatial dispersions.

Figure 3.4(a) plots the empirical spatial dispersions d_{ij}^2 versus geographic distances pictured in Figure 3.3(a), and Figure 3.4(b) plots d_{ij}^2 versus distance h_{ij} represented in the D-image of Figure 3.3(b). The difference in magnitude of the vertical scatters of these two figures demonstrates the improvement in fit of this nonstationary dispersion model compared with a simple stationary isotropic variogram model. Figure 3.4(a) includes a scatterplot smooth (not a legitimate variogram), while Figure 3.4(b) includes the fitted variogram model $\gamma(|h_{ij}|)$ for the D-image representation. In this case it is a simple Gaussian variogram (i.e. the finite measure μ in the expression for the general mixture of Gaussians requires only a single support point). The nugget is estimated at about 0·14. This represents the part of the total variability due

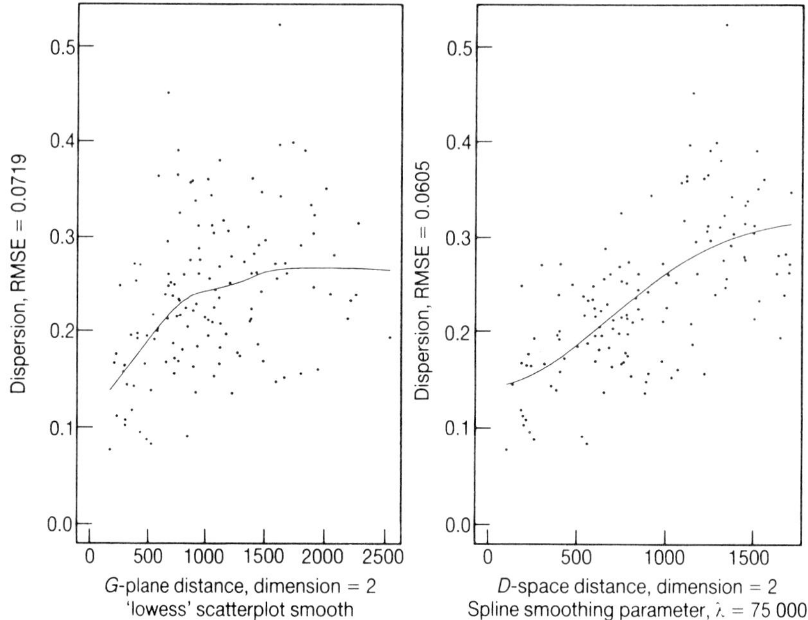

Figure 3.4 Plots of empirical spatial dispersion d_{ij}^2 versus distance in the G-plane (geographic map) and in the D-plane for 17 sites of the UAPSP network. Figure 3.4a includes a simple scatterplot smoother while Figure 3.4b includes a fitted Gaussian variogram.

to measurement error and small scale spatial variability. Data were collected using co-located samplers for 1983–84 (Pollack *et al.*, 1989). The observed variance of the difference in H^+-deposition between the two samplers at the Uvalda, GA site, was 0·051, corresponding to a nugget effect of 0·10, only a little smaller than our estimate which is based on all stations and all years.

Our analysis thus far has produced a model for the spatial dispersion function. To address our design questions we need a model for the spatial covariance function. This we obtain from the spatial dispersion model together with a model for the site variances:

$$\hat{c}_{ij} = \tfrac{1}{2}(\hat{c}_{ii} + \hat{c}_{jj} - \hat{d}_{ij}).$$

For these data there appears to be no significant spatial heterogeneity in the variances of the deseasonalized, detrended log hydrogen deposition. Thus we fitted a constant residual variance, $\hat{c}_{ii} = \hat{c}$ to generate a model for the spatial covariance structure. The resulting fitted covariance matrices are positive-definite.

The purpose of this illustrative analysis is to determine the least important station (in the two senses suggested in Section 3.2). In addition, we show how to replace it with the most informative station among those used in the much denser NADP network (NADP, 1984). The basis of this analysis is the fitted covariance model, $\hat{c}_{ij}$, applied first to the UAPSP network from which it was estimated, and then to all 97 sites in the union of the UAPSP network with 80 sites from the NADP network. We first compute the multiple correlation of each station in the UAPSP network with the remainder of the network. Then we compute all canonical correlations for each subset of 16 out of 17 UAPSP sites with the NADP sites. The corresponding information numbers, for deletion of each UAPSP site, are given in Table 3.1. In addition to the information numbers, estimates of standard errors are given. These were computed by a jackknife procedure, leaving out one four-weekly observation at a time for the entire network and refitting the model. We also jackknifed the standard errors of differences in information relative to the site with the highest information number. Those stations that differ from the highest information station by at least two standard errors are marked with asterisks.

The results were quite similar for the analysis using only the UAPSP sites and for the analysis using the augmented network of UAPSP and NADP sites. In both cases Scranton, PA, was found most expendable.

Cressie *et al.* (1990) found results similar to ours. They used kriging with a quadratic mean function and a spherical variogram applied to two years of annual hydrogen ion deposition from a larger set of UAPSP sites, some of which were subsequently moved or closed down. Their criterion was to minimize the kriging prediction variance when leaving one station out, and resulted in Zanesville, OH, being the most expendable site among those included in our study. The difference between Zanesville and Scranton is relatively small (about 0·2 standard errors) in both our computations.

Table 3.1 Information measures for individual UAPSP stations

Site	Information from UAPSP network only	Standard error	Information from augmented network	Standard error
Scranton, PA	0·254	0·060	2·555	0·788
Zanesville, OH	0·250	0·067	2·551	0·800
Clearfield, KY	0·241	0·064	2·543	0·795
Big Moose, NY	0·238	0·065	2·551	0·798
Turners Falls, MA	0·238	0·058	2·543	0·788
Rockport, IN	0·232	0·068	2·537	0·798
Underhill, VT	0·232	0·063	2·549	0·800
Fort Wayne, IN	0·215	0·057	2·519*	0·784
Selma, AL	0·207	0·068	2·537	0·791
Winterport, ME	0·200	0·063	2·531	0·792
Clinton, MS	0·193	0·061	2·537	0·799
Alamo, TN	0·180*	0·052	2·497*	0·776
Raleigh, NC	0·169	0·057	2·484	0·764
Gaylord, MI	0·167	0·055	2·499	0·751
Uvalda, GA	0·158*	0·049	2·533	0·800
Marshall, TX	0·111*	0·048	2·531	0·790
Lancaster, KA	0·100	0·057	2·485*	0·762

* Sites for which the standardized difference from the highest information site (using jackknifed standard error of difference) is larger than 2·0.

Clearly, these different approaches need not agree in general; differences will depend on the nature of the spatial covariance structure of four-weekly deposition and the distribution of sites in the combined networks. In fact, while naïve approaches are likely to agree on the least important site, typically in the middle of the network, a misspecified covariance model may lead to different conclusions about the regions for which a network provides the least information.

In order to replace a UAPSP site with a more informative one from the NADP network (located within the geographic boundaries of the UAPSP network), we would replace Scranton by Lake Dubay, WI. However, the information calculations show that a number of different substitutions, using Lake Dubay, Big Springs, or Lake Geneva to replace Scranton, Zanesville, Big Moose or Underhill, result in nearly equal values of information. Generally, the UAPSP network has the least information about events in the northwestern part of its coverage, and stations more centrally located are most profitably replaced by NADP sites near the northwestern boundary. It is worth noting that, although the intuitive feeling is correct that stations on the network boundary would be expected to replace stations in the middle of the network, the method proposed in this work points out which part of the network boundary is most uncertain.

Acknowledgements

This work had partial support from the United States Environmental Protection Agency through a cooperative agreement with SIMS and the University of Washington, and from the Electrical Power Research Institute through contract 8010-01.

References

Bernardo, J.M. (1979). Expected information as expected utility. *Ann. Statist.*, **7**, 686–90.

Böhning, D. (1982). Convergence of Simar's algorithm for finding the maximum likelihood estimate of a compound Poisson process. *Ann. Statist.*, **10**, 1006–8.

Caselton, W.F. and Zidek J.V. (1984). Optimal monitoring network designs. *Statist. and Prob. Lett.*, **2**, 223–7.

Cressie, N. (1990). The origins of kriging. *Math. Geol.*, **22**, 239–52.

Cressie, N., Gotway, C.A. and Grondona, M.O. (1990). Spatial prediction from networks. *Chem. Intell. Lab. Sys.*, **7**, 251–71.

Green, R.H. (1979). *Sampling Designs and Statistical Methods for Environmental Biologists*, Wiley, New York.

Guttorp, P. (1990). *Statistical Analysis of Biological Monitoring Data*. SIMS Tech. Rep. No. 147, Dept. of Statistics, University of Washington.

Haas, T.C. (1990). *Using Spatial Covariance Estimation, Estimation Error, or Information Theory to Define Optimality in Monitoring Network Design*. Technical Report, University of Wisconsin, Milwaukee.

Hastie, T. and Tibshirani, R. (1986). Generalized additive models. *Statist. Sci.*, **1**, 297–309.

Journel, A.G. and Huijbregts, C.J. (1978). *Mining Geostatistics*. Academic Press, New York.

Lindley, D.V. (1956). On a measure of information provided by an experiment. *Ann. Math. Statist.*, **27**, 986–1005.

Mardia, K.V., Kent, J.T. and Bibby, J.M. (1979). *Multivariate Analysis*, Academic Press, New York.

Matheron, G. (1971). *The Theory of Regionalized Variables and Its Applications*, Cahiers du Centre de Morphologie Mathématique de Fontainebleau, No. 5. Paris: Ecole Nationale Supérieure des Mines de Paris.

Millard, S.P., Yearsley, J.R., and Lettenmaier, D.P. (1985). Space–time correlation and its effect on methods for detecting aquatic ecological change, *Canad. J. Fish. Aquat. Sci.*, **42**, 1391–1400.

NADP (1984). *Instruction Manual, NADP/NTN Site Selection and Installation*, Natural Resource Ecology Laboratory, Colorado State University, Fort Collins.

Pollack, A.K., Hudischewskyj, A.B., Stoeckenius, T.S., and Guttorp, P. (1989). *Analysis of Variability of UAPSP Precipitation Chemistry Measurements*, Draft Final Report SYSAPP-89/041. Systems Applications, Inc., San Rafael.

Sampson, P.D. and Guttorp, P. (1990). *Nonparametric Estimation of Nonstationary Spatial Covariance Structure*, SIMS Technical Report No. 148, Dept. of Statistics, University of Washington.

Stensland, G.J., Whelpdale, D.M., and Oehlert, G. (1986). Precipitation chemistry. Chapter 5 in *Acid Deposition: Long-Term Trends*, National Academy Press, Washington.

UAPSP (1987). *Quality Assurance Plan for External Audits of the Utility Acid Precipitation Study Program*. UAPSP report 110, Washington.

Warrick, A.W. and Myers, D.E. (1987). Optimization of sampling locations for variogram calculations. *Water Resour. Res.*, **23**, 496–500.

Chapter 4

Spatial covariance estimation for monitoring data

C. Loader and P. Switzer

We discuss the role of spatial covariances particularly with regard to estimates of precision for interpolation. Emphasis is placed on the heterogeneous case where precision can vary locally. We describe procedures for modeling heterogeneous covariances and their estimation from repeated observations at fixed sites. Our methods avoid parametric assumptions that are frequently made for this problem. An example shows the implementation of suggested procedures using monthly rainfall acidity data from a monitoring network in the eastern United States obtained over a two-year period.

4.1 Introduction and summary

Let $\{Z(x); x \in \Xi\}$ be a real-valued spatial field. In our example, $Z(x)$ represents the concentration of a pollutant in rainfall, and Ξ will be a two-dimensional geographic region. We observe Z only at monitored locations $x_1, \ldots, x_p$, for n replications of the process. The replications are denoted by $Z_1(x), \ldots, Z_n(x)$. To motivate the discussion of spatial covariance, suppose we want to estimate or interpolate the values of the random function at unmonitored locations, to estimate the interpolation error or to choose good locations for adding or deleting monitoring locations.

The estimation of interpolation errors, for example, will require some degree of prior specification of properties of the random function. This can be achieved by regarding $Z(x)$ as a realization of a random field and by changing the problem to one of estimation of expected values of interpolation errors or mean square error. It is convenient to use the notation

$$\mu(x, j) = E(Z_j(x)), \qquad x \in \Xi, \quad j = 1, \ldots, n$$

to denote the possibly replication dependent mean function and

$$\sigma(x, y) = E(Z_j(x) - \mu(x, j))(Z_j(y) - \mu(y, j)), \qquad x, y \in \Xi$$

to denote the covariance function of the random field, which we assume to be replication independent. The mean field $\mu(x, j)$ should capture the smooth and regular part of the spatial field $Z_j(x)$, while $\sigma(x, y)$ describes the variability of the residual field. When the interpolation estimates are linear combinations of the monitored values the mean squared errors can be derived explicitly as a function of $\sigma(x, y)$ and the linear estimates used. To estimate the interpolation precision, we first need to estimate the covariance function.

The usual method in geostatistics literature is to specify the covariance function up to a small number of unknown parameters, frequently restricting attention to homogeneous models. For example, a stationary variogram $\gamma(x-y) = E(Z(x) - Z(y))^2$, which does not depend on x, may be used. These assumptions may conceal important spatially specific information. When we have multiple realizations, the need for strong parametric assumptions seems less compelling.

The multiple realizations of the process may exhibit serial correlation, but we will not address this question in detail. Instead, we remove a smooth time trend, and ignore any serial correlation which remains after removal of the trend.

For our example of rainfall acidity, we have 24 monthly values at each of 19 sites for the years 1982 and 1983. For any pair of monitoring sites we can calculate the empirical covariance from the corresponding pair of time series. However, site pairs which are separated by the same spatial vector will inevitably have different empirical covariances, reflecting local properties of the field.

Spatial heterogeneity creates special problems for estimation of interpolation precision because mean interpolation errors depend not only on site-pair covariances but also on covariances between sites and the interpolation points, for which empirical estimates are lacking. Furthermore, the empirical site-pair covariances may themselves be subject to sampling variability when the time series are short. Therefore, some degree of parametric modeling is required which at the same time respects the apparent heterogeneity. Proposals for such modeling are described. Basically, a parametric covariance model is forced on the available empirical covariances and modified covariance estimates are obtained by shrinking the empirical covariances toward the parametric covariances.

Our example exhibits different variability characteristics in different parts of the region, with differences in both the correlation length and the preferred correlation direction.

4.2 Covariance estimation for monitored locations

Let Σ be the $p \times p$ matrix of covariances between the p observed locations $x_1, \ldots, x_p$. In this section we assume the process has known mean $\mu(x)$, which without loss of generality we can take to be 0. The natural estimate of Σ is

$$\hat{\Sigma} = \frac{1}{n} \sum_{j=1}^{n} Z_j Z_j' \tag{4.1}$$

where $Z_j = (Z_j(x_1), \ldots, Z_j(x_p))'$; $1 \leqslant j \leqslant n$. Of course, this tells us nothing about the covariances involving unobserved locations. Furthermore, although we do not wish to make parametric assumptions about $\sigma(x, y)$, it is usually reasonable to make some smoothness assumptions. For example, if the realizations of $Z(x)$ are continuous, the covariance function must also be continuous. Especially if n is small, the estimate $\hat{\Sigma}$ may not reflect this smoothness.

An appealing alternative is to use a shrinkage model, in which the data are allowed to select between a parametric model and the empirical estimate $\hat{\Sigma}$. This suggests an estimate of Σ of the form

$$\hat{\Sigma} = \lambda\hat{\Sigma} + (1-\lambda)C \tag{4.2}$$

for some λ, $0 < \lambda < 1$. The parameter λ is chosen from the data.

To put this in a more formal setting, we show how the estimate (4.2) may be derived as a Bayes estimate with the additional assumption that $Z(x)$ is a Gaussian process. More mathematical details may be found in Anderson (1984).

Let Σ have an inverted Wishart prior distribution, with density

$$w^{-1}(\Sigma \mid \tilde{C}, m) = \frac{|\tilde{C}|^{m/2}|\Sigma|^{(m+p+1)/2}}{2^{mp/2}\Gamma_p(\frac{1}{2}m)} \exp\left(-\tfrac{1}{2}\operatorname{tr}(\Sigma^{-1}\tilde{C})\right) \tag{4.3}$$

where $\tilde{C} = (m-p-1)C$, and $\Gamma_p(t) = \pi^{p(p-1)/4}\prod_{j=1}^{p}\Gamma(t-\frac{1}{2}(j-1))$. The conditional distribution of $n\hat{\Sigma}$ given Σ is a Wishart distribution, with density

$$w(n\hat{\Sigma} \mid \Sigma, n) = \frac{|n\hat{\Sigma}|^{(n-p-1)/2}}{2^{np/2}|\Sigma|^{n/2}\Gamma_p(\frac{1}{2}n)} \exp\left(-\tfrac{1}{2}\operatorname{tr}(\Sigma^{-1}n\hat{\Sigma})\right) \tag{4.4}$$

The posterior mean of Σ is

$$E(\Sigma \mid \hat{\Sigma}) = \frac{1}{n+m-p-1}(n\hat{\Sigma} + \tilde{C}) = \lambda\hat{\Sigma} + (1-\lambda)C \tag{4.5}$$

where $(n+m-p-1)\lambda = n$.

For the empirical Bayes approach, we use the data to estimate the parameter m of the prior distribution of Σ. The marginal density of $n\hat{\Sigma}$ is given by

$$M(n\hat{\Sigma} \mid C, n, m) = \frac{\Gamma_p(\frac{1}{2}(n+m))}{\Gamma_p(\frac{1}{2}n)\Gamma_p(\frac{1}{2}m)} \frac{|\tilde{C}|^{m/2}|n\hat{\Sigma}|^{(n-p-1)/2}}{|n\hat{\Sigma} + \tilde{C}|^{(n+m)/2}} \tag{4.6}$$

The maximum likelihood estimate of m is the value of m that maximizes (4.6), or equivalently maximizes

$$\frac{\Gamma_p(\frac{1}{2}(m+n))}{\Gamma_p(\frac{1}{2}m)}(m-p-1)^{-np/2}\left|I + \frac{nC^{-1/2}\hat{\Sigma}C^{-1/2}}{m-p-1}\right|^{-(m+n)/2} \tag{4.7}$$

where $C^{-1/2}$ is the symmetric matrix satisfying $(C^{-1/2})^2 = C^{-1}$. The advantage of (4.7) over (4.6) is computational. The eigenvalues $\lambda_1, \ldots, \lambda_p$ of $C^{-1/2}\hat{\Sigma}C^{-1/2}$ are real, and

$$\left| I + \frac{nC^{-1/2}\hat{\Sigma}C^{-1/2}}{m-p-1} \right| = \prod_{i=1}^{p} \left[1 + \frac{n\lambda_i}{m-p-1} \right],$$

which simplifies the computation of the determinant when evaluating (4.7) for many values of m. The maximizing value of m is used in (2.5) to obtain the estimate $\tilde{\Sigma}$ of Σ.

4.3 Extension to unmonitored locations

We want to estimate the covariance vectors $\sigma(x) = (\sigma(x, x_1), \ldots, \sigma(x, x_n))'$, for an unobserved $x \in \Xi$. Let $c(x)$ be the estimate from the parametric model. Then Switzer (1989) suggests an estimate of $\sigma(x)$, which modified for our shrinkage estimate of Σ becomes

$$\tilde{\sigma}(x) = \tilde{\Sigma}C^{-1}c(x). \tag{4.8}$$

We note when using (3.1), if $x = x_i$ for some i, the estimate $\tilde{\sigma}(x)$ is the ith column of $\tilde{\Sigma}$, so the extension is faithful to the covariance estimates for the observed locations. We want the extended covariance function to be positive definite. In general, we let x be a vector of q unobserved locations, and $\Sigma_{\text{new}}(x)$ be the covariance matrix between these locations. Let $c(x)$ be the $p \times q$ matrix of parametric estimate of covariances between locations, using the same parametric model used to estimate C in Section 4.2. For positive definiteness we require for all $\lambda_1 \in \mathscr{R}^p$ and $\lambda_2 \in \mathscr{R}^q$,

$$\begin{aligned} 0 &\leqslant (\lambda_1', \lambda_2') \begin{bmatrix} \tilde{\Sigma} & \tilde{\Sigma}C^{-1}c(x) \\ c(x)'C^{-1}\tilde{\Sigma} & \Sigma_{\text{new}}(x) \end{bmatrix} \begin{bmatrix} \lambda_1 \\ \lambda_2 \end{bmatrix} \\ &= (\lambda_1' - \lambda_2' c(x)'C^{-1})\tilde{\Sigma}(\lambda_1 - C^{-1}c(x)\lambda_2) \\ &\quad + \lambda_2'(\Sigma_{\text{new}}(x) - \tilde{\sigma}(x)'\tilde{\Sigma}^{-1}\tilde{\sigma}(x))\lambda_2. \end{aligned} \tag{4.9}$$

A necessary and sufficient condition for (4.9) to be satisfied for all λ_1 and λ_2 is that

$$\Sigma_{\text{new}}(x) - \tilde{\sigma}(x)'\tilde{\Sigma}^{-1}\tilde{\sigma}(x) \tag{4.10}$$

be non-negative definite. Estimating $\tilde{\Sigma}_{\text{new}}(x)$ for multiple sites simultaneously appears to be difficult; we suggest a sequential approach where one site is added at a time. Suppose we have two additional sites x and y. We want to compare the estimates obtained by adding x first with the estimates obtained by adding y first. We have

$$[I \quad C^{-1}c(x)] = [C^{-1} \quad 0] \begin{bmatrix} C & c(x) \\ c(x)' & c(x, x) \end{bmatrix}$$

so

$$[\tilde{\Sigma} \quad \tilde{\sigma}(x)]\begin{bmatrix} C & c(x) \\ c(x)' & c(x,x) \end{bmatrix}^{-1} = \tilde{\Sigma}[I \quad C^{-1}c(x)]\begin{bmatrix} C & c(x) \\ c(x)' & c(x,x) \end{bmatrix}^{-1}$$
$$= \tilde{\Sigma}[C^{-1} \quad 0]. \tag{4.11}$$

From (4.11), we see the estimate $\tilde{\sigma}(y)$ will be the same whether or not we have added x to the model first. This shows that if we let the permutation group S_q operate on the new locations $x_{p+1}, \ldots, x_{p+q}$, then the estimates of covariance between observed and unobserved locations generated sequentially are invariant with respect to elements of S_q of the form $(j \quad j+1)$, $j = 1, \ldots, q-1$. Since these elements generate S_q (Lang (1984), p. 57), it follows that we get the same estimate of covariance between observed and unobserved locations regardless of the order in which new locations are added.

By adding new locations to the model sequentially we only need to consider additional methods to estimate the diagonal elements of $\Sigma_{\text{new}}(x)$; estimates of the off-diagonal elements are obtained automatically. However, the order invariance will not extend to the estimates of covariance between unobserved locations, which will depend on the estimates of diagonal elements of $\Sigma_{\text{new}}(x)$. There are several possibilities for estimating the diagonal elements of $\Sigma_{\text{new}}(x)$. The simplest is to just choose a constant, for example, the mean of the diagonal elements of $\tilde{\Sigma}$. When there is some spatial structure to the variance, it may be preferable to smooth the diagonal elements of $\tilde{\Sigma}$. In either case, we may get problems with positive-definiteness if there is a strong correlation between locations.

The problem of estimating spatially heterogeneous covariances may also be approached by rescaling procedures. The basic idea is to map locations $x \in \mathscr{R}^2$ to new locations $f(x) \in \mathscr{R}^k$ such that spatial covariances in the image space are stationary and isotropic. Such rescaling procedures may be described as follows:

(1) Pick a strictly monotone isotropic covariance function $\tilde{C}(h)$, where h is distance in the image space, e.g. $\tilde{C}(h) = e^{-h}$. This choice is arbitrary and can affect the subsequent covariance estimate.

(2) If c_{ij} are the empirical covariances (or correlations) between monitoring sites x_i and x_j, then

$$\tilde{C}^{-1}(c_{ij}) = \|f(x_i) - f(x_j)\|, \qquad i, j = 1, \ldots, p \tag{4.12}$$

where $\|.\|$ denotes distance in $\mathscr{R}^k$, f is the desired mapping and $\tilde{C}^{-1}$ is the inverse function of $\tilde{C}$. If we choose the image space dimension, k, to be less than $p-1$, then we will not in general be able to choose the transformation f such that (4.12) holds exactly for all i and j. Some approximation will be needed. See, for example, Sampson and Guttorp (1990), who consider mappings from $\mathscr{R}^2$ to $\mathscr{R}^2$ for the spatial variogram rather than covariance estimates.

(3) For an unmonitored location x_0, we interpolate to obtain $f(x_0)$. If the interpolation is linear, we will obtain

$$f(x_0) = \sum_{i=1}^{p} w_i(x_0) f(x_i)$$

where $w_i(x_0)$ is a $k \times k$ matrix.

(4) The estimated covariance is

$$\hat{c}_{0j} = \tilde{C}(\|f(x_0) - f(x_j)\|).$$

Using such procedures readily gives covariance estimates between unmonitored locations. However, the choice of $\tilde{C}$ and f is somewhat arbitrary.

4.4 Interpolation and estimation of the mean

In the preceding section we assume $E(Z(x)) = 0$. This will usually mean an estimate of $E(Z(x))$ has been subtracted before estimating the covariance, and n is reduced to compensate for this estimation. In this section, we consider estimation of $\mu(x, j) = E(Z_j(x))$ and $\hat{Z}(x) = E(Z(x) \mid Z(x_1), \ldots, Z(x_p))$. We also discuss the precision of these estimates.

We assume a generalized additive model (Hastie and Tibshirani, 1987) for $\mu(x, j)$,

$$\mu(x, j) = \mu + X(x) + T(j) \tag{4.13}$$

where x and T are suitably smooth functions, subject to the constraints

$$\sum_{i=1}^{p} X(x_i) = 0 \quad \text{and} \quad \sum_{j=1}^{n} T(j) = 0.$$

Assuming an orthogonal design (i.e. observations are made at the same sites at each time) we can estimate X and T by applying a suitable smoother to the marginal effects of the array $Z_j(x_i)$; $i = 1, \ldots, p$; $j = 1, \ldots, n$. For the purposes of this paper, we use the loess smoother (Cleveland and Devlin, 1988), but many other smoothers or parametric regression models could be used.

The MLE of $Z_j(x)$ for unobserved x if $\mu(x, t)$ and $\sigma(x, y)$ were known is

$$\hat{Z}_j(x) = \mu(x, j) + \sigma(x)'\Sigma^{-1}(Z_j - \mu(\cdot, j)) \tag{4.14}$$

where $\mu(\cdot, j) = (\mu(x_1, j), \ldots, \mu(x_p, j))'$. Using the estimators proposed for μ and σ, (4.14) becomes

$$\tilde{Z}_j(x) = \hat{\mu}(x, j) + c(x)'C^{-1}(Z_j - \hat{\mu}(\cdot, j)). \tag{4.15}$$

We note that if $x = x_1$, then $c(x)'C^{-1} = (1, 0, \ldots, 0)$, so (4.15) reproduces the observed data. Also, (4.15) depends only on the parametric model used to derive C, and not on $\hat{\Sigma}$.

The prediction error is defined by

$$\eta(x, t) = E((Z_t(x) - \tilde{Z}_t(x))^2) \tag{4.16}$$

From standard probability theory and the assumption of no serial correlation, we have

$$\begin{aligned}\eta(x, t) &\geqslant \operatorname{var}(Z_t(x) \mid Z_t) \\ &= \sigma(x, x) - \sigma(x)'\Sigma^{-1}\sigma(x). \end{aligned} \tag{4.17}$$

Note that (4.16) is the prediction error of $\hat{Z}_j(x)$. We might hope the prediction error could be estimated from (4.17). However, this makes no allowance for variability in $\hat{\mu}(x, t)$, and our example in Section 4.5 will show that this may increase prediction error substantially.

We derive the covariance function of $\hat{\mu}(x, t)$ for fixed t. The marginal effects of the array $Z_j(x_i)$ are given by

$$Z\cdot(x_i) = \frac{1}{n}\sum_{j=1}^{n} Z_j(x_i) - \hat{\mu} = X(x_i) + \frac{1}{n}\sum_{j=1}^{n} \varepsilon_j(x_i) - \hat{\mu} \tag{4.18}$$

and

$$Z_j(\cdot) = \frac{1}{p}\sum_{i=1}^{p} Z_j(x_i) - \hat{\mu} = T(j) + \frac{1}{p}\sum_{i=1}^{p} \varepsilon_j(x_i) - \hat{\mu}, \tag{4.19}$$

where

$$\hat{\mu} = \frac{1}{np}\sum_{i=1}^{p}\sum_{j=1}^{n} Z_j(x_i) = \mu + \frac{1}{np}\sum_{i=1}^{p}\sum_{j=1}^{n} \varepsilon_j(x_i). \tag{4.20}$$

Let $l_i(x)$, $i = 1, \ldots, p$ be the smoothing weights for estimating $X(t)$ and $m_j(t)$, $j = 1, \ldots, n$ be the smoothing weights for estimating $T(t)$. Then $\hat{X}(x) = \Sigma_{i=1}^{p} l_i(x) Z\cdot(x_i)$ and $\hat{T}(t) = \Sigma_{j=1}^{n} m_j(t) Z_j(\cdot)$. We assume the smoothing procedure used to estimate $X(x)$ and $T(t)$ will reproduce constants. This implies $\Sigma_{i=1}^{p} l_i(x) = \Sigma_{j=1}^{n} m_j(t) = 1$ for all x and t. Then from (4.18)–(4.20) we get

$$\begin{aligned}\hat{\mu}(x, t) = \mu &+ \sum_{i=1}^{p} l_i(x) X(x_i) + \sum_{j=1}^{n} m_j(t) T(j) \\ &+ \sum_{i=1}^{p}\sum_{j=1}^{n}\left[\frac{l_i(x)}{n} + \frac{m_j(t)}{p} - \frac{1}{np}\right]\varepsilon_j(x_i),\end{aligned} \tag{4.21}$$

giving (with our assumption $\operatorname{cov}(\varepsilon_j(x_i), \varepsilon_{j'}(x_{i'})) = 0$ if $j = j'$),

$$\begin{aligned}\operatorname{cov}(\hat{\mu}(x, t), \hat{\mu}(x', t)) &= \sum_{j=1}^{n}\sum_{i=1}^{p}\sum_{i'=1}^{p}\left[\frac{l_i(x)}{n} + \frac{m_j(t)}{p} - \frac{1}{np}\right] \\ &\quad \times \left[\frac{l_{i'}(x')}{n} + \frac{m_j(t)}{p} - \frac{1}{np}\right]\sigma(x_i, x_{i'}) \\ &= \frac{1}{n} l(x)\Sigma l(x')' + \frac{1_p'\Sigma 1_p}{p^2}\sum_{j=1}^{n}\left(m_j(t) - \frac{1}{n}\right)^2\end{aligned} \tag{4.22}$$

where $l(x) = (l_1(x), \ldots, l_p(x))$ and 1_p represents a vector of 1s of length p.

We can now derive a more accurate formula for the prediction error (4.16). Assuming there is no bias, (4.16) is the variance of $Z_t(x) - \tilde{Z}_t(x)$. Then

$$\eta(x, t) = \sigma(x, x) - 2\operatorname{cov}(Z_t(x), \tilde{Z}_t(x)) + \operatorname{var}(\tilde{Z}_t(x)). \tag{4.23}$$

Neglecting variability in fitting the parametric model, we use (4.19) to get

$$\begin{aligned}\operatorname{cov}(Z_t(x), \tilde{Z}_t(x)) &= \operatorname{cov}(Z_t(x), \hat{\mu}(x, t)) \\ &\quad + c(x)'C^{-1}\operatorname{cov}(Z_t(x), Z_t - \hat{\mu}(\cdot, t)) \\ &= \left[\frac{l(x)'}{n} + \left(m_t(t) - \frac{1}{n}\right)\frac{1_p'}{p} + c(x)'C^{-1}(I - A_t)\right]\sigma(x)\end{aligned} \tag{4.24}$$

where

$$A_t = \frac{1}{n}L + \left[\frac{m_t(t)}{p} - \frac{1}{np}\right]J.$$

Here, L is the $p \times p$ matrix with $L_{i,j} = l_j(x_i)$ and J is the $p \times p$ matrix of 1s. Similarly,

$$\begin{aligned}\operatorname{var}(\tilde{Z}_t(x)) &= \operatorname{var}(\hat{\mu}(x, t)) + c(x)'C^{-1}\operatorname{var}(Z_t - \hat{\mu}(\cdot, t))C^{-1}c(x) \\ &\quad + 2c(x)'C^{-1}\operatorname{cov}(\hat{\mu}(x, t), Z_t - \hat{\mu}(\cdot, t)) \\ &= \frac{1}{n}l(x)'\Sigma l(x) + \frac{1_p'\Sigma 1_p}{p^2}\sum_{j=1}^{n}\left(m_j(t) - \frac{1}{n}\right)^2 \\ &\quad + c(x)C^{-1}\Bigg[(I - A_t)\Sigma(I - A_t') - A_t\Sigma A_t' \\ &\quad + \frac{L\Sigma L'}{n} + \frac{J\Sigma J}{p^2}\sum_{j=1}^{n}\left(m_j(t) - \frac{1}{n}\right)^2\Bigg]C^{-1}c(x) \\ &\quad + 2c(x)'C^{-1}\Bigg[\Sigma\left(\frac{l(x)}{n} + \left(m_t(t) - \frac{1}{n}\right)\frac{1_p}{p}\right) \\ &\quad - \frac{L\Sigma l(x)}{n} - \frac{J\Sigma 1_p}{p^2}\sum_{j=1}^{n}\left(m_j(t) - \frac{1}{n}\right)^2\Bigg].\end{aligned} \tag{4.25}$$

It is easily verified from (4.23)–(4.25) that $\eta(x_i, t) = 0$; $i = 1, \ldots, p$. An estimate of $\eta(x, t)$ is found by replacing $\sigma(x)$ by $\tilde{\sigma}(x)$ in (4.24), Σ by $\tilde{\Sigma}$ in (4.25) and estimating $\sigma(x, x)$ as suggested in Section 4.3. After some algebra, we get

$$\hat{\eta}(x,t) = \tilde{\sigma}(x,x) - c(x)'C^{-1}\tilde{\Sigma}C^{-1}c(x) + \frac{1}{p^2}\sum_{j=1}^{n}\left(m_j(t) - \frac{1}{n}\right)^2$$
$$+ \frac{1}{n}(l(x) - c(x)'C^{-1}L)\tilde{\Sigma}(l(x)' - L'C^{-1}c(x)). \tag{4.26}$$

4.5 Example: acid rain data

We have monthly average sulfate concentrations at 19 sites in the Eastern United States, measured over a two-year period. A log transformation has been applied to the data. The sites are shown in Figure 4.1.

Both components of (4.13) are estimated using the loess smoother. Local quadratic fitting is used to estimate $T(t)$ and bivariate local linear fitting is used to fit $X(x)$. In Figure 4.2, we show M_f plots (Cleveland and Devlin, 1988) for both cases. We select the smoothing parameters $f = 18/24$ for estimating $T(t)$ and $f = 10/19$ for estimating $X(x)$. Figure 4.3 plots the sulfate concentrations for each site, together with the smoothed time trend. Figure 4.4 is a plot of estimated contours for $X(x)$. The highest sulfate concentrations occur in the mid-west and mid-Atlantic areas, while the lowest sulfate concentrations are in the Great Plains States (the high concentrations in the Altantic Ocean are presumably spurious extrapolations).

We use the parametric model $C(x, y) = ae^{-\lambda h}$, where h is the grand circle

Figure 4.1 Acid rain sulfate concentrations; 19 recording locations.

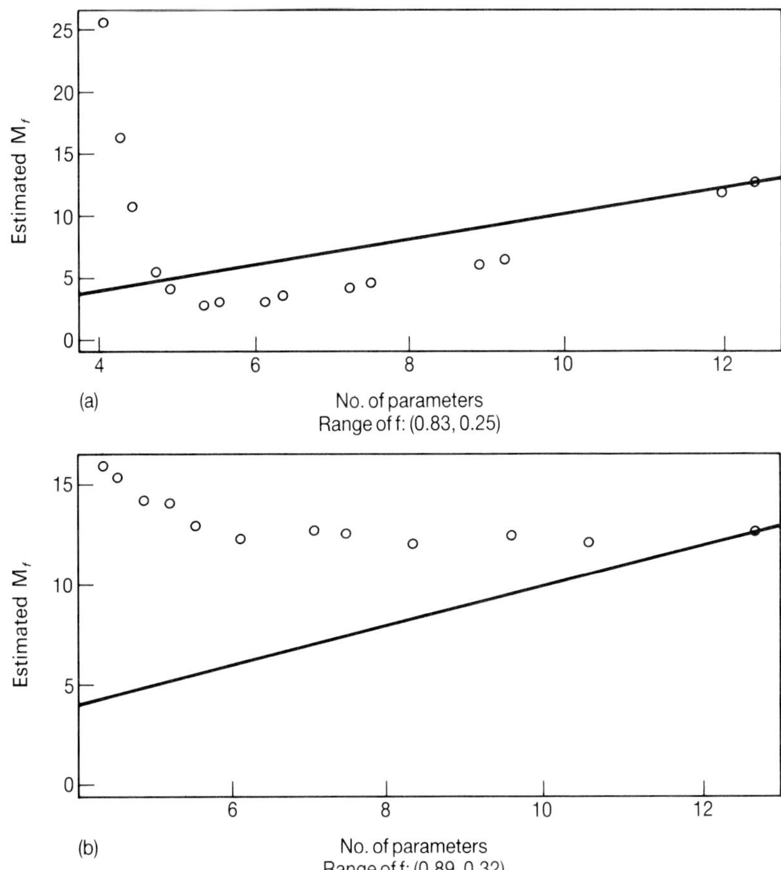

Figure 4.2 M_f plots to select the smoothing parameters to estimate $T(t)$ and $X(x)$.

distance between x and y. Estimates of a and λ are found by least squares applied to the $p(p+1)/2$ variances and covariances betwen sites. Figure 4.5 shows the column of $\hat{\Sigma}$ for each site, and the lines represent the fitted model.

The shrinkage parameter found using (4.5) and (4.6) was $\lambda = 0{\cdot}6$. In Figure 4.6 we plot estimated contours of $\sigma(x, y)$ over the eastern half of the United States, using (4.8).

We use cross-validation to assess how well the estimates of accuracy are performing. We delete each of the sites in turn, and use the remaining 18 sites to estimate the sulfate concentrations at the deleted sites. We compute the site bias,

$$\beta(x) = \frac{1}{n} \sum_{j=1}^{n} (Z_j(x) - \tilde{Z}_j(x)) \tag{4.27}$$

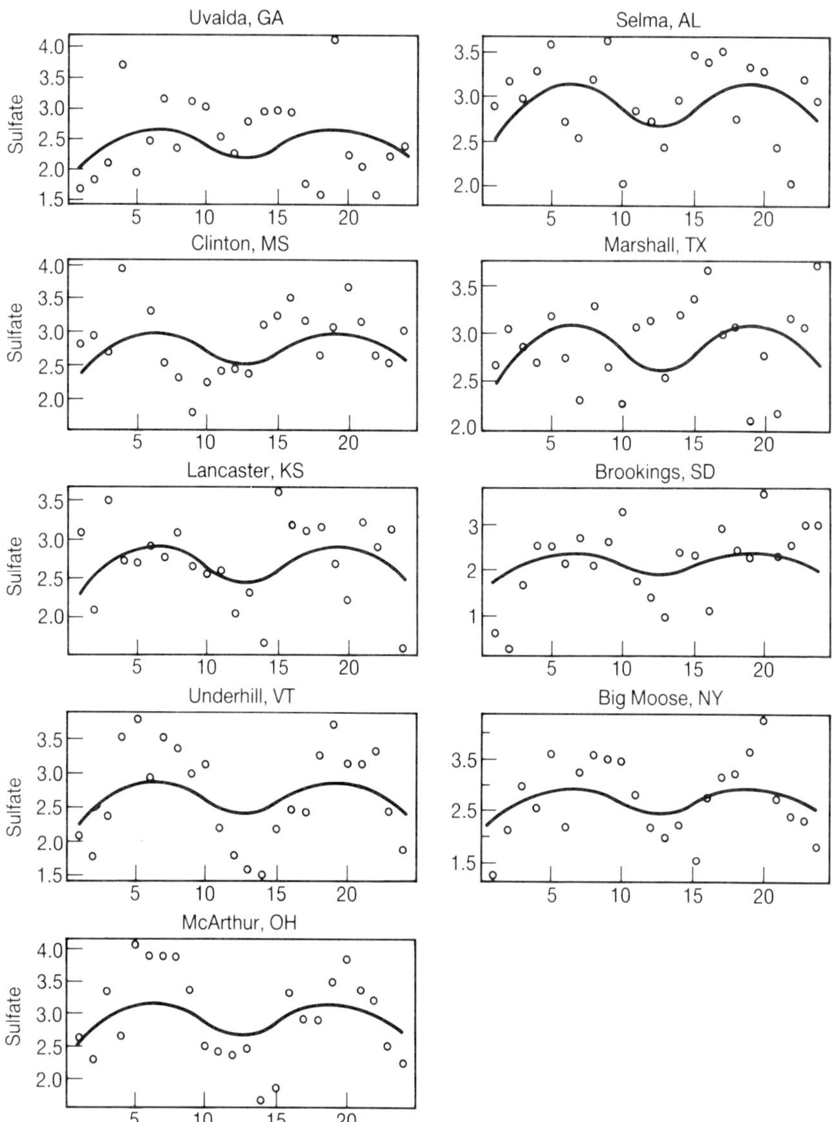

Figure 4.3 Sulfate concentrations and smoothed trends.

and variance,

$$v(x) = \frac{1}{n} \sum_{j=1}^{n} [Z_j(x) - \tilde{Z}_j(x) - \beta(x)]^2. \tag{4.28}$$

The expected values are estimated by splitting (4.26) and summing over t.

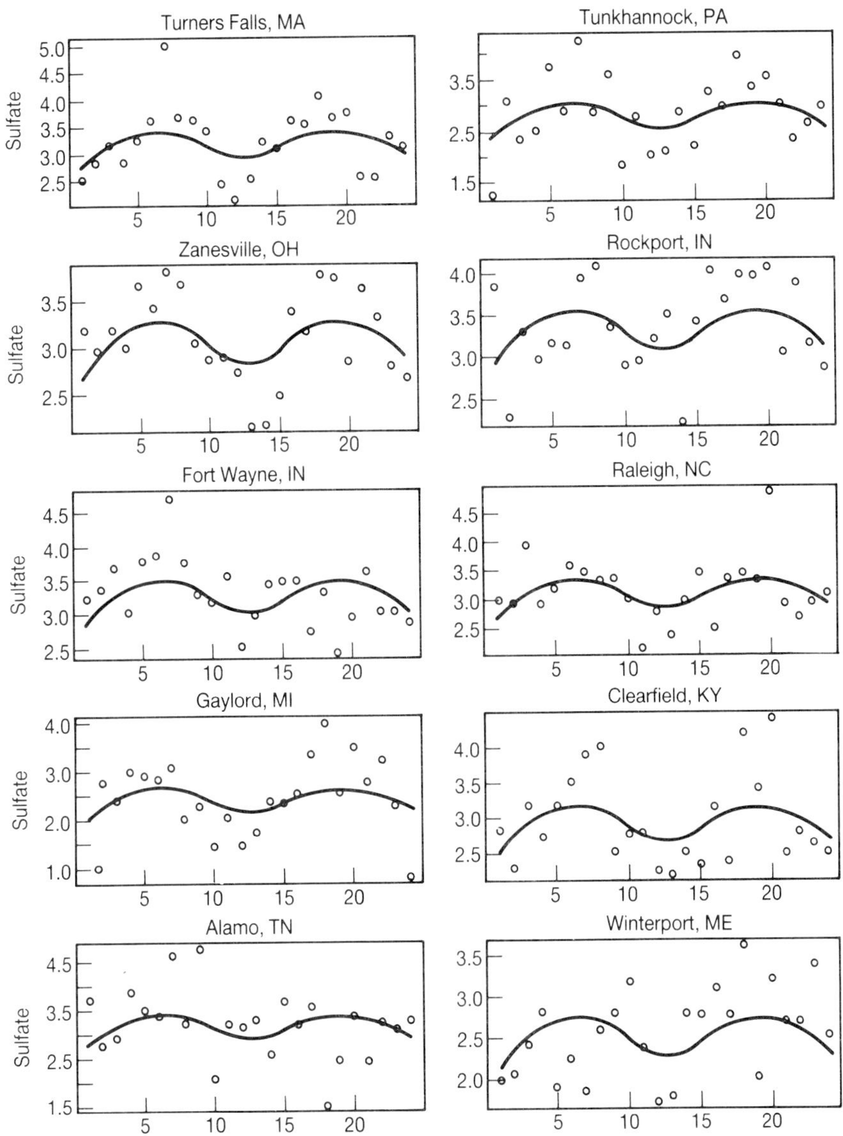

Figure 4.3 continued

This gives

$$E(\beta(x)^2) = n\tilde{\sigma}(x, x) - c(x)'C^{-1}\tilde{\Sigma}C^{-1}c(x) + \frac{1}{p^2}\sum_{t=1}^{n}\sum_{j=1}^{n}\left(m_j(t) - \frac{1}{n}\right)^2$$

and

$$E(v(x)) = (l(x) - c(x)'C^{-1}L)\tilde{\Sigma}(l(x)' - LC^{-1}c(x)).$$

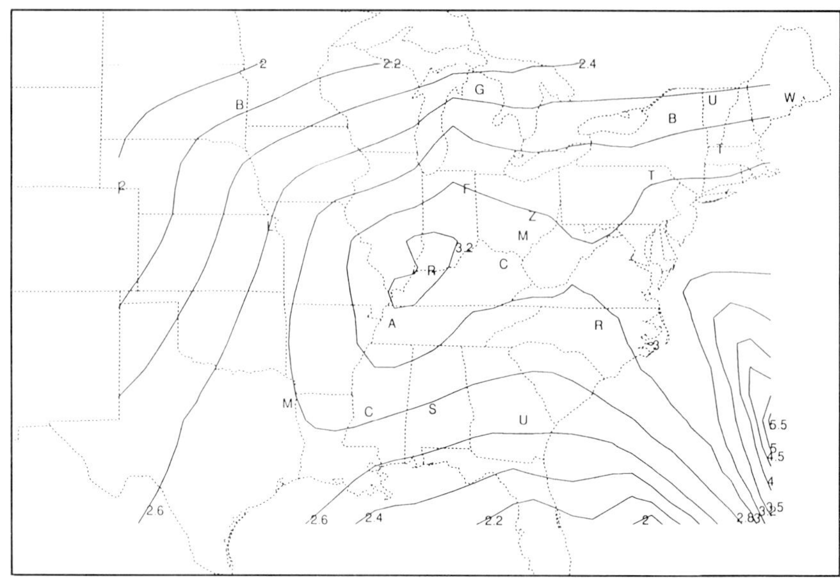

Figure 4.4 Estimated values of $\mu + X(x)$.

Table 4.1 Cross-validated prediction errors and estimates

Site	$\beta(x)^2$	$v(x)$	$E\beta(x)^2$	$Ev(x)$	(4.17)
Turners Falls, MA	0·1445	0·2057	0·0156	0·1993	0·2074
Tunkhannock, PA	0·0873	0·2470	0·0164	0·2924	0·3034
Zanesville, OH	0·0019	0·0958	0·0025	0·0481	0·0499
Rockport, IN	0·0031	0·2061	0·0141	0·2233	0·2317
Fort Wayne, IN	0·1498	0·2290	0·0120	0·2187	0·2269
Raleigh, NC	0·3545	0·2080	0·0236	0·2457	0·2532
Gaylord, MI	1·0999	0·4184	0·0492	0·3646	0·3767
Clearfield, KY	0·1225	0·1823	0·0101	0·2062	0·2146
Alamo, TN	0·0009	0·4928	0·0205	0·4126	0·4288
Winterport, ME	0·3621	0·2744	0·0281	0·2479	0·2561
Uvalda, GA	0·2542	0·4279	0·0238	0·3658	0·3789
Selma, AL	0·0584	0·1885	0·0130	0·2197	0·2277
Clinton, MS	0·0156	0·3156	0·0144	0·2701	0·2804
Marshall, TX	0·1404	0·2462	0·0268	0·2668	0·2747
Lancaster, KS	0·0573	0·2407	0·0178	0·2543	0·2617
Brookings, SD	0·0588	0·5707	0·0446	0·4374	0·4519
Underhill, VT	0·0008	0·2509	0·0112	0·2083	0·2170
Big Moose, NY	0·0019	0·2423	0·0130	0·2669	0·2780
McArthur, OH	0·0227	0·1531	0·0076	0·1726	0·1800

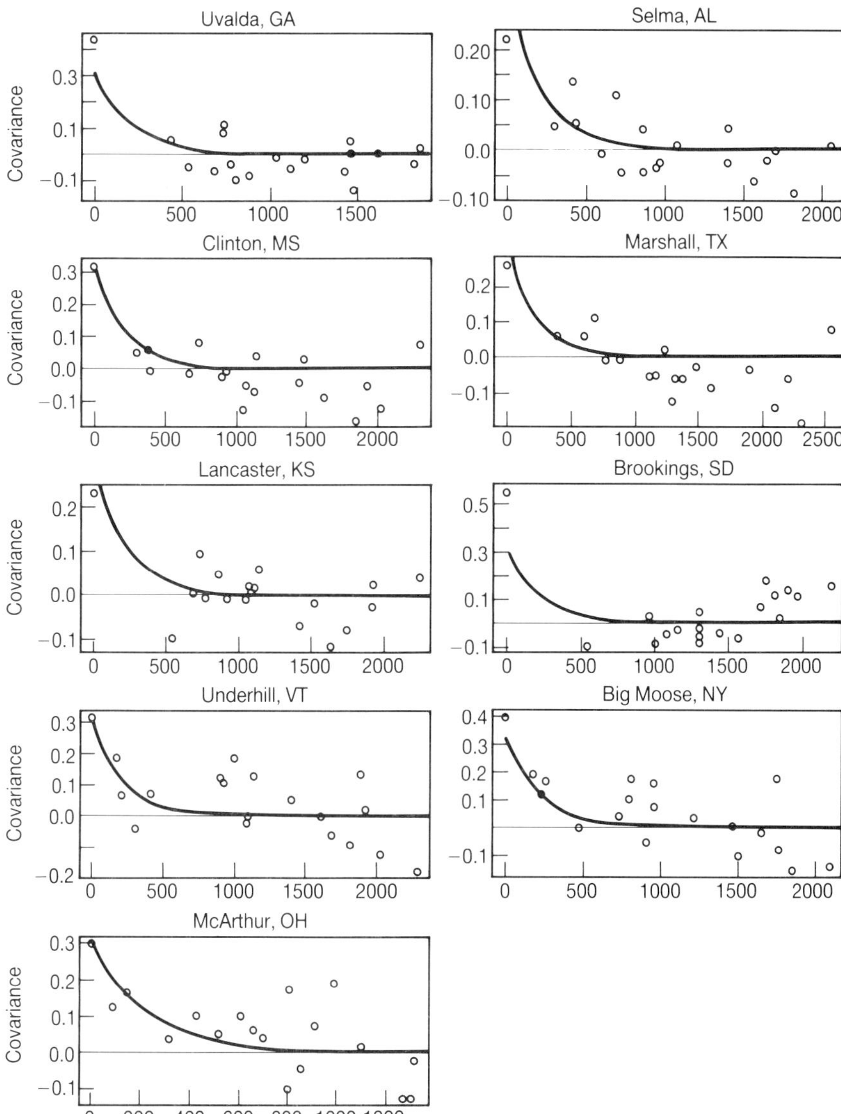

Figure 4.5 Raw covariance estimates (x axis is distance in km).

The results are given in Table 4.1. We include estimates of (4.17) for comparison.

From Table 4.1, we see the site biases are in many cases substantially larger than their estimated values. This means there is bias in the estimation of $X(x)$, so the neighborhoods used for the loess procedure are too large. Due to the small number of sites involved, it is not feasible to reduce the

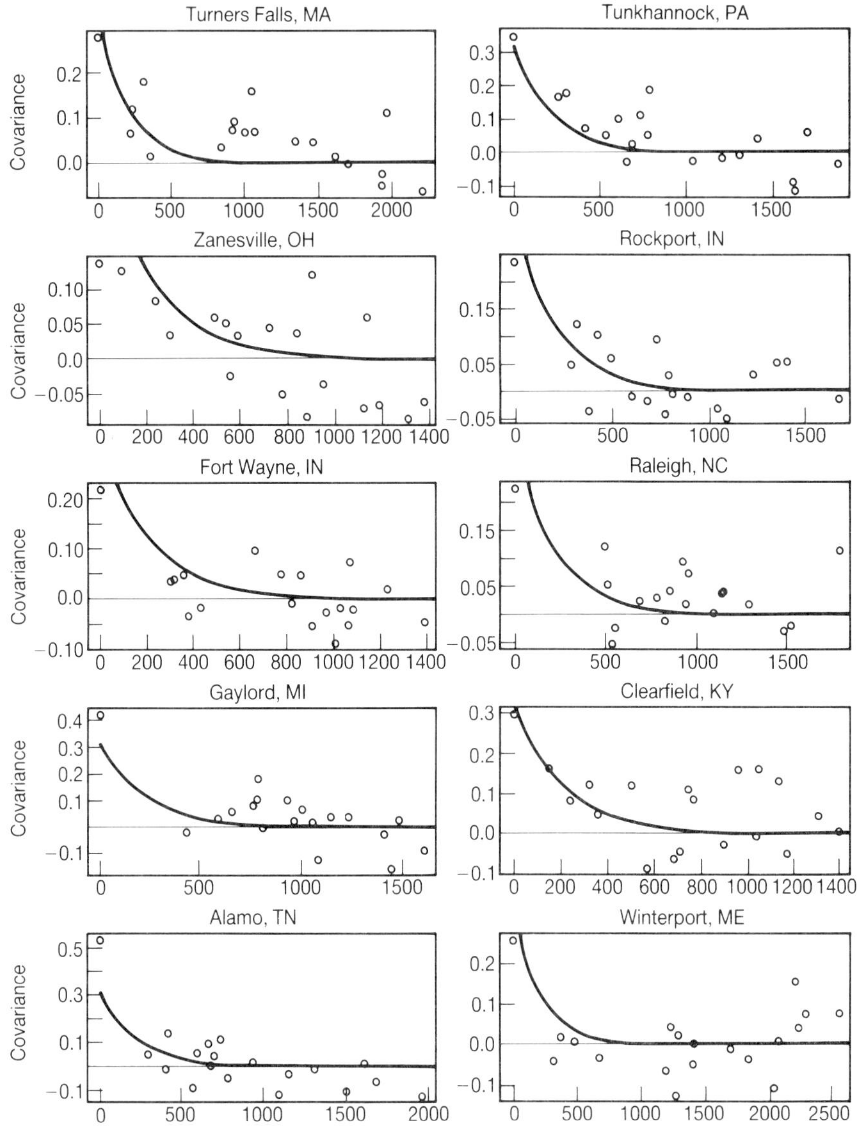

Figure 4.5 continued

neighborhood size further; to get better estimates, we would require data from additional sites.

The estimates derived using (4.17) are adjusted by a factor of 4, to allow for the model fitting with 24 times and 18 sites. For most sites, estimates of variability have performed well. In Figure 4.7 we plot the interpolated values and actual data for each of the 19 sites. In most cases we see a strong correlation between the two, indicating the fitting is performing well.

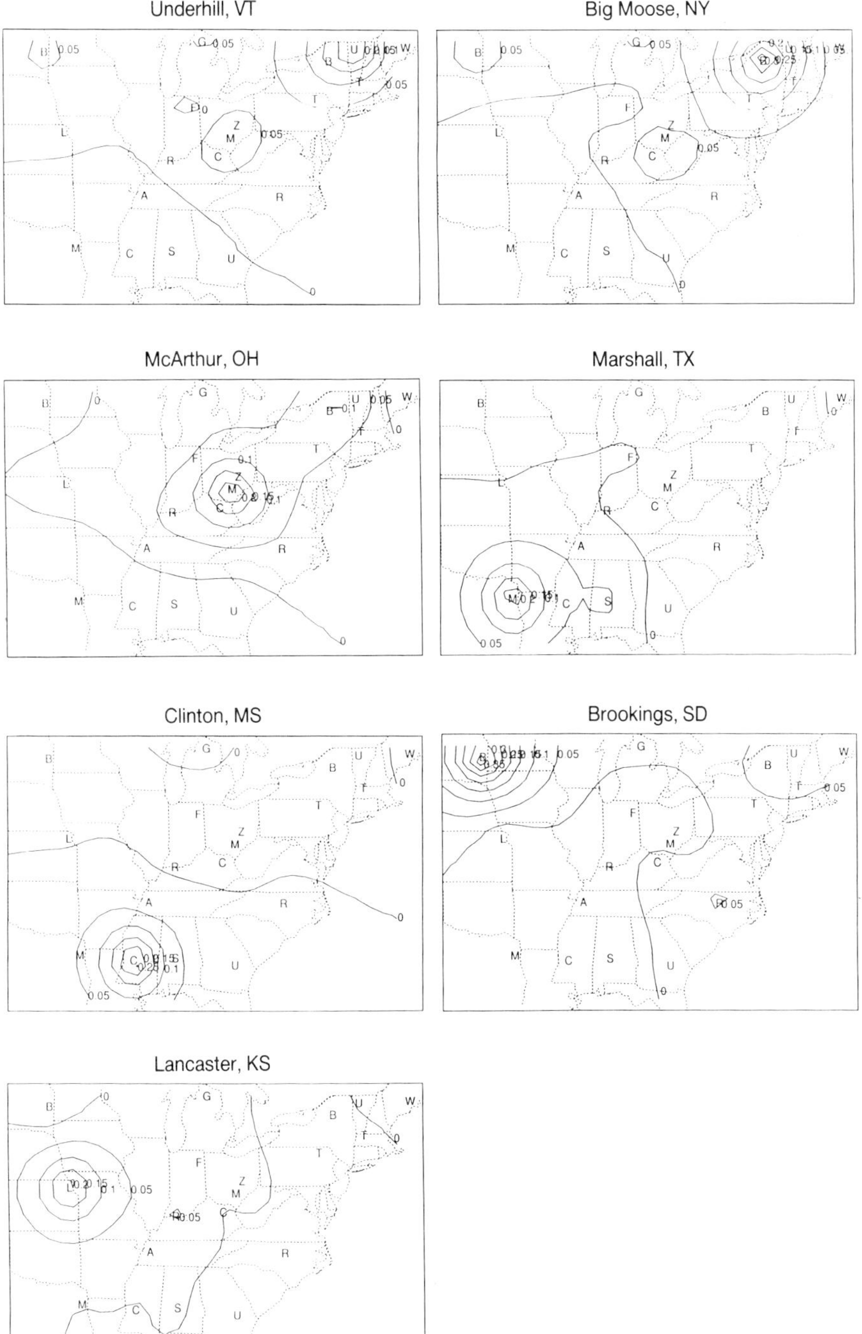

Figure 4.6 Interpolated covariance.

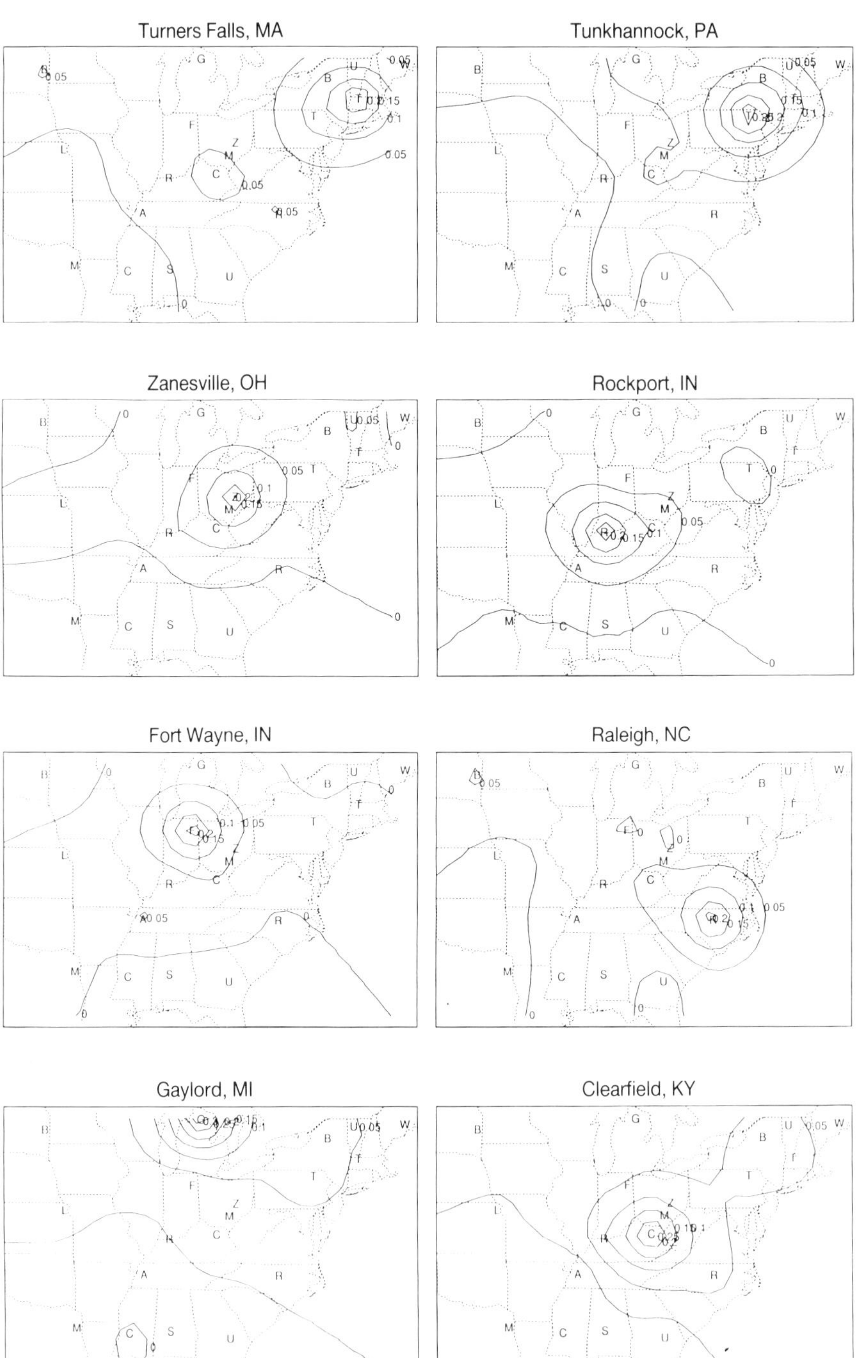

Figure 4.6 continued

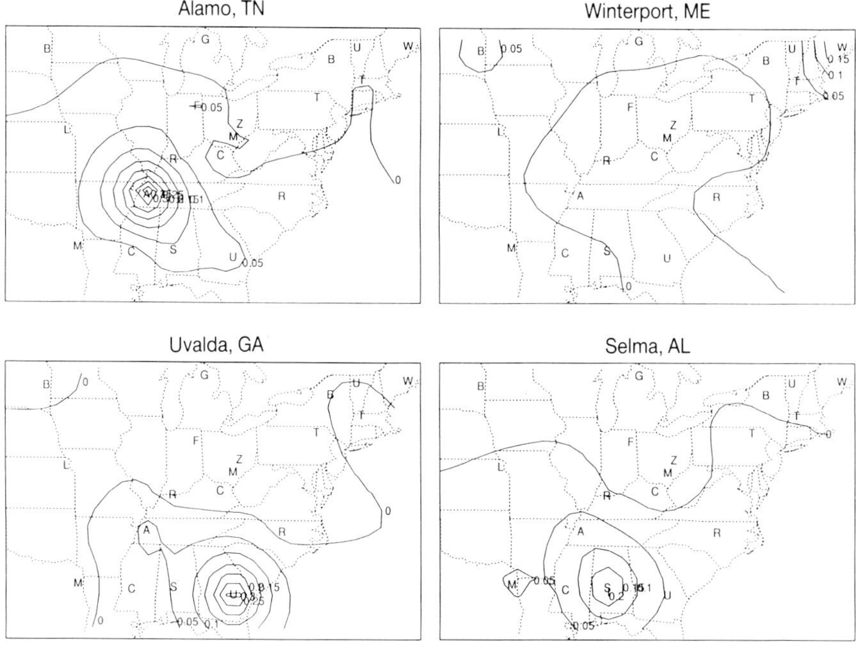

Figure 4.6 continued

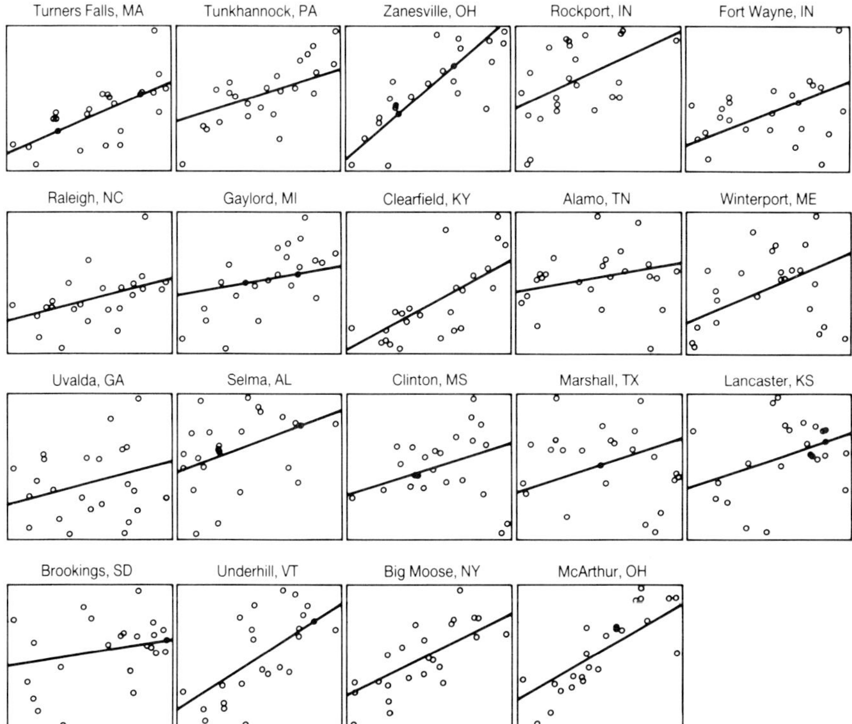

Figure 4.7 Cross validation.

References

Anderson, T.W. (1984). *An Introduction to Multivariate Statistical Analysis* (2nd edn), Wiley, Chichester.

Cleveland, W.S. and Devlin, S.J. (1988). Locally weighted regression: an approach to regression analysis by local fitting. *J. Amer. Statist. Assoc.*, **83**, 596–610.

Hastie, T.J. and Tibshirani, R. (1987). Generalized additive models: some applications. *J. Amer. Statist. Assoc.*, **82**, 371–86.

Lang, S. (1984). *Algebra* (2nd edn), Addison-Wesley, Reading, Mass.

Sampson, P. and Guttorp, P. (1990). *Nonparametric Estimation of Nonstationary Spatial Covariance Structure.* SIMS technical report 148, University of Washington.

Switzer, P. (1989). Non-stationary spatial covariances estimated from monitoring data. In *Geostatistics*, Armstrong, M. (ed.), Kluwer Academic Publishers, pp. 127–38.

Chapter 5

Modeling daily precipitation – progress and problems

D. A. Woolhiser

5.1 Introduction

Precipitation has a major influence on virtually all human activity. Agricultural operations and many engineering activities are strongly influenced by weather phenomena. Examples of applications where precipitation information is essential include: design of irrigation systems (Palmer *et al.*, 1982), design of agricultural or urban drainage systems, design of earthen covers for landfills or storage sites for low-level nuclear waste (Lane, 1984), selection of farm or construction machinery (Von Bargen, 1967), evaluation of erosion hazard and the effects of erosion on agricultural productivity through simulation modeling (Williams and Renard, 1985). New developments in modeling growth and yield of major crops such as corn, wheat, soybeans and cotton will increase the need for precipitation simulation models so that farmers will be able to obtain estimates of the distribution of crop yields as the growing season progresses.

Extensive climatic data, including precipitation, have been collected for many years all over the world with the greatest density of stations in the developed nations. These data were previously available in cumbersome printed form, but are now quite widely available on computer-compatible media. The information content of these data can be summarized by standard statistical analyses and the results presented in tabular or graphical form. For example, extensive analyses were performed by the regional climate committees of State Agricultural Experiment Stations in the United States during the period 1955–67 and the results were presented in a series of reports (Shaw *et al.*, 1960; Dethier and McGuire, 1961; Feyerherm *et al.*, 1965; Gifford *et al.*, 1967).

Individuals from a wide range of disciplines have been interested in the stochastic structure of the precipitation process for a long time. Katz (1984) pointed out that in work published in 1852, Quetelet examined runs of consecutive wet and dry days in Brussels and noted the phenomenon of persistence. To take this dependence into account he developed a special case of what is now called a two-state Markov chain. Many more models have been developed since then, with a major stimulus being provided by the advent of digital computers.

5.2 Stochastic models of daily precipitation

Many models have been developed to describe the daily precipitation process. Precipitation is, of course, an intermittent process in continuous time but precipitation data commonly available are the accumulated depths over a period of one day. Although a substantial literature exists on continuous time models of precipitation they are beyond the scope of this paper. At least in the United States, the time of day that the gauges are read (for standard gauges) may vary from station to station and could also change with time. As we shall see, this methodological factor may be quite significant in modeling. The most common approach has been to develop models describing the occurrence (wet–dry) process and to describe the distribution of rainfall amounts on wet days independently.

The seasonal variation of precipitation throughout the year is an important factor in the construction of models. Several approaches have been used to deal with seasonality: assume that parameters vary as step functions for each month, assume that parameters vary as step functions for seasons which may include two or more months, or use polynomials or Fourier series to provide daily variation of parameters. A comparison of two of these approaches will be presented in a later section.

5.2.1 Occurrence models

Rainfall occurrence can be viewed as a sequence of random variables $\{X(t)\}$; $t = t_1, t_2, \ldots, t_T$, where

$$\begin{aligned} X(t) &= 1 \text{ if rainfall is equal to or greater than a threshold, } d \\ &= 0 \text{ if rainfall is less than } d. \end{aligned} \tag{5.1}$$

The most commonly used model of precipitation occurrence is the two-state Markov chain of first or higher order. Gabriel and Neumann (1962) used a first-order stationary Markov chain to describe rainfall occurrence in Tel Aviv. Non-stationary Markov chains have been used by several investigators (Caskey, 1963; Feyerherm and Bark, 1965; Heerman *et al.*, 1968; Woolhiser and Pegram, 1979; Coe and Stern, 1982, Stern and Coe, 1984). The appropriate order for Markov chain models has been investigated by Chin (1977), Gates and Tong (1976), and Eidsvik (1980). The results varied with the climatic characteristics of the precipitation stations investigated, with the statistical tests used and with the length of record. A first-order chain is adequate for many locations but second or higher order may be required at other locations or during some times of the year.

The alternating renewal process (ARP) for wet and dry sequences has been investigated by Green (1964), Buishand (1977) and Roldan and Woolhiser (1982). Buishand (1977) found that an alternating renewal model with a truncated negative binomial distribution (TNBD) for the length of wet and dry sequences provided an excellent fit for rainfall data from the Netherlands. He assumed that the parameters of the TNBD were constant within four seasons and noted that in dry climates the sample size for wet sequences may be very small and the parameters may be unreliable. Roldan and Woolhiser (1982) compared the ARP with truncated geometric distribution of wet sequences and TNBD dry sequences with a first-order Markov chain. For five US stations studied, the first-order Markov chain was superior to the alternating renewal process according to the Akaike information criterion (Akaike, 1974). Roldan and Woolhiser (1982) used Fourier series to describe the seasonal variation of parameters with the parameters for a specific wet or dry sequence determined by the starting day of the sequence. One of the disadvantages of the ARP is that seasonality is difficult to handle.

Foufoula-Georgiou and Lettenmaier (1987) developed a Markov renewal model for rainfall occurrences in which the times between daily rainfall occurrences were sampled from two different geometric probability distributions. The transition from one inter-arrival type to the other was governed by a Markov chain. They derived maximum likelihood and method of moments estimators for the four parameters and assumed that the parameters were constants within each of five seasons.

Smith (1987) introduced a family of models termed Markov–Bernoulli processes that might be used for rainfall occurrence. The process consists of a sequence of Bernoulli trials with randomized success probabilities described by a first-order two-state Markov chain. At one extreme the model is a Bernoulli process, at the other a Markov chain. This process has some very interesting properties including invariance of the process under random thinning, but parameter identification is quite difficult.

The two-state discrete autoregressive moving average (DARMA) model was first used as a model for rainfall occurrence by Buishand (1977) and more recently by Chang *et al.* (1984) and Delleur *et al.* (1989). Buishand found that an alternating renewal process was superior to the DARMA model for data from the Netherlands but that the DARMA model looked more promising in tropical and monsoon areas. Chang *et al.* (1984) and Delleur *et al.* (1989) used four seasons for two stations in Indiana and found that the first-order autoregressive (Markov chain) or the second-order moving-average model were appropriate for different seasons. Buishand (1977) pointed out an important factor, namely that properties of the rainfall in New Delhi cannot be preserved by a model with constant parameters – stochastic parameters are required. This observation may be generally valid for regions with monsoon climates.

5.2.2 Distribution of amounts

Let $Y(t)$ denote the amount of precipitation on day t and let $U(t) = Y(t) - d$, where d is the smallest amount of rainfall observed or some other threshold for distinguishing between a dry or wet day. It has been observed that the distribution of $U(t)$ is highly skewed and that it exhibits seasonal variability. In a thorough analysis of data from the Netherlands, Buishand (1977) showed that there was evidence for different distributions of rainfall depending on whether it is a solitary rainy day (type I), a rainy day bounded by a rainy day on one side (type II) or a rainy day bounded on both sides by rainy days (type III). Buishand (1977) also showed that there was a small but significant correlation between precipitation amounts on successive wet days. He observed that the discrimination between different types of rainfall amounts has little effect on the distribution of 30-day total rainfall.

Introducing dependencies of the type observed by Buishand (1977) into a nonstationary model leads to difficulties in parameter identification and models with a large number of parameters. In efforts to obtain parsimonious models, investigators have frequently assumed that the daily rainfall, $Y(t)$, is independent of the occurrence process, $X(t)$, and is also independent of rain on previous days $Y(t-1)$, $Y(t-2)$, . . . An intermediate approach is to assume serial independence of $Y(t)$ but to allow $Y(t)$ to depend on $X(t-1)$, $X(t-2)$, . . . – the chain-dependent process as described by Katz (1977a, b). Stern and Coe (1984) describe techniques for estimating parameters for chain-dependent shifted gamma distributions. Woolhiser and Roldan (1982) found that according to the Akaike Information Criterion (AIC), (Akaike, 1974) the independent mixed exponential distribution was superior to chain dependent gamma distributions for five US stations.

The distribution of precipitation depths on wet days is highly skewed and a shifted gamma distribution has been most commonly used (Ison *et al.*, 1971; Buishand, 1977; Katz, 1977b; Richardson and Wright, 1984; Stern and Coe, 1984).

The mixed exponential has been utilized for the distribution of $U(t)$ by Smith and Schreiber (1974), Woolhiser and Pegram (1979), and Woolhiser and Roldan (1982). Richardson (1982) compared the exponential, mixed exponential and gamma distributions for several US stations and found the mixed exponential to be superior to the other two.

Stidd (1953) and Nicks (1974) used normalizing transforms of the type often used to transform gamma-like distributions to normal. Other distributions which have been used include the kappa (Mielke, 1973) and the Weibull (Zucchini and Adamson, 1984). It appears unlikely that a single distribution will provide a good fit to daily rainfall data for all climatic regions.

5.3 Parameter identification and model selection

One striking feature of the precipitation process in most parts of the world is its seasonal variability. This is demonstrated in Figure 5.1 where the mean number of wet days per 14-day period is plotted for four US stations with quite different climates. If we reduced the number of days in each period to five days or to one day we would find that the pattern would become more erratic due to sampling variability, although the seasonal variations would still be evident.

Parameter identification methods are inextricably related to the particular model used and to the techniques used to represent seasonal variation of parameters. If it is assumed that parameters vary as step functions defined

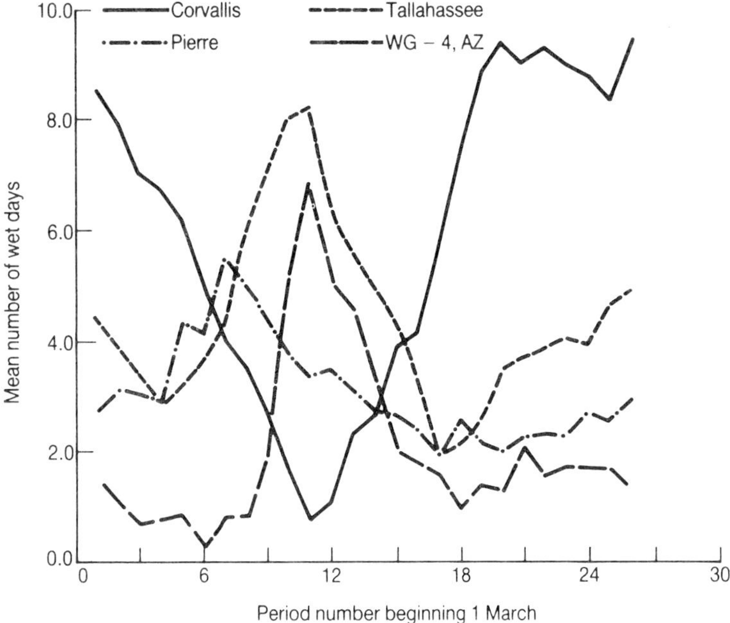

Figure 5.1 Mean number of wet days for 14-day periods for four US stations.

on a monthly or seasonal basis, the techniques are quite straightforward. Maximum likelihood (ML) techniques are typically used to estimate the transition probabilities of Markov chains and ML techniques or method of moments (MOM) are used to estimate parameters for the distribution of daily precipitation and for the distribution of wet and dry periods for alternating renewal models. Buishand (1977) recognized that the small precipitation amounts may be quite unreliable and used ML estimators which included the number of observations below a threshold but not the amount of each observation in this class. Chang *et al.* (1984) used the autocorrelation function to guess the form of a DARMA occurrence model and to obtain initial trial values of the parameters. The final values of the parameters were estimated by an iterative least-squares procedure using the estimated and theoretical autocorrelation functions.

The step function representation of parameters is not intuitively satisfying because we would not expect the precipitation characteristics on adjacent days to vary substantially. The seasonal variation of parameters has been fitted with polynomials (Stern, 1980) but the Fourier series representation has been the most popular (Feyerherm and Bark, 1965; Fitzpatrick and Krishnan, 1967; Ison *et al.*, 1971; Buishand, 1977; Woolhiser and Pegram, 1979; Stern and Coe, 1984; and Zucchini and Adamson, 1984).

Early investigators used ordinary least-squares methods to estimate the Fourier coefficients to describe the seasonal variation of model parameters. These methods are undesirable because the data points represent statistical estimates of parameters and these estimates have unequal variances because of varying sample size. Furthermore, there is no statistically sound procedure to test the significance of individual harmonics. Woolhiser and Pegram (1979) wrote the likelihood equations for a first-order Markov chain and a mixed exponential distribution of amounts as functions of Fourier coefficients and used numerical methods to estimate the coefficients that maximized the likelihoods. They used likelihood ratio tests to determine if harmonics were significant. Stern and Coe (1984) improved upon these techniques in two ways. First they used the logit transform

$$g(t) = \log\{p(t)/[1 - p(t)]\} \tag{5.2}$$

to transform transition probabilities, $p(t)$, which are bounded by 0 and 1 into $g(t)$ which is bounded by plus and minus infinity. They then describe $g(t)$ by a Fourier series and write the likelihood function as the sum of the logs of the likelihoods of binomial distributions. Since the binomial distribution is a member of the exponential family of distributions, the model is a generalized linear model (Nelder and Wedderburn, 1972) and the package GLIM (Baker and Nelder, 1978) can be used for fitting the model. Stern and Coe (1984) used a shifted gamma distribution for the distribution of rainfall amounts and fitted a Fourier series to the log of the mean of the distribution

$$g(t) = \log[\mu(t)]. \tag{5.3}$$

If $g(t)$ is linear in the parameters then this is also an example of a generalized linear model and the GLIM package can again be used for fitting.

Zucchini and Adamson (1984) also used the logit transform for the transition probabilities, but used a Newton–Raphson iteration procedure to estimate the parameters. The procedures for parameter estimation using the logit transform are faster than those used by Woolhiser and Pegram (1979) and the parameters of the harmonics are estimated simultaneously. The logit transform leads to superior log likelihoods in climates where a period of very little rain is followed rather quickly by a wet season. Woolhiser and Pegram (1979) had to introduce a penalty function to provide a constraint on the transition probabilities. In Arizona I have found that this constraint is activated and the optimization scheme cannot adequately fit the dry–dry transition probabilities. This problem is illustrated in Figure 5.2. Note the rapid change in p_{00} from a value near to 1·0 in June to less than 0·6 in July. The Woolhiser and Pegram (1979) method could not accommodate this rapid variation as the penalty function became active and higher harmonics were rejected. With the logit transform the variation was fitted quite closely and substantially higher log likelihoods were obtained. This advantage has not been noted in comparisons made for several stations in more humid areas of the United States. In fact the algorithm described by Woolhiser and Roldan (1986) led to higher likelihoods in a majority of the cases examined, but did require substantially more computer time.

The problem of model selection has been addressed by Gates and Tong (1976), Chin (1977), Eidsvik (1980), Katz (1981), and Zucchini and Adamson (1984). Gates and Tong (1976) used the Akaike information criterion (AIC), Akaike (1974) to determine the appropriate order of Markov chains. Chin

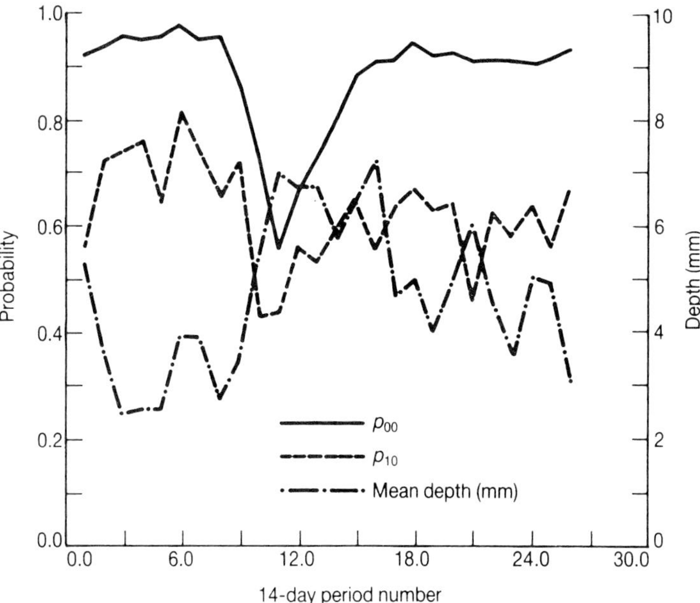

Figure 5.2 Markov chain-mixed exponential model parameters for Walnut Gulch 4, Arizona. 14-day periods beginning 1 March.

(1977) used the AIC technique to identify the appropriate order for Markov chains for more than 100 stations in the United States and found that higher than first-order chains were sometimes required. Eidsvik (1980) used a similar approach for several rainfall stations in Norway. He found that the required order of the chain increased with record length and that with large sample size the AIC minima are not well defined, suggesting significant uncertainty. Katz (1981) showed that the AIC has a substantial probability of overestimating chain order. He then used the Bayesian information criterion (BIC) proposed by Schwarz (1978) and the AIC to estimate the correct order using simulated data. He found that the BIC had a tendency to underfit and proposed a modification to the BIC to correct this problem.

I have calculated the log likelihood functions of the Markov chain and the mixed exponential model as a function of the number of parameters for the Fourier series representation and for two-step function representations (14- and 28-day periods). I analyzed data for the same stations studied by Roldan and Woolhiser (1982). The results are presented graphically for one station in Figures 5.3 and 5.4. The likelihood functions, of course, increase as the number of parameters increase, but after the first 8–10 parameters the rate of increase is slow. The AIC shows a minimum with the Fourier series representation. The results for the other stations were consistent with Figures 5.3 and 5.4.

The representation of seasonality is relevant in selecting the order of Markov chains, for Diggle (1984) has pointed out that stochastic dependence

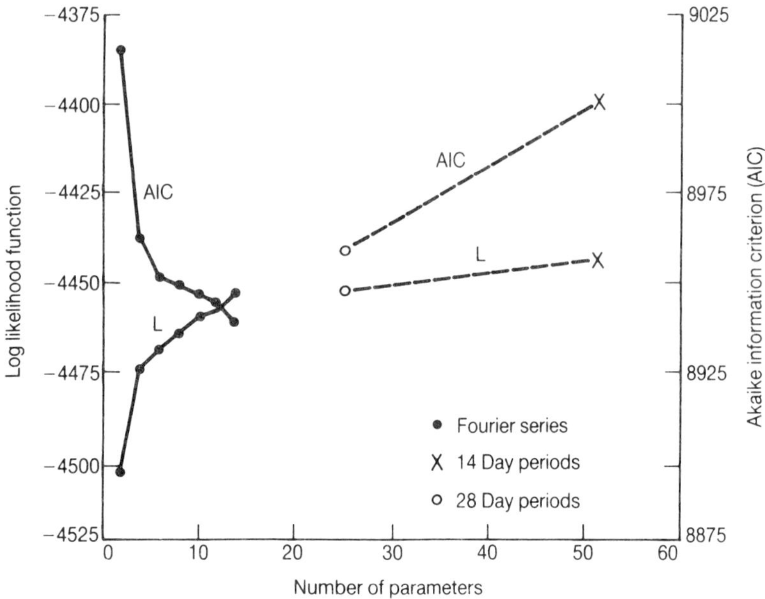

Figure 5.3 Log likelihood and AIC versus number of parameters – Markov chain.

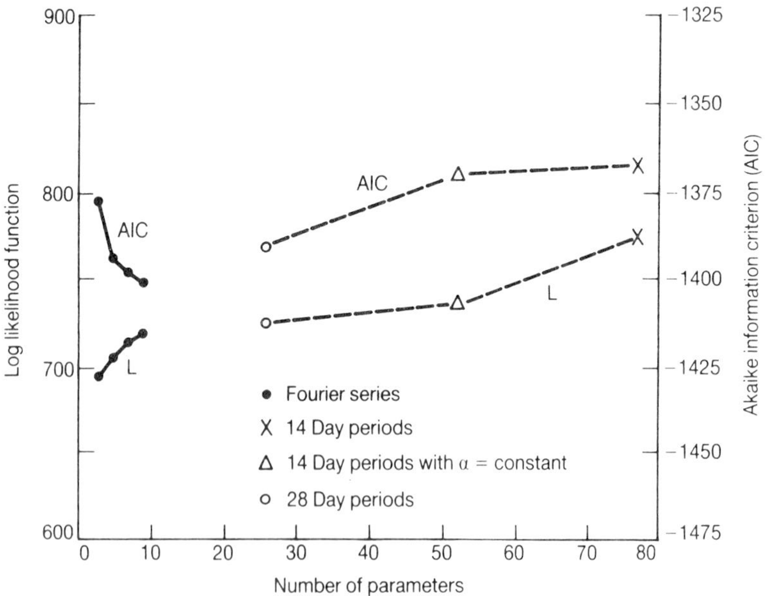

Figure 5.4 Log likelihood and AIC versus number of parameters – mixed exponential distribution.

and deterministic trends may be indistinguishable in practice. It should also be noted that the choice of model should depend to a great extent on the proposed use of the model. A model with a relatively simple structure that can be used in practice (agricultural decision making for example) can be superior to a more complicated model that is unlikely to be used. The important question then becomes: 'How sensitive are the decisions to be made to the precipitation model structure?' Pickering *et al.* (1989) examined this question by comparing three stochastic weather models with historical data as input to a nutrient loss simulation model.

5.4 Some problems with precipitation data

Persons who have collected precipitation data in the field and have some training in meteorology and statistics are usually quite aware of problems that may arise that affect the quality of the data. Unfortunately the collectors of field data often do not have training in meteorology and statistics.

Systematic errors and nonhomogeneities may be introduced into rainfall data by changing the type or location of a rain gauge or by construction of buildings nearby or the growth of trees. These factors should be noted in the station history and the user should always be aware of this potential problem.

Frequently there is substantial diurnal variability in the occurrence of rainfall as shown in Figure 5.5. In the United States precipitation data are gathered by a variety of organizations. The National Weather Service (NWS) has a relatively sparse network of first-order stations where an observer is always present. These data are very reliable and midnight is used as the day delimiter. The cooperative observer network is much denser but the observation time varies, usually being around 8.00 a.m. or 5.30–6.00 p.m. The Agricultural Research Service (ARS), the US Geological Survey (USGS) and others often operate rain gauges in conjunction with hydrological research. These data are obtained in analog or digital form by weighing recording gauges or in digital form with tipping-bucket gauges.

It has been well documented that tipping-bucket gauges under register during very intense rainstorms. The weighing-recording gauges with analog output on a chart may underestimate the number of days with small amounts of rainfall if the chart drum revolves daily or more frequently because the thickness of the pen trace may hide small rises. Unless rain gauges are shielded, all will show an underestimate of precipitation when it is accompanied by wind.

Woolhiser and Roldan (1986) found that the time of observation was a significant factor introducing 'noise' into both the occurrence process as defined by a first-order Markov chain and in the distribution of rainfall depth. They attributed this to diurnal variation of rainfall and to evaporation of small rainfall amounts for gauges read in the afternoon.

Data from cooperative stations usually show a reduced number of wet days and greater mean daily amounts as compared to data from nearby first-order stations. Part of this is due to observation time, but part is due to the

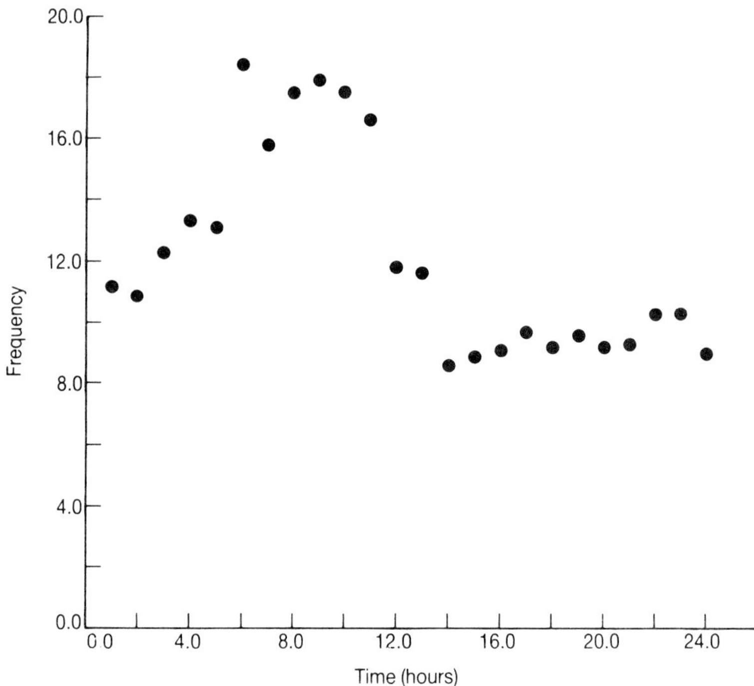

Figure 5.5 Hourly frequencies of precipitation, Hastings, Nebraska.

fact that the observer has missed making an observation on one day and the total accumulation is recorded on the following day. The effect of these methodological factors and errors is primarily reflected in the occurrence process and in the smaller amounts of rainfall. Histograms of rainfall depth are usually erratic, reflecting the rather small sample size, and in many cases it is noted that observers favor certain depths. For example some observers will record an excess of rainfall at 0·05 inches (1·25 mm) and a deficit at 0·04 (1·00 mm) and 0·06 inches (1·50 mm).

These practical problems have a bearing on model selection and on the testing criteria. For example, it may well be a waste of time to develop a complex model to fit the distributions of runs of wet and dry days precisely, because these statistics are very sensitive to the observation time and show wide variations among adjacent stations.

5.5 Application to practical problems

Agricultural applications of daily precipitation models fall into two categories: (1) short-run problems such as irrigation management, pest management and planning field operations and (2) long-range problems, including irrigation system design, drainage design, farm planning and examination of the national impact of erosion and flood-control policy. For the short-run problems, the precipitation models can be used in a simulation mode along with the models of irrigation scheduling (Hubbard and Wilhite, 1987), pest monitoring (Welch, 1984) and crop yield (Jones and Kiniry, 1986). With initial conditions known, several equally likely precipitation sequences can be simulated and distributions of various functionals such as time to next irrigation, time to pesticide application or crop yield can be obtained. The consequences of various management decisions can then be evaluated in a probabilistic sense.

Because these models require more than just precipitation as input, multivariate models including such variables as temperature and solar radiation must be used (see Richardson and Wright, 1984).

Stern and Coe (1984) demonstrate the use of rainfall models to obtain distributions of soil water content as a function of time and the distributions of dry spells. Zucchini and Adamson (1984) identified parameters for 2550 climatic stations in South Africa and used simulation procedures to examine the spatial characteristics of climatic indices, optimal planting time for maize and wheat and characteristics of drought. Because the parameters of rainfall models provide a very concise description of precipitation climatology, a combination of these parameters along with temperature and soils information should prove useful in determining the adaptability of plant species.

Some of the best examples of long-run or policy problems that can be examined using precipitation models include the evaluation of potential nonpoint pollution from agriculture through the use of the CREAMS model (Knisel, 1980) and the evaluation of the effects of erosion on crop productivity (Williams and Renard, 1985).

Many engineering activities are weather dependent and daily rainfall models can be useful in estimating the probability that a project may be disrupted by rain. Weather records or rainfall models are also useful in developing design criteria for drainage works, small dams, urban storm drainage structures and in the evaluation of the trafficability of unpaved areas.

5.6 Some new approaches

5.6.1 Conditioning models on monthly amounts

Wilks (1989) has developed an interesting precipitation model in which he estimated parameters for a Markov chain and gamma distributions of daily amounts separately for months in the lower 30%, middle 40% and upper 30% of the climatological distributions of total monthly precipitation. These classes correspond to those used by the Climate Analysis Center (CAC) of the National Oceanic and Atmospheric Administration for 30–90-day forecasts. Simulations of daily precipitation sequences can then be conditioned on the monthly forecasts.

Long-term simulations can be carried out by using probability mixtures of the conditional parameter sets which reproduce the observed probabilities of transitions among dry, near-normal, and wet months. These transition probabilities are represented by a three-state, first-order Markov chain. Wilks (1989) used generalized likelihood ratio tests to show that the increase from four to ten parameters per month is justified by the data. Distributions of monthly rainfall totals simulated using the conditionally derived suites of parameters exhibited upper and lower tails that were closer to the observed distributions than those obtained by unconditional parameters. It would be interesting to try this approach for stations with a monsoon climate.

5.6.2 Southern oscillation index

Simulation studies using daily precipitation models with annually periodic parameters typically result in underestimation of the variance of monthly and annual precipitation (Buishand, 1977; Zucchini and Adamson, 1984; Woolhiser *et al.*, 1988). This may be due to changes in data-collection techniques during the period of record, real long-term trends, or the assumption of annual periodicity may be incorrect because of large-scale meteorological circulation patterns that do not exhibit annual periodicities. One such phenomenon that has attracted recent scientific interest is the Southern Oscillation (SO). The Southern Oscillation is 'a coherent variation of barometric pressures at interannual intervals that is related to weather phenomena on a global scale, particularly in the tropics and subtropics' (Enfield, 1989). The SO is the atmospheric counterpart to El Niño, the warm, southward-flowing current along the coast of southern Ecuador and northern Peru that appears at irregular intervals. The SO is quantitatively described by the Southern Oscillation Index (SOI), the time series of the anomalies of atmospheric pressure differences between Papeete (Tahiti) and Darwin (Australia).

Several recent studies have documented statistical relationships between the El Niño–Southern Oscillation, commonly referred to as ENSO, and weather patterns and precipitation in the western United States and South America (Caviedes, 1975, 1984; Redmond and Koch, 1991). These studies have typically involved regression analyses between ENSO and annual or seasonal precipitation for groups of stations. These results prompt the question: 'Is it possible to incorporate the information in the SOI into a stochastic daily precipitation model?' In the remainder of this section I will describe one possible approach to answering this question.

We will assume that the Markov chain–mixed exponential model (MCME) (Woolhiser and Pegram, 1979) is appropriate for the stations investigated. In this model precipitation occurrence is described by a first-order Markov chain and the mixed exponential distribution is used for the distribution of daily rainfall, given that rain occurs.

Let

$$\begin{aligned} X(t) &= 0 \qquad \text{if day } t \text{ is dry, } t = t_1, \ldots, t_T \\ &= 1 \qquad \text{if day } t \text{ has rain over a threshold, } d. \end{aligned} \tag{5.4}$$

We assume that $\{X(t)\}$ is a first-order Markov chain with transition probabilities

$$P_{ij}(t) = P[X(t) = j \mid X(t-1) = i], \qquad i, j = 0, 1, \tag{5.5}$$

Let $Y(t)$ be the amount of precipitation on day t when $X(t) = 1$. We assume that $Y(t)$ is serially independent and is independent of $X(t-1)$. Let the random variable $U(t) = Y(t) - d$ be distributed as a mixed exponential (ME)

$$f_t(u) = \alpha(t)/\beta(t) \exp(-u/\beta(t)) + (1 - \alpha(t))/\delta(t) \exp(-u/\delta(t)) \tag{5.6}$$

where $0 < u < \infty$, d is a threshold (in the United States normally 0·01 inch (0·25 mm)), $0 < \alpha(t) < 1$, $0 < \beta(t) < \delta(t)$. The mean, $\mu(t)$, is given by

$$\mu(t) = \alpha(t)\beta(t) + [1 - \alpha(t)]\delta(t). \tag{5.7}$$

The model is nonhomogeneous so the parameters $p_{00}(t)$, $p_{10}(t)$, $\alpha(t)$, $\beta(t)$, and $\mu(t)$ are written in the polar form of a finite Fourier series

$$G_i(t) = G_{i0} + \sum_{k=1}^{m_i} [C_{ik} \sin(2\pi tk/365 + \phi_{ik})] \tag{5.8}$$

where $i = 1, 2, \ldots, 5$, $G_i(t)$ is the value of the ith parameter on day t, m_i is the maximum number of harmonics, G_{i0} is the mean, C_{ik} = amplitude of the kth harmonic and ϕ_{ik} = phase angle of the kth harmonic for the ith parameter. Instead of using the Markov chain transition parameters directly we use the logit transform as demonstrated by Stern and Coe (1984) and Zucchini and Adamson (1984):

$$g_{ij}(t) = \log\{p_{ij}(t)/[1 - p_{ij}(t)]\}. \tag{5.9}$$

The Fourier series are fit to the logits and the transition probabilities are obtained by the inverse transform

$$p_{ij}(t) = \exp[g_{ij}(t)]/\{1 + \exp[g_{ij}(t)]\}. \tag{5.10}$$

To incorporate the effect of the SO let us suppose that the periodic parameters are perturbed by a lagged linear function of the SOI

$$G_i'(t) = G_i(t) + b_i S(t - \tau_i) \tag{5.11}$$

where b_i and τ_i are parameters to be estimated from the data and $S(t)$ is the SOI on day t. Both of the parameters of the Markov chain and the mean, $\mu(t)$, of the mixed exponential distribution were assumed to be affected by the SOI.

The Fourier coefficients for the logits were estimated by maximum likelihood techniques as described by Zucchini and Adamson (1984) and coefficients for the parameters of the mixed exponential distribution were estimated by numerical maximum likelihood as described by Woolhiser and Roldan (1986). Then the characteristics of the likelihood response surface for the Markov chain and the mixed exponential were investigated by varying the parameters b_i and τ_i. A monthly SOI series was used, so $S(t)$ is represented as a step function.

Data from stations in Arizona, Idaho and Oregon were analyzed in this preliminary study. Some characteristics of the data are shown in Table 5.1. Note that there is a large range in the mean annual precipitation and in the mean number of wet days per year.

Results of perturbing the periodic logits of the transition probabilities are shown in Table 5.2. The increase in log likelihood resulted in a minimum

Table 5.1 Precipitation stations analyzed

Station	Years of record	Annual precipitation (mm)	Mean number wet days
Arizona			
Phoenix	1949–81	177	34·39
Prescott	1953–81	478	68·97
Tucson	1948–81	277	49·88
Walnut Gulch 4	1955–87	305	53·61
Idaho			
Boise	1940–89	304	91·38
Grangeville	1940–84	600	121·82
Reynolds Cr. 116	1962–81	469	105·45
Reynolds Cr. 163	1962–87	1128	132·15
Oregon			
Corvallis	1948–87	1078	156·68
Crater Lake	1947–87	1732	142·02
Bonneville Dam	1950–87	1945	171·87

Table 5.2 Effect of SOI on log likelihood functions for occurrence of rainfall – first-order Markov chain

Station	Unperturbed L	Perturbed L	b_i	Lag (days)
Arizona				
Phoenix	−3306·560	−3300·482*	+0·120	95
Prescott	−4240·921	−4236·580*	+0·100	95
Tucson	−4277·916	−4269·211*	+0·142	95
Walnut Gulch 4	−4140·479	−4129·321*	+0·147	93
Idaho				
Boise	−3696·153*	−3695·075	+0·04	330
Grangeville	−9576·560	−9573·180*	−0·04	104
Reynolds Cr. 116	−3767·502*	−3766·384	+0·053	150
Reynolds Cr. 163	−5199·465	−5196·788*	+0·060	0
Oregon				
Corvallis	−7320·965*	−7320·580	−0·02	90
Crater Lake	−7487·811*	−7487·811	0·00	0
Bonneville Dam	−7126·041*	−7124·312	−0·06	30

* Minimum AIC

AIC for six stations, with the Arizona stations being most strongly affected. The signs of the coefficients are fairly consistent with previous studies, with a negative SOI leading to more rainfall in the Southwest and the opposite effect in the Pacific Northwest.

Results of perturbing the seasonally varying mean of the ME are shown in Table 5.3. The perturbed mean precipitation resulted in the minimum AIC for all stations and the signs of the coefficients are consistent with expectations except for Boise, ID. Again the Arizona stations exhibit the strongest

Table 5.3 Effects of perturbing the mean of the ME distribution on the log likelihood function

Station	Unperturbed L	Perturbed L	b_i (mm)	Lag (days)
Arizona				
Phoenix	876·973	876·180*	−0·43	90
Prescott	950·163	955·856*	−0·60	60
Tucson	1184·797	1187·926*	−0·39	75
Walnut Gulch 4	1092·404	1096·816*	−0·45	89
Idaho				
Boise	2166·453	2170·052*	−0·19	0
Grangeville	4162·459	4164·689*	+0·13	60
Reynolds Cr. 116	1795·836	1803·659*	+0·39	101
Reynolds Cr. 163	705·685	714·536*	+0·40	180
Oregon				
Corvallis	2889·396	2894·771*	+0·20	104
Crater Lake	−1188·772	−1185·422*	+0·254	94
Bonneville Dam	−638·453	−635·349*	+0·254	115

* Minimum AIC

effects. The most common lag is about 90 days, which raises interesting possibilities of conditioned simulations of rainfall based upon the SOI for the past 90 days. However, this analysis is preliminary and we cannot reach strong conclusions based on the evidence presented. For example, although the Markov chain–mixed exponential appears to fit well for the Arizona and Reynolds Creek data at least a second-order Markov chain may be required for the Oregon stations. A more thorough analysis using more stations and higher-order Markov chains is presently under way.

5.6.3 Elevation effects

It is often desirable to have daily precipitation data in mountainous regions but frequently data are only available for valley stations and usually these data do not represent conditions at higher elevations. Therefore it would be useful if relationships could be developed between precipitation model parameters and characteristics of higher elevation sites such as elevation or annual precipitation so that simulation model parameter sets could be estimated. Hanson *et al.* (1989) utilized data from a network of rain gauges in southwest Idaho, United States, to determine if such relationships could be found. Fourier coefficients describing the seasonal variability for the parameters of the MCME model were estimated for each station and regression relationships were obtained between means, amplitudes and phase angles and annual precipitation. Highly significant relationships were found between the logits of the mean p_{00}, p_{10}, the amplitudes of the first harmonic of both p_{00} and p_{10} and annual precipitation, P_a. For the distribution of amounts, the logit of the weighting parameter, α, showed a linear decrease with P_a, the mean $[\alpha\beta + (1 - \alpha)\delta]$ and the amplitude of the first harmonic increased linearly with P_a, and the phase angle of the first harmonic exhibited a nonlinear decrease with P_a. The regression relationships were used to estimate a complete set of parameters for four sites and the statistical properties of 50 years of simulated data were compared with historical data at the same sites. Mean monthly precipitation and number of wet days were closely preserved. There appeared to be a slight tendency for the simulated data to have a lower variance than the historical data. The procedure appears promising but it must be tested over a larger area and information must be provided so that the user can evaluate the potential errors involved.

5.7 Discussion

Models to describe the daily precipitation process are well developed and a great deal of progress has been made recently in developing techniques for parameter estimation – particularly when the seasonal variation is described by Fourier series. Various practical problems related to data collection may affect the models chosen and may limit the degree of fit that can be obtained. Increasing use of crop yield models and models simulating runoff, erosion and chemical transport will lead to a greater demand for models for simulating precipitation and other weather variables.

A shortcoming of existing models is an inability to maintain the variance of simulated monthly and annual totals. Conditioning model parameters on monthly amounts (Wilks, 1989) or perturbing periodic parameters with the SOI both result in better agreement between the variance of simulated and observed annual total precipitation. These approaches should be investigated more thoroughly and the statistical problems and the practical impact should be assessed.

Acknowledgement

I wish to thank T. O. Keefer for his assistance in the SOI analysis.

References

Akaike, H. (1974). A new look at the statistical model identification. *IEEE Trans. Autom. Control*, **19** (6), 716–23.

Baker, R.J. and Nelder, J.A. (1978). *The GLIM System*, NAG, Oxford.

Buishand, T.A. (1977). Stochastic modeling of daily rainfall sequences. *Meded. Landbouwhogesch. Wageningen*, **77** (3), 211 pp.

Caskey, J.E. (1963). A Markov chain model for the probability occurrence in intervals of various lengths. *Mon. Weather Rev.*, **91**, 298–301.

Caviedes, C.N. (1975). El Niño 1972: Its climatic, ecological and human implications. *Geograph. Rev.*, **65**, 493–509.

Caviedes, C.N. (1984). El Niño 1982–83. *Geograph. Rev.*, **74**, 268–90.

Chang, T.J., Kavvas, M.L., and Delleur, J.W. (1984). Daily precipitation modeling by discrete autoregressive moving average processes. *Water Resources Res.*, **20**, 565–80.

Chin, E.H. (1977). Modeling daily precipitation occurrence process with Markov chain. *Water. Resour. Res.*, **13** (6), 949–56.

Coe, R. and Stern, R.D. (1982). Fitting models to daily rainfall data. *J. Appl. Met.*, **21**, 1024–31.

Delleur, J.W., Chang, T.J., and Kavvas, M.L. (1989). Simulation models of sequences of dry and wet days. *J. Irrig. and Drainage Engr.*, **115** (3), 344–57.

Dethier, B.E. and McGuire, J.K. (1961). The climate of the Northeast. Probability of selected weekly precipitation amounts in the Northeast region of the U.S. *Agronomy Mimeo 61–4*. Cornell Univ. Agr. Exp. Stn., Ithaca, NY.

Diggle, P.J. (1984). Discussion of paper by Stern and Coe. *J. Roy. Statist. Soc.* A, **147**, Part 1, 27–8.

Eidsvik, K.J. (1980). Identification of models for some time series of atmospheric origin with Akaike's information criterion. *J. Appl. Met.*, **19** (4), 357–69.

Enfield, D.B. (1989). El Niño, past and present. *Rev. Geophys.*, **21** (1), 159–87.

Feyerherm, A.M. and Bark, L.D. (1965). Statistical methods for persistent precipitation patterns. *J. Appl. Met.*, **4**, 320–8.

Feyerherm, A.M., Bark, L.D., and Burrows, W.C. (1965). Probabilities of sequences of wet and dry days in South Dakota. *North Central Region Res. Pub. 161* and *Tech. Bull. 139h*. Kansas Agric. Exp. Stn., Manhattan, KS.

Fitzpatrick, E.A. and Krishnan, A. (1967). A first order Markov model for assessing rainfall discontinuity in Central Australia. *Arch. Met. Geophys. Biokl.*, **B13**, 270–86.

Foufoula-Georgiou, E. and Lettenmaier, D.P. (1987). Markov renewal model for rainfall occurrence. *Water Resour. Res.*, **23** (5), 875–84.

Gabriel, K.R. and Neumann, J. (1962). A Markov chain model for daily rainfall in Tel Aviv. *Quart. J. Roy. Met. Soc.*, **88**, 90–5.

Gates, P. and Tong, H. (1976). On Markov chain modelling to some weather data. *J. Appl. Met.*, **15**, 1145–51.

Gifford, R.O., Ashcroft, G.L., and Magnuson, M.D. (1967). Probability of selected precipitation amounts in the Western Region of the U.S. *Western Regional Res. Pub.* and *T-8*. Nevada Agric. Exp. Stn., Reno, NV.

Green, J.R. (1964). A model for rainfall occurrence, *J. Roy. Statist. Soc.*, Series B, **26**, 345–53.

Hanson, C.L., Osborn, H.B., and Woolhiser, D.A. (1989). Daily precipitation simulation model for mountainous areas. *Trans. ASAE*, **32** (3), 865–73.

Heerman, D.F., Finkner, M.D., and Hiler, E.A. (1968). Probability of sequences of wet and dry days for eleven Western States and Texas. *Colorado Agric. Exp. Stn. Tech. Bull.*, 117.

Hubbard, K.G. and Wilhite, D.A. (1987). A demonstration and evaluation of the use of climate information to support irrigation scheduling and other agricultural operations. *CAMaC Prog. Report* 87–4. Center for Agricultural Meteorology and Climatology, Univ. of Nebraska, Lincoln.

Ison, N.T., Feyerherm, A.M., and Bark, L.D. (1971). Wet period precipitation and the gamma distribution. *J. Appl. Met.*, **10** (4), 658–65.

Jones, C.A. and Kiniry, J.R. (eds) (1986). *CERES-MAIZE: A Simulation Model of Maize Growth and Development*, Texas A & M University Press, College Station, TX, 194 pp.

Katz, R.W. (1977a). An application of chain-dependent processes to meteorology. *J. Appl. Prob.*, **14**, 598–603.

Katz, R.W. (1977b). Precipitation as a chain-dependent process. *J. Appl. Met.*, **16** (7), 671–6.

Katz, R.W. (1981). On some criteria for estimating the order of a Markov chain. *Technometrics*, **23** (3), 243–9.

Katz, R.W. (1984). Discussion of the paper by Dr. Stern and Mr. Coe. *J. Roy. Statist. Soc.* A., **147**, Part 1, 29.

Knisel, W.G. (ed.) (1980). *CREAMS: A Field Scale Model for Chemicals, Runoff and Erosion from Agricultural Management Systems*, U.S. Dept. of Agric., Conservation Research Report No. 26, 640 pp.

Lane, L.J. (1984). Surface water management: a user's guide to calculate a water balance using the CREAMS model. *LA-10177-M. Manual UC-70B*, Los Alamos National Laboratory, Los Alamos, N.M.

Mielke, P.W. (1973). Another family of distributions for describing and analyzing precipitation data. *J. Appl. Met.*, **10** (2), 275–80.

Nelder, J.A. and Wedderburn, R.W.M. (1972). Generalized linear models. *J. Roy. Statist. Soc.* A, **135**, 370–84.

Nicks, A.D. (1974). Stochastic generation of the occurrence, pattern, and location of maximum amount of daily rainfall. *Proc. Symp. Statistical Hydrology*, Misc. Publ. No. 1275, U.S. Dept. of Agriculture, ARS. 154–71.

Palmer, W.L., Barfield, B.J., and Hann, C.T. (1982). Sizing farm reservoirs for supplemental irrigation of corn. Part I: Modeling reservoir size yield relationships. *Trans. ASAE*, **25**, 372–6.

Pickering, N.B., Stedinger, J.R., and Haith, D.A. (1989). Weather Input for Non-Point source Pollution Models. *J. Irrig. and Drainage Engr.*, ASCE, **114** (4), 674–90.

Redmond, K.T. and Koch, R.W. (1991). Surface climate and streamflow variability in the Western United States and their relationship to large scale circulation indices. *Water Resour. Res.*, **27** (9), 2381–99.

Richardson, C.W. (1982). A comparison of three distributions for the generation of daily rainfall amounts. In Singh, V.P. (ed.), *Statistical Analysis of Rainfall and Runoff*, pp. 67–78, Proc. Int. Symp. on Rainfall–Runoff Modeling, Water Resour. Publ., 700 pp.

Richardson, C.W. and Wright, D.A. (1984). *WGEN: A Model for Generating Daily Weather Variables*. U.S. Dept. of Agric., Agric Res. Svc., ARS-8, 83 pp.

Roldan, J. and Woolhiser, D.A. (1982). Stochastic daily precipitation models 1. A comparison of occurrence models. *Water. Resour. Res.*, **18** (5), 1451–9.

Schwarz, G. (1978). Estimating the dimension of a model. *Ann. Statist.*, **6**, 461–4.

Shaw, R.H., Barger, G.L., and Dale, R.F. (1960). *Precipitation Probabilities in the North Central States*, Bull. 753, Missouri Agric. Exp. Stn., Columbia, MO.

Smith, J.A. (1987). Statistical modeling of daily rainfall occurrences. *Water Resour. Res.*, **23** (5), 885–93.

Smith, R.E. and Schreiber, H.A. (1974). Point processes of seasonal thunderstorm rainfall, 2. Rainfall depth probabilities. *Water Resour. Res.*, **10** (3), 418–23.

Stern, R.D. (1980). Analysis of daily rainfall at Samaru, Nigeria, using a simple two-part model. *Arch. Met. Geophys. Biokl.*, Ser. B, **28**, 123–35.

Stern, R.D. and Coe, R. (1984). A model fitting analysis of daily rainfall data. *J. Roy. Statist. Soc.* A, **147**, Part 1, 1–34.

Stidd, C.K. (1953). Cube root normal precipitation distributions. *Amer. Geophys. Union Trans.*, **34** (1), 31–5.

Von Bargen, K. (1967). A systems approach to harvesting alfalfa hay, *Trans. ASAE*, **10** (3), 318–19.

Welch, S.M. (1984). Developments in computer-based IPM extension delivery systems. *Ann. Rev. Ent.*, **29**, 359–81.

Wilks, D.S. (1989). Conditioning stochastic daily precipitation models on total monthly precipitation. *Water Resour. Res.*, **23** (6), 1429–39.

Williams, J.R. and Renard, K.G. (1985). Assessments of soil erosion and crop productivity with process models (EPIC). In *Soil Erosion and Crop Productivity*, Follett, R.F. and Stewart, B.A. (eds), American Society of Agronomy and Crop Science Society of America, Madison, WI, pp. 67–103.

Woolhiser, D.A. and Pegram, G.G.S. (1979). Maximum likelihood estimation of Fourier coefficients to describe seasonal variation of parameters in stochastic daily precipitation models. *J. Appl. Met.*, **18** (1), 34–42.

Woolhiser, D.A. and Roldan, J. (1982). Stochastic daily precipitation models 2. A comparison of distribution of amounts. *Water Resour. Res.*, **18** (5), 1461–8.

Woolhiser, D.A. and Roldan, J. (1986). Seasonal and regional variability of parameters for stochastic daily precipitation models: South Dakota, U.S.A. *Water. Resour. Res.*, **22** (6), 965–78.

Woolhiser, D.A., Hanson, C.L., and Richardson, C.W. (1988). *Microcomputer Program for Daily Weather Simulation*, U.S. Dept. of Agric., Agric. Res. Svc., ARS-75, 49 pp.

Zucchini, W. and Adamson, P.T. (1984). *The Occurrence and Severity of Droughts in South Africa*, Dept. of Civil Engr., Univ. of Stellenbosch and Dept. of Water Affairs, WRC Report No. 91/1/84, 198 pp.

Chapter 6

Aging functions and their nonparametric estimation in point process models of rainfall

M. J. Phelan

We treat the role of aging functions in point process models of rainfall and in the analysis of turbulence in rainfall fields. We do so in a class of models featuring a Le Camian representation of rainfall fields being generated by convective rain systems. One such function in the model specifies, for example, the lifetime distribution for the convective cells supporting the system, in which case the aging function is also called a cumulative hazard function. We consider the problem of nonparametric estimation of the aging functions from data drawn from a typical observational scheme and a method of tracking such cells over time and space. As the scheme yields only partial observation of the cells, we propose an Aalen-type estimator of the aging function for its ability to handle censored data. In the present work we develop the martingale property and the large-sample properties of the proposed estimator.

Keywords and phrases: aging functions, nonparametric estimation, point processes, rainfall fields, turbulence.

6.1 Introduction

The convective rain cell is an important geophysical construct in meteorological descriptions of precipitative phenomena. Together with stratiform precipitation, the notion of convective precipitation informs empirical treatments of a range of systems including rain bands in extratropical cyclones, mid-latitude thunderstorms, and tropical cloud clusters, see for example Houze (1981) and Houze and Hobbs (1982). For our purposes, a convective rain cell refers to a localized region of high-intensity rainfall embedded within a connecting region of stratiform precipitation of lower, more-or-less uniform intensity. This structure essentially defines a cloud cluster in Houze and Hobbs (1982) and the surface rainfall trace from such a cluster is shown in Figure 6.1. The data depicted there originates from radar observations of rain falling over the tropical Atlantic Ocean.

The observed structure of precipitating cloud systems has led many researchers in hydrology to consider models of rainfall fields based on point

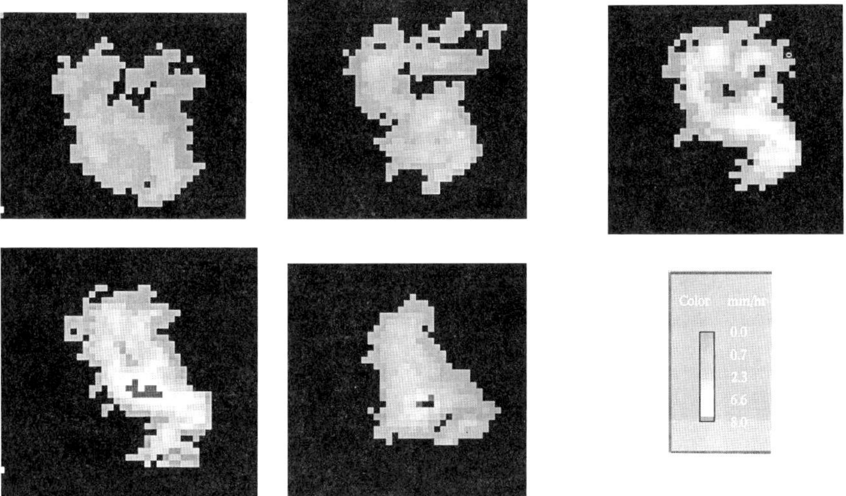

Figure 6.1 Surface rainfall trace from a cloud cluster observed during the GATE experiment over a 5-hour period. The observed pattern reflects the underlying structure of convective systems. The enclosed area is about 120 km square and each pixel represents a 4 km square.

processes. Such efforts include, for example, the work of Smith and Karr (1983), Waymire *et al.* (1984), Rodriguez-Iturbe *et al.* (1986–88), and Cox and Isham (1988). The point process in these models generates the spatiotemporal occurrence of convective rain cells. Typically, the rain cells are endowed with a set of distinguishing characteristics or marks, including a duration, water content or rain volume, dispersion, and velocity.

As is seen in Figure 6.1, precipitating cloud systems produce highly structured patterns of surface rainfall. These patterns mirror the observed structure of the cloud system itself, such as the pattern of convective and stratiform precipitation produced by the cloud clusters described above. In the models referred to above, surface rainfall fields are thus represented by a smoothing transformation of the point process generating the rain cells. The result is a spatial stochastic process describing the temporal evolution of the distribution of surface rainfall intensity. Stochastic descriptions of precipitation such as those originate with the work of Le Cam (1961), although it was not until after the Global Atmospheric Research Program's Atlantic Tropical Experiment (GATE) of 1974 that these descriptions came into widespread use.

The WGR model, proposed by Waymire *et al.* (1984), provides an example of a Le Camian representation of convective rainfall fields. Using data from the GATE experiment, Phelan and Goodall (1990) assessed the ability of the model to represent tropical rainfall fields. In analyzing the rainfall trace from a particular cloud cluster, we found that the model fitted reasonably well, but that the age of the system mattered. Specifically, the representational ability of the model apparently deteriorates during the dissipative phase of the system. We were thus led to consider the role of aging and, in particular, aging functions in models of rainfall fields.

We begin by treating three empirical results concerned with aging in the analysis of rainfall fields. These motivate three problems in modeling and inference from such fields, including the one considered here. We then introduce a class of Le Camian representations of rainfall fields and we provide a number of examples. Our treatment features the role of aging functions in the study of turbulence and parametrizations of the model. We describe a typical observational scheme and consider the problem of nonparametric estimation of the aging function. This turns out to be equivalent to estimating the distribution of the rain cell's characteristics. Our approach is to implement an Aalen-type estimator for its ability to handle censored data. Here we develop its large-sample properties and suggest a number of applications to the analysis of rainfall fields.

6.2 Aging in rainfall fields

The process of aging is of course integral to all transient phenomena. Here we present three empirical results that illustrate this point in the analysis of rainfall fields. We treat specifically the role of aging of convective rain cells in the propagation of storm systems, turbulence in rainfall fields, and the geometry and kinematics of stochastic representations of such fields. This motivates three problems in the modeling and inference for rainfall fields, including the one of present concern.

6.2.1 Propagation of cloud systems

The aging of convective rain cells in precipitating cloud systems apparently serves the persistence and propagation of the system. Houze (1981), for example, describes a mechanism for this service for tropical cloud clusters, which he defines roughly to be a cluster of convective cells supporting a canopy of stratiform cloud. Propagation of the system is thought to be the product of new cells developing at the leading edge of the system. That is, it is argued that the individual rain cell undergoes a life history consisting generically of three stages: a developing stage, a mature stage, and finally a dissipative stage. Stage one is characterized by the presence of updrafts alone, stage two by the coexistence of updrafts and downdrafts, and finally stage three by the presence of downdrafts alone. In tropical cloud clusters, as in mid-latitude thunderstorms, the occurrence of downdrafts in an aging rain cell may encourage the vertical displacement of upward-moving air to its level of free convection, thereby inducing the birth of a new rain cell. The system then propagates as this 'wave' of convection advances across the sky. In an earlier simulation study, Ogura and Takahashi (1971) investigated the dynamics of this phenomenon and obtained results consistent with the empirical record just described.

In our introduction, we drew attention to a class of stochastic models of precipitation having Le Camian representations of the rainfall fields. In our survey of such models, we find them to be reasonable descriptions of many features of actual rainfall fields, but each fails to account for the propagative mechanism described above. Generally speaking, they model the evolution of cloud systems along the lines of classical spatial birth-and-death processes, wherein individual cells are born in a Poisson process, live for a period of time while advancing across space, and then die. There is no deliberate linkage of the death of one cell to the birth of another as described above, nor is there much of an effort to model the life cycle of a rain cell.

The discussion above presents a problem in the modeling of precipitating cloud systems; namely that of developing a stochastic description of such systems having a propagative mechanism that is tied to the life cycle of convective cells. We expect such a development to improve the predictive capabilities of the model, particularly when predicting features of precipitative events associated with turbulence or the spatial distribution of rainfall; as we discuss below, both of these are affected by the process of aging of convective rain cells. In a working paper, we are addressing this problem by introducing what we call 'propagating coverage processes' that handle propagation in rainfall fields in the manner sought after above.

6.2.2 Turbulence and Taylor's hypothesis

The idea behind Taylor's hypothesis for rainfall fields originates with the study of turbulence in wind tunnels found in Taylor (1937). Briefly, Taylor analyzed the wind-velocity field in two ways. In the first he used the Eulerian coordinate system and derived the spectral decomposition of the field at a fixed point in space. In the second he used the Lagrangian coordinate system and derived the spectral decomposition of the field from a point moving in the stream of the wind flow. Taylor showed that if the small-scale fluctuations or turbulent velocities are negligible relative to the large-scale flow or stream velocity, then these two spectra are the same. The restricting condition here occurs in turbulent flows of high Reynolds number and no shearing, and the identity between spectra is known as Taylor's hypothesis.

Zawadski (1973) analyzed the role of Taylor's hypothesis in his statistical treatment of precipitation patterns. He observed that the expected duration of a convective rain cell may define a cut-off for the approximate validity of the hypothesis for rainfall fields. This empirical result has an interesting connection to aging in convective rain cells, so we reconstruct Zawadski's observation here.

We argue informally in describing this result. Let T denote a subset of $\mathfrak{R}$ and an index set of times. Let S denote a subset of $\mathfrak{R}^2$ and an index set of geographical locations on the surface of the Earth. Finally, let $R = (R(t, x))$, $t \in T$, $x \in S$ denote a random field over $T \times S$. Suppose that a storm passes over the region S and that the system moves at velocity v_0, where v_0 is a fixed vector in $\mathfrak{R}^2$. We assume that the process R models the surface rainfall intensity due to the storm as it passes over S in the horizon T, so that $R(t, x)$ denotes the rate of rain falling at location x at time t. For each (t, x) in $T \times S$, let $q(t, x, \tau)$ and $p(t, x, \tau)$ denote the lag-τ temporal and spatial autocovariance functions satisfying

$$q(t, x, \tau) = \text{cov}\,(R(t, x), R(t + \tau, x))$$

and

$$p(t, x, \tau) = \text{cov}\,(R(t, x), R(t, x - \tau v_0)),$$

for every τ in $\Re$, where cov denotes the covariance operator. Observe that these functions measure autocovariance in the Eulerian and the Lagrangian sense, respectively, so the Taylorian hypothesis for rainfall fields is equivalent to the equality of p and q.

Zawadski (1973) investigated the Taylorian hypothesis for rainfall fields by estimating p and q from rainfall generated by a mid-latitude thunderstorm. He assumed that they were free of t and x, so that his estimators depend only on the lag τ. The results are displayed graphically in Figure 6.2, where we have reproduced part of Figure 15 in Zawadski (1973). Observe that the two estimators of autocovariance are approximately equal for lags less than about 40 minutes, while they diverge at longer lags. On the basis of this result, Zawadski (1973) concluded for the approximate validity of the hypothesis at short lags. It turns out that the cut-off, 40 minutes, was also the average duration or lifetime of the convective rain cells observed during the generating storm. It thus appears that the aging, or in this case the death, of convective cells affects the structure of turbulence in rainfall fields.

This result establishes empirically a connection between the structure of turbulent flows and stochastic descriptions of precipitation. The further connection between this and the phenomenon of aging adds to our interest in the result. We propose that it be reinvestigated over a broader range of storm systems, hypotheses on turbulence in fluid flows, and stochastic descriptions of rainfall fields. We expect this reinvestigation to benefit from advancements made in our understanding of turbulence, the analysis of time series and spatial covariance in random fields, and the measurement of convective activity in precipitating storm systems; see for example Frisch and Orszag (1990), Brillinger (1988), Sampson and Guttorp (1990), and Rosenfeld (1987).

6.2.3 Geometry of stochastic representations

Waymire *et al.* (1984) developed a stochastic model of rainfall fields featuring a Le Camian representation of the rainfall intensity process. Gupta and Waymire (1987) showed mathematically that the covariance function of the process achieves the approximate validity of Taylor's hypothesis as observed empirically by Zawadski (1973). The details of this model, known as the WGR model, are given below, where we briefly illustrate this important result. Here we describe a statistical assessment of the ability of the model to *represent* actual rainfall fields.

We remarked earlier that a Le Camian representation of a rainfall field is determined by a smoothing transformation of the point process generating the rain cells. In the examples below, we specify this transformation by a particular choice of smoothing kernel, which entails the implicit choice of a particular 'geometry' to underly the storm system. For example, the WGR model associates a spherically symmetric Gaussian kernel to each rain cell,

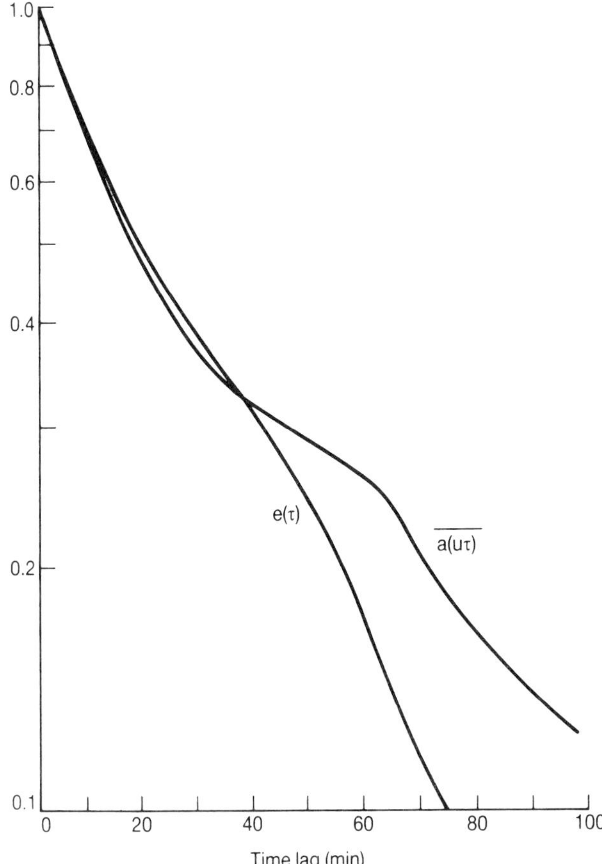

Figure 6.2 Partial reproduction of Figure 15 in Zawadski (1973) depicting the approximate validity of the Taylorian hypothesis for rainfall fields. In Zawadski's notation, e denotes the Eulerian covariance function, a denotes the Lagrangian covariance function.

which determines the spatial distribution of its water content. This Gaussian-based 'geometry' of the storm system is illustrated in Figure 6.3, where we represent selected contours of constant rainfall intensity being generated by a Poisson number of rain cells distributed randomly in the plane.

Phelan and Goodall (1990) assess a generalization of the WGR model for its ability to represent actual (GATE) rainfall fields. Our generalization differs from the original model in allowing the rain cells to have more random characteristics, including velocities, aging rates, water contents, and dispersions. Otherwise, our model preserves both the geometry and the kinematics of the original, since our principal aim was to assess these two features of the chosen Le Camian representation. We briefly describe this work here and summarize our conclusions below.

Consider the problem of fitting the generalized WGR model to the five-

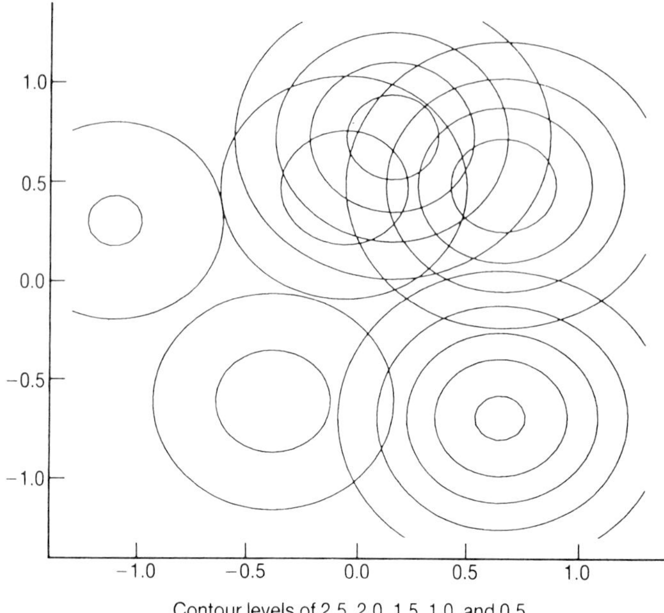

Figure 6.3 Selected contours of constant rainfall intensity as simulated from the WGR model.

hour rainfall trace depicted in Figure 6.1. As described above, we view the trace as the product of the precipitative activity of a cluster of convective cells supporting the system. Their activity is most readily seen at hour 3 in Figure 6.1. Notice the S-shaped ridge of high-intensity rainfall snaking through the image. This localizes the rainfall activity of six convective cells as assembled along the ridge. The cells are identified by their relative maxima in rainfall rate which are then tracked over the five-hour period. The details of cell identification and tracking are in Phelan and Goodall (1990).

Given the cells as identified above, Phelan and Goodall (1990) fit the generalized WGR model to the observed rainfall trace using nonlinear least squares. The results yield least-squares estimates of the vector of rain-cell characteristics that were then used to generate least-squares fits to the rainfall trace as shown in Figure 6.4(a). These images are thus model-based representations of the observed field. The residuals are shown in Figure 6.4(b). They provide an assessment of the ability of the model to represent actual rainfall fields. We expect a good fit to yield spatially and temporally homogeneous residuals resembling Gaussian noise.

Phelan and Goodall (1990) conclude that the model fits reasonable well, but that age matters. That is, consider the evolution of the spatial pattern of residuals in Figure 6.4(b). During hours 1, 2, and to a lesser extent hour 3, the residuals manifest little structure, thus suggesting a reasonable fit. In contrast, during hours 4 and 5, the residuals manifest increasing spatial inhomogeneity. This suggests a deteriorating fit. Phelan and Goodall (1990)

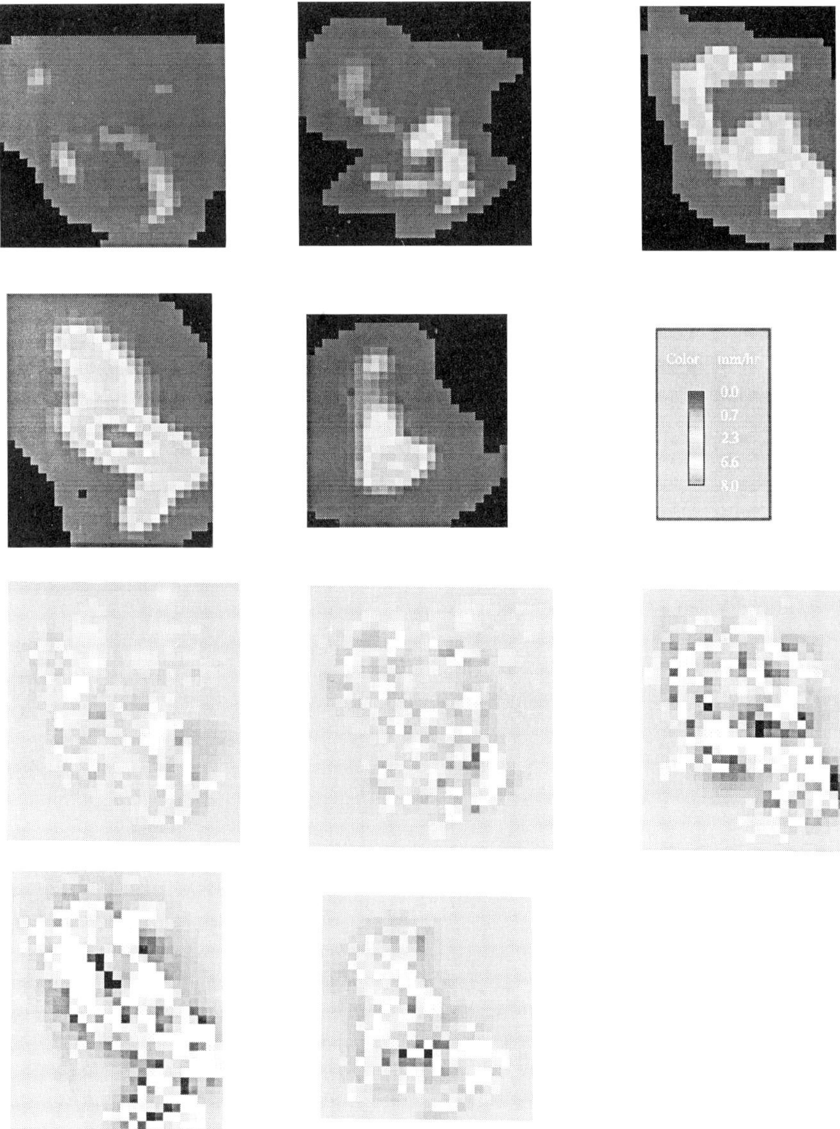

Figure 6.4 (a) Least-squares estimate of the surface rainfall trace as estimated from the observations in Figure 6.1 for the WGR model. (b) The residuals from the least-squares estimates of Figure 6.4a.

argue that the three earlier hours cover the formative and mature phase of the system, whereas the two later hours cover the dissipative phase. During dissipation, the rain cells lose momentum, age rapidly, and expire. The fit deteriorates apparently as the convective activity in the system dies off.

This result sparked the present interest in the phenomenon of aging in rainfall fields. We pursue this interest here by investigating the role of aging functions in stochastic models of rainfall. In doing so, we propose the problem of nonparametric estimation of aging functions as a first step in describing the life history of convective rain cells. The broad aim is to incorporate such descriptions into future models of precipitation.

6.3 Aging functions in rainfall models

The previous section outlines three empirical observations regarding the role of aging of convective rain cells in the propagation of storm systems, turbulence, and stochastic representations of rainfall fields. Here we specify a class of point-process models featuring a Le Camian representation of rainfall fields being generated by a convective storm system. Our purpose is to consider a parametrization of the model in terms of aging functions with particular regard for the one specifying the life distribution of the convective cells supporting the system. In doing so, we return to the problem of Taylor's hypothesis and examine the role of the life distribution in its approximate validity for rainfall fields.

6.3.1 Convective rain cells and Poisson random measures

Convective rain cells refer to a geophysical construct associated with the exchange of latent heat from the surface of the Earth to the atmosphere. We imagine warm columns of moisture-laden air rising into the upper atmosphere where the moisture condenses, reaches negative bouyancy, and subsequently falls as precipitation. This sets up a precipitating cell of cycling updrafts and downdrafts called a convective rain cell. Convection is the principal source of rainfall in the tropics, for example, and it produces rainfall fields such as those depicted in Figure 6.1.

To model the occurrence of convective rain cells, we introduce a class of point processes based on Poisson random measures. Let T denote a subset of $\Re$ and an index set of times. Let S denote a subset of $\Re^2$ and an index set of geographical locations on the surface of the Earth. Finally, for n an integer, let $(E, \mathscr{E})$ denote the product space $T \times S \times \Re^n_+$ endowed with its Borel sets. Now with reference to Definition 1.1 in Karr (1986), let N denote a point process on $(E, \mathscr{E})$ defined on a complete probability space $(\Omega, \mathscr{F}, \mathscr{P})$. The purpose of N is to generate the spatiotemporal occurrence of convective rain cells in $T \times S$ and endow them with a vector of rain-cell characteristics in $\Re^n_+$. That is, fix ω and suppose that (t, x, r) is an atom or a rain cell of $N(\omega)$. We then say that (t, x) specifies the placement of the rain cell in time and space and r specifies the n-vector of its characteristics. Such characteristics include a water content, a lifetime or duration, a scale or dispersion, and a speed along an ambient wind velocity field. We give two examples below.

Typically, one assumes that rain cells are generated as points in a Poisson process or as points in a cluster point process obtained thereof. Thus, N is taken to be a Poisson random measure of a Cox process on $(E, \mathscr{E})$, where the latter is also called a doubly stochastic Poisson process. Rainfall models of this sort are found in, for example, Waymire *et al.* (1984), Smith and Karr (1985), Rodriguez-Iturbe *et al.* (1987, 1988), and Cox and Isham (1988).

For the present purposes and to fix ideas, we consider the following Poisson random measure as our probability model for N. Let φ denote the probability distribution on $\Re^n_+$ satisfying

$$\begin{aligned}\varphi(\mathrm{d}r) &= \varphi(\mathrm{d}r_1, \ldots, \mathrm{d}r_n)\\ &= \mathrm{d}r_1 \cdots \mathrm{d}r_n \exp\left(-\sum_1^n r_i\right), \qquad (r_i \geqslant 0, i = 1, \ldots, n),\end{aligned} \tag{6.1}$$

so that φ is the distribution of n independent, exponential random variables. Consider the following definition.

Definition 6.1 Let N denote a Poisson random measure on $(E, \mathscr{E})$ having mean measure ν, say. We say that N is a *cardinal Poisson random measure* whenever ν satisfies

$$\nu(\mathrm{d}t, \mathrm{d}x, \mathrm{d}r) = \mathrm{d}t\, \mathrm{d}x \varphi(\mathrm{d}r),$$

for every $t \in T$, $x \in S$, and $r \in \Re^n_+$.

According to Definition 6.1, the process N distributes rain cells over time and space according to a Poisson process of unit rate, and it endows them with a set of n independent, exponentially distributed characteristics.

In general, we consider the class of point-process models of rainfall obtained by an elementary transformation of the cardinal Poisson random measure above. In the examples treated below, such transformations arise from a simple change of variables, including a change of rate in the Poisson process, or a change of distribution of the characteristics, or both. In anticipation of this discussion, we recall an important fact here. Let f denote a measurable mapping from E into E and suppose that N is a cardinal Poisson random measure having mean measure ν. Using Definition 1.36 of Karr (1986), the point process Nf^{-1} is the mapping of N by f, where f^{-1} denotes the inverse image of f. In the applications treated below, it turns out that Nf^{-1} is itself a Poisson random measure on $(E, \mathscr{E})$ having mean measure νf^{-1}.

6.3.2 Rainfall fields and Le Camian representations

Rainfall fields generally refer to the rate of intensity of rain falling over a geographic region. Le Camian representations of rainfall fields refer to random fields over $T \times S$ obtained by a smoothing transformation of the point process N. First introduced by Le Cam (1961), they form the basis of the stochastic descriptions of rainfall treated here.

Let $(E, \mathscr{E})$ denote the set $T \times S \times \Re^n_+$ equipped with its measurable sets. For generating rain cells and their characteristics, let N denote a cardinal Poisson random measure on $(E, \mathscr{E})$ having mean measure ν. For each $t \in T$

and $x \in S$, let $k(t, x)$ denote a positive, measurable function defined on E and let $k = \{k(t, x), (t, x) \in T \times S\}$ denote a family of such functions satisfying

$$\int_E \nu(\mathrm{d}z)k^2(t, x, z) < \infty, \tag{6.2}$$

for every (t, x) in $T \times S$. Now consider the spatial stochastic process $R = (R(t, x))$, $t \in T$, $x \in S$, where $R(t, x)$ satisfies

$$R(t, x) = \int_E N(\mathrm{d}z)k(t, x, z). \tag{6.3}$$

The $R(t, x)$ denote the intensity of rain falling on position x at time t, so the process R models the temporal evolution of the rainfall intensity over the geographic region S. We interpret eq. (6.3) as a smoothing transformation of N and refer to k as the corresponding smoothing kernel.

The process R is an example of a Le Camian representation of the rainfall field. By virtue of condition (6.2), R admits a mean function m and a covariance function q, say, satisfying

$$m(t, x) = ER(t, x) = \int_E \nu(\mathrm{d}z)k(t, x, z) \tag{6.4}$$

and

$$q(t, x, s, y) = \operatorname{cov}(R(t, x), R(s, y)) = \int_E \nu(\mathrm{d}z)k(t, x, z)k(s, y, z), \tag{6.5}$$

for every $t, s \in T$ and $x, y \in S$. These facts follow readily from the fact that N is a Poisson random measure and the derivation of Example 1.15 in Karr (1986).

We provide two examples.

(1) *WGR model.* This model is based on the WGR model as developed in Waymire *et al.* (1984). It provides an example of Gaussian-based geometry in a stochastic representation of precipitation. Let $n = 2$ and let T and S denote the sets $\mathfrak{R}$ and $\mathfrak{R}^2$, respectively, so that E denotes the product space $\mathfrak{R} \times \mathfrak{R}^2 \times \mathfrak{R}^2_+$. We assume that N is a cardinal Poisson random measure on $(E, \mathscr{E})$. Now fix ω and suppose that (t, x, w, z) is an atom of $N(\omega)$. We then say that the rain cell has placement (t, x) in time and space, water content w, and duration z. For purposes below, we assume that each rain cell moves with fixed velocity ν_0 in the plane.

Next let σ denote a fixed, positive number. Let k denote the smoothing kernel satisfying: for each $t \in T$ and $x \in S$,

$$k(t, x, u, y, w, z) = w1(u \leqslant t)1(z > t - u)\sqrt{(2/\pi)}\sigma e^{-\sigma d(x, y + (t-u)\nu_0)},$$

for every $u \in T$, $y \in S$, and $w, z \in \mathfrak{R}_+$. Here $d(\cdot, \cdot)$ denotes the *square* of the Euclidean distance between points in S. Finally, using this smoothing kernel, let R denote the rainfall intensity process as defned by eq. (6.3).

For each t and x, $R(t, x)$ models the total intensity of rain falling at (t, x). It is the sum of the contributions of those rain cells born before time t and that remain active at that time. In particular, fix ω and suppose that (u, y, w, z) is an atom or rain cell of $N(\omega)$. By definition, $k(t, x, u, y, w, z)$ is the contribution of that rain cell to the total intensity $R(\omega; t, x)$. Note that, as long as its time of birth u precedes t and its duration z exceeds its current age $t - u$, its contribution is proportional to its water content ω. The proportion decays exponentially with the distance between x and the rain cell's current location $y + (t - u)v_0$. Therefore, the smoothing kernel of Model 1 generates Gaussian contours of constant rainfall intensity such as those rendered in Figure 6.3. ■

(2) *Random-disks model.* This model is developed in Cox and Isham (1988). It envisions the coverage of a precipitating rain cell to be a random disk in the plane. The initial placement of the disk is specified at birth, when the rain cell is characterized by a radius, water content, duration, and velocity. A rain cell becomes active at birth and it remains so for its duration. We assume here that the ensemble of rain cells propagates across the plane along the fixed direction v_0, the individual motions being at uniform rate depending upon that rain cell's speed. In doing so, they deposit their water contents uniformly over their identifying disks and uniformly over their durations. Therefore, the rain cell vanishes.

Let $n = 4$ and let T and S denote the sets $\mathfrak{R}$ and $\mathfrak{R}^2$, respectively, so that E denotes the product space $\mathfrak{R} \times \mathfrak{R}^2 \times \mathfrak{R}_+^4$. We assume that N is a cardinal Poisson random measure on $(E, \mathscr{E})$. Now fix ω and suppose that (t, x, r, s, w, z) is an atom of $N(\omega)$. We then say that the rain cell has placement (t, x) in time and space, radius or dispersion r, speed s, water content w, and duration z.

Next let k denote the smoothing kernel satisfying: for each $t \in T$ and $x \in S$,

$$k(t, x, u, y, r, s, w, z) = w1(u \leqslant t)1(z > t - u)1(d(x, y + (t - u)sv_0) \leqslant r),$$

for every $u \in T$, $y \in S$, and $r, s, w, z \in \mathfrak{R}_+$. Here $d(\cdot, \cdot)$ denotes the Euclidean distance between points in S. Now, using this smoothing kernel, let R denote the rainfall intensity process as defined according to eq. (6.3).

For each t and x, $R(t, x)$ models the total intensity of rain falling at (t, x). According to the model, this is given by the sum of the water contents or intensities of those active rain cells that cover the site x. As we see from the kernel above, the contributing rain cells are determined by a coverage process of random disks migrating in the plane. ■

6.3.3 Aging functions and turbulence

In the models above, we take the characteristics of the convective rain cells to be exponential random variables occurring in a cardinal Poisson random measure. As a modest generalization, we let these have more-or-less arbitrary, but continuous distributions. We develop the generalization in terms of a transformation of the underlying Poisson random measure. In doing so, we fix ideas initially and develop the transformation with respect to the durations of the rain cells. Along the way, we treat the connection between these and Taylor's hypothesis for turbulence in rainfall fields.

The life history of a convective cell determines the length-of-life or duration of a convective rain cell. Here we characterize the life distribution of such durations in terms of aging functions. Consider the following definition.

Definition 6.2 Let X denote a positive random variable modeling the duration of a convective rain cell. Let $A = (A(t)), t \geqslant 0$ denote a strictly increasing, continuous function satisfying $A(0) = 0$ and $A(t)$ increases to infinity with t. We say that X has *aging function* A whenever the life distribution of X satisfies

$$P(X > t) = \exp(-A(t)), \qquad (t \geqslant 0),$$

so that the survival curve of X decreases exponentially in $A(t)$.

We refer to A of Definition 6.2 as the *aging function* parametrizing the life distribution of X. If X has the exponential distribution with unit parameter, then its aging function equals t for every $t \geqslant 0$. On the other hand, if X has the Pareto distribution with shape parameter b and scale parameter c, then its aging function A satisfies

$$A(t) = b \ln(1 + ct/b) \qquad (t \geqslant 0). \tag{6.6}$$

In the statistical context of survival analysis, aging functions are called cumulative-hazard functions. We interpret A as the intrinsic clock that measures the age of a rain cell from its duration. In this we refer to the well-known transformation identifying the duration with a time-change of its age or a unit-parameter exponential random variable. That is, if X has aging function A, then $A(X)$ is a unit-parameter exponential random variable. Conversely, if W is a unit-parameter exponential random variable and A is an aging function, then $A^{-1}(W)$ has aging function A, where A^{-1} denotes the functional inverse of A. We say that $A(X)$ measures the age or intrinsic duration of the rain cell, whereas $A^{-1}(W)$ measures its duration as a time change of its age. As the clock in this transformation, A measures the cumulative rate of aging of the rain cell.

We next define a generalization of our rainfall model in terms of a transformation of the underlying cardinal Poisson random measure. The transformation involves a change of rate in the Poisson process coupled with a change in distribution for the rain cell characteristics. For example, let c denote a positive number and let $A, B, \ldots, C$ denote n aging functions as defined in Definition 6.2. Consider the transformation f satisfying

$$f(t, x, u, v, \ldots, w) = (t/c, x, A^{-1}(u), B^{-1}(v), \ldots, C^{-1}(w)) \tag{6.7}$$

for every $t \in T$, $x \in S$, and $(u, v, \ldots, w) \in \Re^n_+$. In the following definition, we refer to this measurable mapping of E into itself as a scale.

Definition 6.3 Let M denote a random measure on $(E, \mathscr{E})$. Let N denote a cardinal Poisson random measure on $(E, \mathscr{E})$ and let f denote the transformation defined above by eq. (6.7). We say that M is a *cardinal Poisson process at scale* f whenever M is the mapping of N under the change of variable f.

According to Definition 6.3, M is stochastically equivalent to a cardinal Poisson random measure up to a change of scale. We noted earlier that it is given by the random measure Nf^{-1}, where f^{-1} denotes the inverse image of f. Moreover, if M has mean measure μ and N has mean measure ν, then μ satisfies

$$\begin{aligned}\mu(\mathrm{d}t, \mathrm{d}x, \mathrm{d}r) &= \nu f^{-1}(\mathrm{d}t, \mathrm{d}x, \mathrm{d}r)\\ &= c\, \mathrm{d}t\, \mathrm{d}x \varphi \circ (A, B, \ldots, C)(\mathrm{d}r)\\ &= c\, \mathrm{d}t\, \mathrm{d}x A(\mathrm{d}u)\mathrm{e}^{-A(u)} B(\mathrm{d}v)\mathrm{e}^{-B(v)} \ldots C(\mathrm{d}w)\mathrm{e}^{-C(w)} \qquad (6.8)\end{aligned}$$

for every $t \in T$, $x \in S$ and $r = (u, v, \ldots, w) \in \Re^n_+$. The process M thus distributes rain cells over time and space according to a Poisson process of rate c, and it endows them with a set of n independent characteristics having respective aging functions $A, B, \ldots, C$. We work with this parametrization of the model below.

We give an example of this parametrization in the context of each of the models introduced above. At the same time, we revisit the approximate validity of Taylor's hypothesis for rainfall fields as described empirically by Zawadski (1973). In a mathematical treatment of this result, we demonstrate the role of the aging function of the rain cells' durations.

(a) Taylor's hypothesis in the WGR model

Consider the WGR model as introduced above. Recall that a cardinal Poisson process generates the occurrence of rain cells having random, exponentially distributed water contents and durations.

Let α denote a fixed, positive number. Let Φ denote the distribution function of a Gaussian random variable having mean 0 and variance 1. We introduce two aging functions A and B satisfying

$$A(t) = -\ln(1 - \Phi(\ln(t))) \qquad \text{and} \qquad B(t) = \alpha t \qquad (t \geqslant 0).$$

Note that A is the aging function of a so-called lognormal random variable, whereas B is that of an exponential random variable with parameter α.

Next we introduce the transformation f satisfying

$$f(t, x, w, z) = (t, x, \exp(\Phi^{-1}(1 - \mathrm{e}^{-w})), z/\alpha)$$

for every $t \in T$, $x \in S$, and $w, z \in \Re_+$. Here the functional inverse of A transforms w, the functional inverse of B transforms z. The function Φ^{-1} is thus the functional inverse of Φ.

Let N denote the Poisson random measure generating the rain cells and their characteristics. We consider the mapping of N by f and denote the resulting random measure by M. According to Definition 6.3, M is a cardinal Poisson process at scale f having mean measure μ satisfying

$$\mu(\mathrm{d}t, \mathrm{d}x, \mathrm{d}w, \mathrm{d}z) = \mathrm{d}t\, \mathrm{d}x\, \mathrm{d}w(\phi(\ln(w))/w)\, \mathrm{d}z\, \alpha \mathrm{e}^{-\alpha z}$$

for every $t \in T$, $x \in S$ and $w, z \in \Re_+$. Here ϕ denotes the density belonging to Φ. Accordingly, in this rescaled WGR model, rain cells are distributed over time and space as points in a Poisson process of unit rate, and they are endowed with lognormally distributed water contents and exponentially distributed durations having parameter α.

We show that the Taylorian hypothesis for rainfall fields is approximately valid for the WGR model as demonstrated in Gupta and Waymire (1987). First observe that all of the rain cells move at the fixed velocity v_0 so that there are no small-scale fluctuations in the velocity field. This means that the kinematics in the model satisfy a condition commensurate to Taylor's condition on turbulent velocities as cited above as 'sufficient' for the hypothesis in his analysis of wind-velocity fields. To demonstrate our claim, it thus remains to compare the Lagrangian to the Eulerian autocovariance function as determined from the rainfall field.

Let $R = (R(t, x))$, $t \in T$, $x \in S$ denote the rainfall field in the rescaled WGR model. For each t and x, $R(t, x)$ satisfies

$$R(t, x) = \int_{T \times S \times \Re_+^2} M(\mathrm{d}u, \mathrm{d}y, \mathrm{d}w, \mathrm{d}z) k(t, x, u, y, w, z),$$

where k denotes the smoothing kernel specified earlier for the WGR model. The covariance function belonging to R is given by eq. (6.5) upon substituting the measure μ given above for ν. Therefore, for $\tau > 0$, we have

$$\begin{aligned} q(t, x, t + r, x) &= \operatorname{cov}(R(t, x, R(t + \tau, x)) \\ &= \int_{T \times S \times \Re_+^2} \mu(\mathrm{d}u, \mathrm{d}y, \mathrm{d}w, \mathrm{d}z) \\ &\quad \times k(t, x, u, y, w, z) k(t + \tau, x, u, y, w, z) \\ &= \mathrm{e}^{-\alpha\tau} \int \mu(\mathrm{d}u, \mathrm{d}y, \mathrm{d}w, \mathrm{d}z) \\ &\quad \times k(t, x, u, y, w, z) k(t, x - \tau v_0, u, y, w, z) \\ &= \mathrm{e}^{-\alpha\tau} \operatorname{cov}(R(t, x), R(t, x - \tau v_0)) \\ &= \mathrm{e}^{-\alpha\tau} q(t, x, t, x - \tau v_0), \end{aligned}$$

where these equalities easily follow from either definition, direct substitution, or elementary algebraic manipulation. Here $q(t, x, t + \tau, x)$ measures the lag-τ autocovariance of the rainfall field in the Eulerian sense, whereas $q(t, x, t, x - \tau v_0)$ measures the same in the Lagrangian sense. Observe that for $\alpha\tau$ negligible or $\tau \ll 1/\alpha$ we have

$$q(t, x, t + \tau, x) = q(t, x, t, x - \tau v_0) + O(\tau),$$

where $O(\tau)/\tau$ remains bounded and $O(\tau)$ tends to zero with τ. This equation, of course, asserts the approximate validity of Taylor's hypothesis for 'small' lags. The cut-off is $1/\alpha$ or the expected duration of the rain cells, confirming the empirical observation of Zawadski (1973) as described above. Here we emphasize that the cut-off is determined by the aging rate of the rain cells.

(b) Taylor's hypothesis in the random-disks model

Consider the random-disks model as introduced above. Recall that a cardinal Poisson process generates the occurrence of rain cells characterized by random disks in the plane. Each rain cell propagates at random speed along a fixed velocity, while it uniformly distributes its water content over the identifying disk for a period of time given by its duration.

Let a, b, c, and d denote fixed, positive numbers. We introduce two aging functions A and B satisfying

$$A(t) = \int_0^t \mathrm{d}s \left(\int_0^\infty \mathrm{d}u \, \mathrm{e}^{-bu} (1 + u/s)^{a-1} \right)^{-1}$$

and

$$B(t) = \mathrm{d} \ln (1 + ct/d) \qquad (t \geqslant 0).$$

Note that A is the aging function of a Gamma random variable with shape parameter a and scale parameter b, whereas B is that of a Pareto random variable with scale parameter c and shape parameter d.

Next we introduce the transformation f satisfying

$$f(t, x, r, s, w, z) = (t, x, r, A^{-1}(s), w, \mathrm{d}(\mathrm{e}^{z/d} - 1)/c)$$

for every $t \in T$, $x \in S$, and $r, s, w, z \in \Re_+$. Here A^{-1} denotes the functional inverse of A and the functional inverse of B transforms the variable z.

Let N denote the Poisson random measure generating the rain cells and their characteristics. We consider the mapping of N by f and denote the resulting random measure by M. Accordingly, M is a cardinal Poisson process at scale f having mean measure μ satisfying

$$\mu(\mathrm{d}t, \mathrm{d}x, \mathrm{d}r, \mathrm{d}s, \mathrm{d}w, \mathrm{d}z) = \mathrm{d}t \, \mathrm{d}x \, \mathrm{d}r \, \mathrm{e}^{-r} \, \mathrm{d}s \, b(bs)^{a-1} \, \mathrm{e}^{-bs} / \Gamma(a) \, \mathrm{d}w \, \mathrm{e}^{-w} \, \mathrm{d}z \, c(1 + cz/d)^{-(d+1)}$$

for every $t \in T$, $x \in S$, and $r, s, w, z \in \Re_+$. Here Γ denotes the gamma function. This transformation defines a rescaled random-disks model. The rain cells are distributed over time and space as points in a Poisson process of unit rate. They are then endowed with exponentially distributed radii and water contents, Gamma distributed speeds, and Pareto distributed durations.

We next investigate the Taylorian hypothesis for rainfall fields in the rescaled random-disks model. Let $R = (R(t, x))$, $t \in T$, $x \in S$ denote the rainfall field. For each t and x, $R(t, x)$ satisfies

$$R(t, x) = \int_{T \times S \times \Re_+^4} M(\mathrm{d}u, \mathrm{d}y, \mathrm{d}r) k(t, x, u, y, r),$$

where M is defined above and k is the smoothing kernel specified earlier for the random-disks model. The notation is simplified here on letting r denote a point in $\Re_+^4$. To investigate the hypothesis, it remains to compare the spatial with the temporal autocovariance function belonging to R.

The covariance function belonging to R is given by eq. (6.5) upon substituting the mean measure μ given above for ν. Therefore, for $\tau > 0$, the lag-τ temporal autocovariance satisfies

$$\begin{aligned} q(t, x, t+\tau, x) &= \operatorname{cov}(R(t, x), R(t+\tau, x)) \\ &= \int_{T\times S\times \Re^4_+} \mu(\mathrm{d}u, \mathrm{d}y, \mathrm{d}r) k(t, x, u, y, r) k(t+\tau, x, u, y, r) \\ &= \int_{-\infty}^{t} \mathrm{d}u\, \mathrm{e}^{-B(t+\tau-u)} \int_0^\infty A(\mathrm{d}s) s^2 \mathrm{e}^{-A(s)} \\ &\quad \times \int_0^\infty \mathrm{d}r\, s\mathrm{e}^{-sr} r^2 h(\tau \|v_0\| / r), \end{aligned}$$

where A and B are the aging functions introduced above and $\|v_0\|$ denotes the length of v_0. The function $\delta \to h(\delta)$ denotes the area of the intersection of two disks of unit radius whose centers are δ apart; it is given by eq. (6.5) in Cox and Isham (1988). Obtained after some simplification, the third equality specifies the Eulerian autocovariance function for the rainfall field R. For comparison, the Lagrangian autocovariance function satisfies

$$\begin{aligned} q(t, x, t, x-\tau v_0) &= \operatorname{cov}(R(t, x), R(t, x-\tau v_0)) \\ &= \int_{-\infty}^{t} \mathrm{d}u\, \mathrm{e}^{-B(t-u)} \int_0^\infty \mathrm{d}r\, \mathrm{e}^{-r} r^2 h(\tau \|v_0\| / r). \end{aligned}$$

Observe that the aging functions A and B dominate the comparison of the two autocovariance functions, where these determine the distribution of the speeds and the durations of the rain cells. This is qualitatively the observation made in Cox and Isham (1988) as raised by their treatment of this comparison.

We give two conditions under which the Taylorian hypothesis is approximately valid for the rainfall field R. First suppose that $a = b$ so that the rain cell speeds have mean 1 and variance $1/b$. Then it is easily demonstrated that $q(t, x, t+\tau, x)$ satisfies

$$q(t, x, t+\tau, x) = \mathrm{e}^{-B(\tau)} Q(t, \tau) q(t, x, t, x-\tau v_0) + O(1/b),$$

where $bO(1/b)$ remains bounded and $O(1/b)$ goes to zero as b tends to infinity. Here $Q(t, \tau)$ satisfies

$$Q(t, \tau) = \left(\int_{-\infty}^{t} \mathrm{d}u\, \mathrm{e}^{-C_\tau(t-u)} \right) \Big/ \left(\int_{-\infty}^{t} \mathrm{d}u\, \mathrm{e}^{-B(t-u)} \right),$$

where C_τ is the aging function satisfying

$$C_\tau(t) = d \ln(1 + ct/(d + c\tau)) \qquad (t \geqslant 0).$$

Therefore, if $\tau \ll 1/c$ and if b is 'large', then $q(t, x, t+\tau, x)$ well approximates $q(t, x, t, x-\tau v_0)$, so that Taylor's hypothesis is approximately valid for this model. Note that for b sufficiently large the kinematics of the model satisfies a condition commensurate to Taylor's condition of having negligible small-scale fluctuations in the velocity field of the storm. Given that condition, the condition $\tau \ll 1/c$ suggests the cut-off phenomenon as observed with the WGR model.

6.4 Nonparametric estimation of aging functions

The previous section treated the role of aging functions in a class of rainfall models. The models are based on Poisson random measures that generate the occurrence of convective rain cells. We used aging functions to parametrize the mean measure of the Poisson process and demonstrated their importance in analyzing turbulence in rainfall fields.

Here we consider the problem of estimating aging functions from observed rainfall fields. Note well that this is in essence equivalent to estimating the distribution of the rain-cell characteristics. We begin by describing a typical observational scheme and the generated data. We then introduce a nonparametric estimator of the aging function in the class of models defined above. We do so in the context of a rescaled-WGR model, where we specify the principal ideas in detail. Our arguments, however, readily carry over to a rescaled random-disks model or any model in the class considered here.

Observational scheme. Imagine that we are positioned at a fixed weather radar station such as the one presently operating off Darwin, Australia. From time to time, storm systems pass in the vicinity of our instruments, which record the activity of the supporting convective cells. In a typical field operation the radar beams scan a spherical volume of the troposphere over a period of time. The aim is to measure rainfall intensity or rain volume rate as a function of reflectivity from the water content in the volume of space.

Here we consider the observed intensity of rainfall reaching the surface of the Earth and the physical properties of the convective cells supporting this precipitation. We specifically assume that our observations include the set of rain-cell characteristics as specified in the models described above. It turns out that Rosenfeld (1987) provides an objective method of determining these characteristics from the weather-radar data described above. The method analyzes the temporal evolution of the rainfall intensity and extracts measures of the desired characteristics of the rain cells, including trajectories, durations, water contents, velocities, and so on. As we see below, some of these random variables are only partially observed during a typical field operation, so we must deal with censored data in solving our estimation problem.

Censoring and the observational scheme in the WGR model. We fix ideas in the context of the WGR model. Let T_0 denote an interval of time in T and let S_0 denote a disk of fixed radius in S. Here T_0 refers to the period of observation and S_0 refers to the field of view of our instruments, so that $T_0 \times S_0$ denotes the coverage window of our observational scheme. Recall that according to the kinematics of the WGR model, the trajectory of a rain cell traces a line segment in S along the fixed velocity v_0. Under our observational scheme, we observe that part of the trajectory contained in the field of view S_0, for every rain cell that enters S_0 during the period of observation. For every one that does so, moreover, this yields a partial observation of the temporal evolution of its rainfall intensity and so a partial observation of its derived characteristics.

Let M denote the Poisson random measure generating the rain cells and their characteristics in $T \times S \times \Re_+^2$ in a possibly rescaled WGR model. We identify M with the countable sequence $(T, X, W, Z) = (T_k, X_k, W_k, Z_k)$, $k \geqslant 0$ of $T \times S \times \Re_+^2$-valued random variables. For the rain cell having label k, (T_k, X_k) specifies its time and place of birth in $T \times S$, while W_k specifies its water content, and Z_k specifies its duration. Here we describe the partial observation of the W_k and the Z_k with respect to our observational scheme.

For each $k \geqslant 0$, let L_k denote the line segment in S traced by the trajectory of the rain cell labeled k. The L_k satisfy

$$L_k = \{x : x \in S, x = X_k + tv_0 \quad \text{for some } t \in [0, Z_k)\}. \tag{6.9}$$

Now provided $\|v_0\| > 0$, we have

$$Z_k = l(L_k)/\|v_0\|,$$

where $l(L_k)$ denotes the length of L_k. We thus observe the duration of the rain cell by observing its trajectory. According to our observational scheme, however, we observe only the subtrajectory $L_k \cap S_0$, which is that part of L_k contained in S_0. Figure 6.5 depicts the four contingencies, depending on the position of the rain cell at birth and at death. This partial observation of the L_k entails the partial observation of the Z_k. In particular, we observe

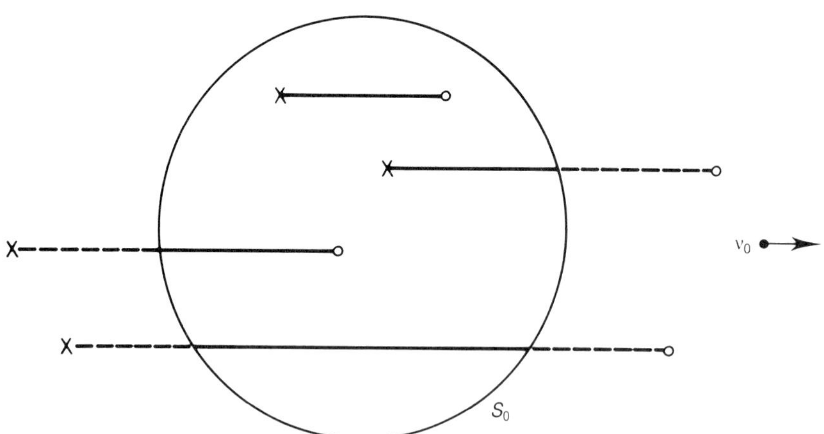

Figure 6.5 Four contingencies in the partial observation of L_k. Here S_0 denotes a disk in the plane, v_0 points in the horizontal direction, x denotes the position of the rain cell at birth, and 0 denotes its position at death.

only the censored duration $\tilde{Z}_k$ satisfying

$$\tilde{Z}_k = l(L_k \cap S_0)/\|v_0\|. \tag{6.10}$$

In this context, the $\tilde{Z}_k$ model the partially observed durations of the rain cells which are available through Rosenfeld's (1987) algorithm.

Turning to the water contents of the rain cells, we find that these are fully observed for every rain cell that enters the field of view. According to the smoothing kernel k in the WGR model, W_k is the ratio of the rain volume or cumulative intensity of the rain cell to its duration Z_k. If the latter two quantities are censored in accordance with our observational scheme, W_k is the ratio of the observed rain volume to a quantity determined by the observed duration $\tilde{Z}_k$, the observed trajectory $L_k \cap S_0$, and the Gaussian kernel in the smoothing kernel k. The observed rain volume, duration, and trajectory are available through Rosenfeld's (1987) algorithm. Thus, W_k in the WGR model is fully observed independently of censoring for every rain cell that enters S_0 in the period T_0.

Finally, we draw attention to one other potential source of censoring. The durations, for example, are observed at most over the period T_0, so that they, but not the water contents, may be further censored. Although this is easily handled by modifying the right-hand side of expression (6.10), we proceed as though T_0 is sufficiently large to make its contribution to censoring negligible. We do so in view of our aim in the next section of developing large-sample properties of our nonparametric estimators as the size of T_0 grows unboundedly. These estimators are introduced presently.

The nonparametric estimators. We work explicity with a rescaled WGR model. In doing so, we assume that the W_k have aging function A and the Z_k have aging function B, where these may take arbitrary functional forms. We therefore propose a nonparametric estimator of A to be drawn from the observed W_k and one of B to be drawn from the censored Z_k. To begin, let I denote the random set of labels satisfying

$$I = \{k : [T_k, T_k + Z_k) \cap T_0 \neq \varnothing, L_k \cap S_0 \neq \varnothing\}, \tag{6.11}$$

where the L_k are defined by eq. (6.9) and the $[T_k, T_k + Z_k)$ denote random subintervals determined by the times T_k and $T_k + Z_k$. Here I is the index set of labels of those rain cells that exist during the observational period and enter our field of view. Now for each $k \geqslant 0$, let δ_k denote the indicator $1(Z_k = \tilde{Z}_k)$, which is one for any rain cell whose trajectory is fully contained in S_0. Finally, the data for estimating A and B consist of the set of observations $\{W_k, \tilde{Z}_k, \delta_k : k \in I\}$.

Now let $N^a = (N_t^a)$, $t \geqslant 0$, $R^a = (R_t^a)$, $t \geqslant 0$, $N^b = (N_t^b)$, $t \geqslant 0$, and $R^b = (R_t^b)$, $t \geqslant 0$ denote the stochastic processes satisfying

$$N_t^a = \sum_{k \in I} 1(W_k \leqslant t) \qquad \text{and} \qquad R_t^a = \sum_{k \in I} 1(W_k \geqslant t)$$

and

$$N_t^b = \sum_{k \in I} 1(\tilde{Z}_k \leqslant t, \delta_k = 1) \qquad \text{and} \qquad R_t^b = \sum_{k \in I} 1(\tilde{Z}_k \geqslant t). \tag{6.12}$$

Finally, we define the stochastic processes $\hat{A} = (\hat{A}_t)$, $t \geqslant 0$ and $\hat{B} = (\hat{B}_t)$, $t \geqslant 0$ satisfying

$$\hat{A}_t = \int_0^t \frac{\mathrm{d}N_t^a}{R_t^a} \quad \text{and} \quad \hat{B}_t = \int_0^t \frac{\mathrm{d}N_t^b}{R_t^b}. \tag{6.13}$$

We propose $\hat{A}$ as the nonparametric estimator of A, and $\hat{B}$ as the estimator of B.

The processes defined at (6.12) give analogues of the survival counting process and the risk process as encountered in the counting process approach to censored survival data. The estimators $\hat{A}$ and $\hat{B}$ are in turn analogs of the so-called Nelson–Aalen estimator of the cumulative hazard function. We refer the reader to Andersen and Borgan (1985) for details.

In general, the mean measure belonging to M is parametrized by n aging functions, say $A, B, \ldots, C$. Each aging function belongs to one characteristic describing a physical property of the convective rain cells. In the above, we describe a typical observational scheme entailing the partial observation of these characteristics, where the details vary with the specific model and the particular characteristic. Nevertheless, given a set of possibly censored observations of the characteristics, we propose the estimators $\hat{A}, \hat{B}, \ldots, \hat{C}$ constructed in the manner illustrated above.

6.5 Properties of the estimator

We treat the properties of the nonparametric estimator introduced above. We show in specific that it is a martingale estimator, and we give conditions under which it is consistent and converges weakly to a Gaussian process. As the former leads naturally to the latter, we begin with the martingale property.

The Martingale property. The estimators $\hat{A} = (\hat{A}_t), t \geqslant 0$ and $\hat{B} = (\hat{B}_t), t \geqslant 0$ are defined at (6.13). In the context of the rescaled-WGR model, we show that $\hat{A}$ is a martingale estimator of the aging function A, and that $\hat{B}$ is the same of B.

The underlying probability space is $(\Omega, \mathscr{F}, \mathscr{P})$. Let $H = (\mathscr{H}_t)$, $t \geqslant 0$ denote the filtration of sub-σ-algebras of $\mathscr{F}$ satisfying

$$\mathscr{H}_t = \sigma(N_s^a, R_s^a, N_s^b, R_s^b, s \leqslant t). \tag{6.14}$$

Here H is the history of the observed durations and water contents of the rain cells. We develop the martingale property with respect to this history.

Proposition Suppose M is a rescaled cardinal Poisson random measure on $T \times S \times \Re_+^2$ having mean measure μ as defined at (6.8). Suppose further that the integral $\int_0^\infty B(\mathrm{d}t)\, \mathrm{e}^{-B(t)} t^2$ is finite, so that the durations have finite second moment. The estimators $\hat{A}$ and $\hat{B}$ are the martingale estimators of A and B, respectively, with respect to the history $H = (\mathscr{H}_t)$, $t \geqslant 0$ as defined at (6.14).

Proof Let $\|I\|$ denote the cardinality of the random set I as defined at (6.11). Also, let d_0 denote the diameter of the disk S_0 and let $\|T_0\|$ denote the same of the interval T_0. For each $z \geqslant 0$, let $D(z)$ denote the set satisfying

$$D(z) = \{t, x : [t, t+z) \cap T_0 \neq \varnothing, \qquad \inf\{s : x + s v_0 \in \bar{S}_0\} \leqslant z\},$$

where $\bar{S}_0$ denotes the closure of S_0. By virtue of the hypotheses of the proposition, we have

$$E\|I\| = \int_{T \times S \times \Re_+^2} \mu(\mathrm{d}t, \mathrm{d}x, \mathrm{d}w, \mathrm{d}z) 1((t, x) \in D(z))$$

$$\leqslant \int_0^\infty B(\mathrm{d}z) \mathrm{e}^{-B(z)} (\pi(d_0/2)^2 + d_0 z)(\|T_0\| + z) < \infty. \qquad (6.15)$$

Now it is readily seen that $N^a = (N_t^a), t \geqslant 0$ and $N^b = (N_t^b), t \geqslant 0$ as defined at (6.12) are counting processes such that EN_t^a and EN_t^b are finite for every t. We therefore introduce the stochastic processes $\tilde{A} = (\tilde{A}_t), t \geqslant 0$ and $\tilde{B} = (\tilde{B}_t), t \geqslant 0$ satisfying

$$\tilde{A}_t = \int_0^t A(\mathrm{d}s) R_s^a \qquad \text{and} \qquad \tilde{B}_t = \int_0^t B(\mathrm{d}s) R_s^b.$$

Clearly, each of these processes is adapted to the history H and, moreover, $\tilde{A}$ and $\tilde{B}$ are predictable. Therefore, by virtue of Definition 5.3 in Karr (1986) or the treatment in Andersen and Borgan (1985), the proposition is proved upon showing that N^a has compensator $\tilde{A}$ and N^b has compensator $\tilde{B}$ relative to the history H. We handle the two cases separately.

(a) *N^a has compensator $\tilde{A}$.* We again identify M with the sequence $(T_k, X_k, W_k, Z_k), k \geqslant 0$ and $\mathscr{I}$ with the σ-algebra generated by I. Consider the $\mathscr{I}$-conditional distribution of the set of random variables $\{W_k; k \in I\}$. Since the sequence $(T_k, X_k, Z_k), k \geqslant 0$ determines I, it follows from the hypothesis of the proposition that the W_k for $k \in I$ are $\mathscr{I}$-conditionally independent random variables having common aging function A. Moreover, the counting process N^a is $\mathscr{I}$-conditionally independent of the processes N^b and R^b. The desired result now follows directly.

(b) *N^b has compensator $\tilde{B}$.* Let $\xi = (\xi_t), t \geqslant 0$ denote an $\mathscr{H}_t$-predictable process and suppose that the random measure ν on $(\Re_+, \mathscr{R}_+)$ is the unique predictable projection belonging to the counting process N^b. We then have

$$E \int_0^\infty N^b(\mathrm{d}t) \xi_t = E \int_0^\infty \nu(\mathrm{d}t) \xi_t.$$

Therefore, by interchanging the order of integration and expectation and by virtue of the uniqueness of the predictable projection, it suffices to demonstrate the equality

$$EN^b(\mathrm{d}t)\xi_t = E\nu(\mathrm{d}t)\xi_t = E\tilde{B}(\mathrm{d}t)\xi_t = B(\mathrm{d}t) ER_t^b \xi_t,$$

for every $t \geqslant 0$, where this equality is understood as one among measures on $(\Re_+, \Re_+)$. On the other hand, by virtue of the fact that R^b and ξ are predictable, it suffices to demonstrate that $E(N^b(\mathrm{d}t)\,|\,\mathscr{H}_{t-})$ equals $B(\mathrm{d}t)R_t^b$ almost surely.

We need some preliminaries. We set our argument in a setting that well represents the observational scheme and the resulting censoring mechanism. For each $t \in T$, $x \in S$, and $z \in \Re_+$, let i_{tz} and l_{xz} denote the sets satisfying

$$i_{tz} = \{s : s \in T, t \leqslant s < t + z\} \qquad \text{and}$$

$$l_{sz} = \{y : y \in S,\ y = x + s\mathrm{v}_0 \text{ for some } s \in [0, z)\}.$$

Now let $(G, \mathscr{G})$ denote the Borel space where G satisfies

$$G = \{(i_{tz}, l_{xz}) : t \in T, x \in S, z \in \Re_+\}.$$

Finally, we introduce the measurable mapping $f : T \times S \times \Re_+^2 \to G$ that satisfies: $(t, x, w, z) \to (i_{tz}, l_{xz})$, for every $(t, x, w, z) \in T \times S \times \Re_+^2$.

Next, let Mf^{-1} denote the mapping of M by f in accord with Definition 1.36 in Karr (1986). This defines a Poisson random measure on $(G, \mathscr{G})$. According to the action of f, Mf^{-1} models the 'life horizons' in T and the trajectories in S of the convective rain cells generated by M. These quantities determine the observational scheme as depicted in part in Figure 6.5.

For each $(i, l) \in G$, let $g(i, l)$ denote the indicator function $1(i \cap T_0 \neq \varnothing, l \cap S_0 \neq \varnothing)$. Also, let $\|l\|$ denote the length of l along v_0 and let c denote the length of v_0. Now, for each $t \geqslant 0$, let ε_{ct} denote the Dirac measure concentrated at ct and let $h(t, l)$ denote the indicator function $1(\|l \cap S_0\| \geqslant ct)$. Finally, by virtue of the definition of Mf^{-1}, we have

$$N^b(\mathrm{d}t) = \int_G Mf^{-1}(\mathrm{d}i, \mathrm{d}l)g(i, l)\varepsilon_{ct}(\|l\|)h(t, l)$$

and

$$R_t^b = \int_G Mf^{-1}(\mathrm{d}i, \mathrm{d}l)g(i, l)h(t, l),$$

for every $t \geqslant 0$. We use these representations below.

We identify Mf^{-1} with the countable sequence $(I, L) = (I_k, L_k)$, $k \geqslant 0$ of G-valued random variables. For each $(i, l) \in G$, recall that we have

$$Mf^{-1}(\mathrm{d}i, \mathrm{d}l)g(i, l)\varepsilon_{ct}(\|l\|)h(t, l) = \sum_{k \geqslant 0} \varepsilon_{(I_k, L_k)}(i, l)g(i, l)\varepsilon_{ct}(\|l\|)h(t, l).$$

In light of the desired result and the representation of $N^b(\mathrm{d}t)$ given earlier, we examine the determinants of each term in the summation above. In particular, for each $k \geqslant 0$, we show that on the set $\{\omega : Z_k(\omega) \geqslant t\}$ the kth term is determined by the random measure $\varepsilon_t(Z_k)$, the placement (T_k, X_k), the point (i, l), and t.

Fix $k \geqslant 0$ and suppress it from the notation. We refer therefore to the pair (I, L) as determined from the triplet (T, X, Z). For each $t, u \in T$ and $x \in S$, let $D(u, x, t)$ denote the set in G satisfying

$$D(u, x, t) = \{(i, l) : i = i_{us},\ l = l_{xs} \qquad \text{for some } s \geqslant t\}.$$

Now an elementary argument reveals that we have

$$\varepsilon_{(I,L)}(i,l)\varepsilon_{ct}(\|l\|)1(Z \geqslant t) = \varepsilon_t(Z)1((i,l) \in D(T,X,t))1(Z \geqslant t),$$

for every $(i, l) \in G$ and $t \geqslant 0$. Finally, by virtue of the definition of M, we observe that given the placements (T, X) and the event $\{Z \geqslant t\}$, the random measure $\varepsilon_t(Z)$ has conditional mean $B(\mathrm{d}t)1(Z \geqslant t)$, for every $t \geqslant 0$.

Finally, let $F = (\mathscr{F}_t)$, $t \geqslant 0$ denote the filtration satisfying

$$\mathscr{F}_t = \sigma(M(B), B \in \sigma(T \times S \times \Re_+ \times [0, t])).$$

Here $\mathscr{F}_t$ is a history of the placements and the durations of those rain cells whose durations lie inside of t. For each $t \geqslant 0$, by virtue of the arguments above, the $\mathscr{F}_{t^-}$-conditional mean of $N^b(\mathrm{d}t)$ depends at most on the configuration of placements and the number of durations lying outside of t. We therefore have

$$\begin{aligned}
E(N^b(\mathrm{d}t) \mid \mathscr{H}_{t^-}) &= E(E(N^b(\mathrm{d}t) \mid \mathscr{F}_{t^-} \vee \mathscr{H}_{t^-}) \mid \mathscr{H}_{t^-}) \\
&= B(\mathrm{d}t)E\Bigg(\int_G \sum_{k \geqslant 0} 1((i,l) \in D(T_k, X_k, t))1(Z_k \geqslant t) \\
&\qquad \times g(i,l)h(t,l) \mid \mathscr{H}_{t^-}\Bigg) \\
&= B(\mathrm{d}t)E\left(\int_G Mf^{-1}(\mathrm{d}i, \mathrm{d}l)g(i,l)h(t,l) \mid \mathscr{H}_{t^-}\right) \\
&= B(\mathrm{d}t)R_t^b,
\end{aligned}$$

for every $t \geqslant 0$. This completes our proof. ■

The Proposition generalizes to contexts other than the rescaled-WGR model. This requires that the observational scheme allows for at least the partial observation of the rain-cell characteristics in the model. For example, the method of proof used in part (b) transfers more or less directly to the random-disks model, even though the rain cells are permitted to have random speeds.

Consistency and asymptotic normality. The martingale property leads naturally to the large-sample properties of the estimators. Here we show that $\hat{A}$ and $\hat{B}$ are consistent and converge weakly to a Gaussian process as the observational period grows unboundedly.

We begin with these preliminaries. Let g denote the function satisfying

$$g(x,z) = 1(l_{xz} \cap S_0 \neq \varnothing) \Big/ \int_S \mathrm{d}y\, 1(l_{yz} \cap S_0 \neq \varnothing),$$

for every $x \in S$ and $z \in \Re_+$, where l_{xz} denotes the rain cell trajectory as defined in part (b) of the proof of the Proposition. Now introduce the function $r = (r(t))$, $t \geqslant 0$ satisfying

$$r(t) = \int_{S \times \Re_+} \mathrm{d}x\, B(\mathrm{d}z)\mathrm{e}^{-B(z)} g(x,z) 1(\|l_{xz} \cap S_0\| \geqslant t \,\|\mathrm{v}_0\|). \qquad (6.16)$$

Finally, let α and β denote functions on $\Re_+$ satisfying

$$\alpha(t) = \int_0^t A(\mathrm{d}s)\mathrm{e}^{A(s)} \qquad \text{and} \qquad \beta(t) = \int_0^t B(\mathrm{d}s)(r(s))^{-1} \qquad (t \geqslant 0). \tag{6.17}$$

Observe that both α and β are increasing functions that are 0 at time zero, so they can be used as 'clocks' in a transformation of time. We do so here and introduce two Gaussian processes $W^\alpha = (W_t^\alpha), t \geqslant 0$ and $W^\beta = (W_t^\beta), t \geqslant 0$ satisfying

$$W_t^\alpha = W_{\alpha(t)}^1 \qquad \text{and} \qquad W_t^\beta = W_{\beta(t)}^2,$$

as time changes of the independent Weiner processes W^1 and W^2. Thus W^α and W^β are independent Gaussian processes with variance functions as defined by (6.17).

For $0 < c \leqslant +\infty$, let $\mathscr{D}_c$ denote the space of right-continuous functions defined on $[0, c)$ and having limits from the left. Also, let a denote the extended real number $\sup\{t : A(t) < \infty\}$, and similarly let b denote $\sup\{t : r(t) > 0\}$, for r defined at (6.16). Finally, let d_0 denote the diameter of the disk S_0 and let $\| T_0 \|$ denote that of the interval T_0.

Theorem Suppose that the hypotheses of the Proposition above are true. Let $\| I \|$ denote the cardinality of the random set I as defined at (6.11). In light of the preliminaries treated above, we have the following results:

(1) *Consistency.* The estimator $\hat{A}$ converges to A uniformly on $[0, a)$ in probability as $\| T_0 \|$ tends to infinity, and $\hat{B}$ does the same to B uniformly on $[0, b \wedge g_0)$,

(2) *Asymptotic normality.* The estimator $\hat{A}$ and $\hat{B}$ converge weakly to independent Gaussian processes. That is, $\| I \|^{1/2}(\hat{A} - A)$ converges weakly to W^α in the space $\mathscr{D}_a$ endowed with the Skorokhod topology as $\| T_0 \|$ tends to infinity, and similarly $\| I \|^{1/2}(\hat{B} - B)$ converges weakly to W^β in $\mathscr{D}_{b \wedge d_0}$. ■

The proof of the theorem rests on the elementary observation that $R_t^a / \| I \|$ converges uniformly to $\mathrm{e}^{-A(t)}$ on $[0, a)$ in probability as $\| T_0 \|$ tends to infinity, and $R_t^b / \| I \|$ does the same to $r(t)$ on $[0, b \wedge d_0)$. Therefore, the demonstration involves standard applications of martingale inequalities and martingale central limit theorems. An expository treatment of such applications in statistics is found, for example, in Phelan (1988). We omit the details here.

We refer the reader to Andersen and Borgan (1985) and Andersen *et al.* (1988) for applications of the results of this section. In this we include the construction of confidence sets, tests of hypotheses, kernel-based smoothing of the estimators, and the estimation of the distribution of the rain-cell characteristics via the Kaplan–Meier or product-limit estimator. (Regarding interpretation, we believe that the argument in Johansen (1978) will carry over to this context.) Phelan (1990) promotes the idea of incorporating these estimators into the method of primary features for estimating parametric models. We leave the details of these applications to a further work.

6.6 Concluding remarks

We have drawn attention to the issue of aging in the analysis of rainfall fields. In particular, we treated aging of convective rain cells in its connection to propagation, turbulence, and stochastic representations of such fields. In doing so, we suggested directions for future research in the modeling and inference for precipitation.

In the present work, we highlighted the role of aging functions in understanding turbulence in rainfall fields and in parametrizations of point process models for the same. We introduced nonparametric estimators of such functions as drawn from data determined by a typical observational scheme.

The application of this work across a wide variety of convective storm systems remains in prospect. We expect to widen our scope to cover the issues raised here with respect to turbulence and stochastic representations.

References

Andersen, P.K. and Borgan, O. (1985). Counting process models for life history data: a review. *Scand. J. Statist.*, **12**, 97–158.

Andersen, P.K., Borgan, O., Gill, R.D., and Keiding, N. (1988). Censoring, truncation, and filtering in statistical models based on counting processes. *Contemp. Math.*, AMS, **80**, 19–60.

Brillinger, D.R. (1988). Some statistical methods for random process data from seismology and neurophysiology. *Ann. Statist.*, **16**, 1–54.

Cox, D.R. and Isham, V.S. (1988). A simple spatial–temporal model of rainfall. *Proc. Roy. Soc. Lond.*, A, **915**, 317–28.

Frisch, U. and Orszag, S.A. (1990). Turbulence: Challenges for theory and experiment. *Physics Today*, January, **43**, 24–32.

Gupta, V.K, and Waymire, E. (1987). On Taylor's hypothesis and dissipation in rainfall. *J. Geophys. Res.*, **92**, 9657–60.

Houze, R.A. (1981). Structure of atmospheric precipitation systems: a global survey. *Radio Sci.*, **16**, 671–89.

Houze, R.A. and Hobbs, P.V. (1982). Organization and structure of precipitating cloud systems. *Adv. Geophys.*, **24**, 225–315.

Hudlow, M.D. and Patterson, P. (1979). *GATE Radar Rainfall Atlas*, NOAA Special Report.

Johansen, S. (1978). The product limit estimator as maximum likelihood estimator. *Scand. J. Statist.*, **5**, 195–9.

Karr, A. (1986). *Point Processes and Their Statistical Inference*, Marcel Dekker, New York.

Le Cam, L.M. (1961). A stochastic description of precipitation. In *Proc. 4th Berkeley Symp. Math. Statist. and Probability*, Neymann, J. (ed.), **3**, 165–86, Berkeley, CA.

Phelan, M.J. (1988). Some applications in statistics of semimartingale weak convergence theorems. *Contemp. Math.*, AMS, **80**, 153–90.

Phelan, M.J. (1990). Point processes and inference for rainfall fields. To appear in *Proc. IMS Symp. Appl. Prob.*, Sheffield.

Phelan, M.J. and Goodall, C.R. (1990). An assessment of a generalized Waymire-Gupta-Rodriguez-Iturbe model for GARP Atlantic Tropical Experiment rainfall. *J. Geophys. Res.*, **95**, 7603–16.

Rodriguez-Iturbe, I., Cox, D.R., and Eagleson, P.S. (1986). Spatial modelling of total storm rainfall. *Proc. Roy. Soc. Lond.* A, **403**, 27–50.

Rodriguez-Iturbe, I., Cox, D.R., and Isham, V. (1987). Some models for rainfall based on stochastic point processes. *Proc. Roy. Soc. Lond.* A, **410**, 269–88.

Rodriguez-Iturbe, I., Cox, D.R., and Isham, V. (1988). A point process model for rainfall: further developments. *Proc. Roy. Soc. Lond.* A, **417**, 283–98.

Rosenfeld, D. (1987). Objective method for tracking convective cells as seen by radar. *J. Atmos. and Oceanic Tech.*, **4**, 422.

Sampson, P.D. and Guttorp, P. (1990). *Nonparametric estimation of nonstationary spatial covariance structure.* SIMS Technical Report No. 148.

Smith, J.A. and Karr, A.F. (1985). Parameter estimation for a model of space-time rainfall. *Water Resour. Res.*, **21**, 1251–7.

Taylor, G.I. (1937). The spectrum of turbulence. *Proc. Roy. Soc. Lond.* A, **164**, 476–90.

Waymire, E., Gupta, V.K., and Rodriguez-Iturbe, I. (1984). Spectral theory of rainfall intensity at the meso-β scale. *Water Resour. Res.*, **20**, 1453–65.

Zawadski, I.I. (1973). Statistical properties of precipitation patterns. *J. Appl. Met.*, **12**, 459–72.

Introduction to statistics in the earth sciences

The use of statistical methodology in earth sciences is ubiquitous. Many quantities of interest are, and can only be, sensed remotely (e.g. the Earth's core, earthquake intensities). To obtain accurate estimates and images much data redundancy is commonplace, and data analysis is often almost totally statistical in character. It is undoubtedly the case that the largest usage of many time series analysis techniques is in seismic signal processing in oil and gas exploration. This section contains a selection of papers which give a good flavor of the wide variety and complexity of statistical methods being brought to bear on current earth science problems. They show an invigorating interplay between the real-world problem and the statistical methods designed for their solution.

One of the most important statistical techniques in regular use in the earth sciences is that of deconvolution. It finds application in all areas where deblurring of a time series (one dimension) or an image (two dimensions) is required. The technique sprang to scientific prominence in 1990 with its use for deblurring images from the Hubble space telescope caused by the spherical aberration in the optical system. In the first paper of this section Scargle looks at the deconvolution of chaotic time series, which are currently of great interest to earth scientists. Despite tailoring his cost function for chaotic time series, Scargle finds that it can also be successful at deconvolving conventional random processes. In oil and gas exploration, a cross-sectional representation of the earth is built up from seismic reflection surveys. The picture comprises thousands of time series plotted vertically next to each other. Here the seismic explosion ('wavelet') acts as a blurring filter. Over the years, billions of such series have been deconvolved to sharpen the cross-sectional representation. With no *a priori* knowledge of the phase of the blurring filter, deconvolution can be accomplished by a two-step approach: spectral whitening followed by a phase correction. The paper by Walden examines the phase correction step, and shows how the performance of two projection indices (which are maximized to find the phase correction) are influenced by data bandwidth and signal-to-noise ratio. The method is illustrated on some real seismic data. Deconvolution is also critical in analysing seismic recordings of earthquakes and explosions: for example improving the quality of the estimated source functions can enable estimation of yields of nuclear explosions. In the frequency domain, the Fourier transforms of individual recordings can be factored into the product of site (receiver) and source frequency response functions. Der *et al.* simultaneously estimate receiver and source functions by applying the EM algorithm. Amongst other analyses, they present results of deconvolutions of data from explosions at Soviet, French, Chinese and US test sites.

Statistical analysis of earthquake data continues in the next two papers. Many processes in nature do not consist of simple sinusoids, but rather are quasi-periodic, having a dominant oscillation that suffers amplitude and phase variation. Park derives the multitaper complex demodulate at the carrier frequency and shows it to be the minimum-size solution for the envelope. Using this multitaper approach he looks at the minimization of other quadratic functionals. His approach allows the modulating functions to have complicated (but physically plausible) structures. Park applies his method to investigate envelopes of coupled long-period seismic oscillations from the great Macquarie Ridge earthquake of 1989, and to examine quasi-periodic changes in the earth's climate. Vere-Jones reviews statistical methods for space or space–time modeling of the data in an earthquake catalog. Of particular interest are models which lead to estimates of intensities and mean magnitudes for graphical display. Among the techniques discussed are kernel-smoothing methods, orthogonal- and spline-function representations, and stochastic-process models. The point-process approach is flexible enough to incorporate causal clustering and stress release models. Vere-Jones stresses the need for a better physical understanding of the empirical observations.

The final two papers are both concerned with resource modeling and estimation. Ripley looks at stochastic models for the distribution of rock types in petroleum reservoirs. Such results can be used for numerical simulation of multi-phase flow through the reservoir. The techniques surveyed by Ripley in his much needed wide-ranging review include Markov random fields, Gibbs point processes, the Gibbs sampler and Metropolis algorithm, among many others. Computational considerations and problems are covered. In the last paper, Conradsen *et al.* apply the spatial interpolation method known as kriging to the chemical analysis of stream sediment samples. A novel aspect of their analysis is the use of classification and regression trees (CART) to produce estimates of the probability of finding heavy minerals (such as gold or casserite) using the stream sediment data as predictors.

Chapter 7

Deconvolution of chaotic and random time series

J. D. Scargle

Chaos is the deterministic but disordered evolution of a physical system that has high sensitivity to arbitrarily small changes in initial conditions. Mathematically it is a special type of random process. Two sample chaotic processes illustrate the characteristics that are important in time-series analysis. Recently developed tools for chaotic time-series analysis include state-space visualization, construction of an observationally accessible state space (the embedding space), plus linear and nonlinear prediction. These tools, combined with a classic representation for stationary chaotic and random processes (the Wold decomposition) give a practical approach to time-domain modeling. Specifically, a novel linear predictive deconvolution procedure, while not achieving the uncorrelated Wold representation, can detect, model, and separate chaos and randomness in time-series data.

7.1 Introduction: the nature of chaos

Physics abounds with initial-value problems. Laws governing the evolution of a system fix its future behavior, given an initial state. The mathematical representation is usually a set of dynamical differential equations, with unique solutions. The evolution is therefore deterministic, and a given initial state generates a unique evolutionary path. In short: identical initial conditions give identical system behavior.

The common textbook examples of such deterministic systems evolve in an orderly way – i.e. the physical variables are smooth functions of time, and two nearby initial states evolve along paths that remain close for all time. In particular, linear systems are always well-behaved in this sense.

On the other hand, even simple nonlinear systems frequently do not have this property. They are *chaotic*. Evolutionary paths from initial states close together, no matter how close, diverge exponentially from each other. This sensitivity to initial conditions (abbreviated SIC) is fundamental to chaos. Frequently the evolution of the physical variables appears very disordered. Chaotic processes appear random. They are in fact random, in ways that will become clear. Furthermore we take all chaotic processes to be stationary; the importance of this assumption will become clear. Chaotic behavior occurs in surprisingly simple systems, as nearly all sets of nonlinear differential equations exhibit chaos for some values of the system parameters.

Most of the characteristics of chaos are evident in the simplest possible chaotic system, the *doubling map* (sometimes called the *one-sided Bernoulli shift*):

$$X_{n+1} = \begin{cases} 2X_n & 0 \leqslant X_n < \frac{1}{2} \\ 2X_n - 1 & \frac{1}{2} \leqslant X_n \leqslant 1. \end{cases} \tag{7.1}$$

This is a discrete-time system, evolving according to a difference (not differential) equation. Repeated application of this map, starting from an initial value in [0, 1], generates an infinite sequence X_n in the same interval. The evolutionary behavior depends on the initial value, but for almost all X_0 the sequence looks random. Each iteration of the map is equivalent to discarding the most significant binary digit of the previous iterate, then shifting the remaining digits one place to the left. Therefore any irrational initial value will produce a nonrepeating sequence with a disordered appearance.

As it stands eq. (7.1) does not quite define a random process, since each realization with a given initial value is identical. Random appearance of the time series does not guarantee randomness, nor does sensitivity to computational details (see below). For X to be a true random process the initial value X_0 must be a random variable.

Also, since X is stationary, one cannot specify the distribution of X_0 arbitrarily. By eq. (7.1) the distribution of X_0 determines that of X_1, which determines that of X_2, and so on. Unless these distributions are all the same the process is not stationary. We thus ask the question: Is there a function $f(x)$ on [0, 1] that is the distribution function of all the random variables X_n, $n = 0, 1, 2, \ldots$? If so, is it unique? The answers are yes, but no – there is more than one. One such f is a δ-function at $X = 0$; for $X_0 = 0$ implies $X_n = 0$ for all n. Another one is a δ-function at 1. These singular orbits do not have the disorder that arises from most initial values. Is there a distribution function that describes these more typical cases? The answer is yes, and it is unique; the reader can verify that it is the uniform distribution on [0, 1].

Many mathematicians have studied the conditions necessary and sufficient for the existence of distribution functions invariant under transformations of the form $X_{n+1} = F(X_n)$. Such distributions are known as *invariant measures*. The above result for eq. (7.1) is typical: one invariant measure describes almost all orbits (this is sometimes called the *physical measure*) and other singular ones for special cases. Consult, e.g. Eckmann and Ruelle (1985) or Ruelle (1989) for more on this topic.

Another simple example of chaos is the *logistic map*, also on [0, 1]:

$$X_{n+1} = \lambda X_n(1 - X_n), \qquad 0 \leqslant \lambda \leqslant 4. \tag{7.2}$$

Figure 7.1 shows a typical realization of this process for $\lambda = 4$. As before the system behavior depends on the initial value. (To see this consider $X_0 = \frac{3}{4}$; also compare with $X_0 = \frac{3}{4} + 10^{-100}$.) It also has an extraordinarily complex dependence on λ that we cannot describe here. For $\lambda = 4$ almost all initial values generate a random sequence – one that is uncorrelated, in contrast

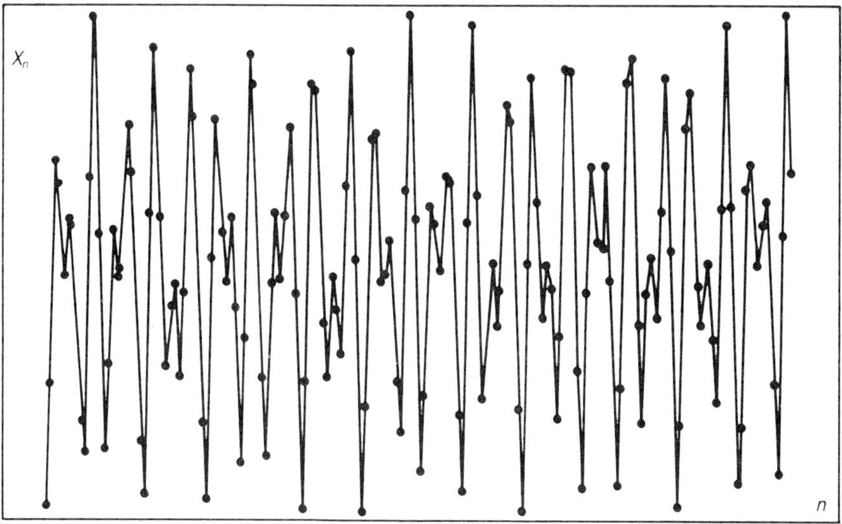

Figure 7.1 Chaotic time series generated by the logistic map, eq. (7.2), with $\lambda = 4$. These data are uncorrelated but highly dependent.

to the Bernoulli shift's exponentially shaped correlation function (Scargle, 1989a).

There is a large literature on the logistic map (e.g. Schuster, 1988; May, 1976; Grebogi *et al.*, 1982). Simple numerical experiments verify that both of the above maps have SIC. There is also a large literature dealing with many similar discrete-time systems. The property necessary to produce chaos is a 'stretch-and-fold' character to the map (think of kneading bread dough or taffy). Chaos requires a nonlinear map.

Similar considerations apply to continuous-time systems described, say, by differential, difference, or differential-difference (Bellman and Cooke, 1963) equations. Sensitivity to initial conditions has many implications for numerical integrations of dynamical equations of motion. Orbits starting from the same initial state calculated with different numerical schemes diverge from each other. This is because different algorithms, or even the same one with different time-steps, produce slightly different errors; then SIC amplifies the differences until they become large. Similarly, the same integration scheme on two computers with different precision or round-off (Quinn and Tremaine, 1990) yields solutions grossly diverging from each other after a finite time. Thus SIC dooms to failure, or at least greatly limits, computation verifications such as convergence with diminishing step size.

These and other considerations lead to the conclusion that it is average, or statistical, behavior of chaotic systems that is important, not details of specific orbits. Even the gross character of computed evolution of the Bernoulli shift is dependent on the precision used. More generally, since a finite computer can represent only a finite number of values, all computed orbits of a bounded map are eventually periodic. Yet almost no real orbit of a chaotic system is periodic; the corresponding initial states are a set of measure zero (e.g. rational numbers in the Bernoulli shift). Almost all initial values (e.g. irrational numbers) produce nonrepeating orbits not accessible to a finite computer.

Note: While chaotic time series often have the grossly disordered appearance of the above examples, this is not always the case. For example a system with two or more states, each of which generates an ordered time series, may exhibit chaos only by occasional but erratic jumps between states.

The existence of deterministic systems that exhibit chaotic pseudo-randomness, behavior akin to indeterminism and ordinarily associated with true randomness, has several implications for data analysis. Time-series data that seem random may, in reality, be chaotic. Such processes can be better understood physically if, using the tools described here or otherwise, one can 'unveil the order hidden in chaos' – e.g. find the map. Chaos and true randomness can be present in the same physical system. We need models and analysis tools to detect chaos and randomness, to study their structure, and to separate them if both are present. In the next sections we describe tools designed for these tasks. It will be seen that tangible distinctions between randomness and chaos usually occur in state space, not in the appearance of the time series.

For a general introduction to chaos consult, for example, Schuster (1988), May (1976), Grebogi *et al.* (1982), Ford (1983; 1986), Wolfram (1985), Abraham and Shaw (1983), Lichtenberg and Lieberman (1983), Bai-Lin (1984), Berge *et al.* (1984), Thompson and Stewart (1986), Moon (1987), Ruelle (1989), Bai-Lin (1989). Mandelbrot (1983) and Feder (1988) discuss interesting connections between fractals, time series, and random walks.

7.2 Tools for chaotic time-series analysis

Researchers in various fields have developed new ways to analyze chaotic time-series data. Here we review only those necessary to develop the deconvolution procedure that is the main topic of this paper. The standard tools of time-series analysis, e.g. power spectrum analysis, have been of limited value in the study of chaotic processes. Accordingly, new tools have been responsible for most recent progress.

It is generally useful to study a physical system in its state space – an abstract space, the coordinates of which are a complete set of independent variables. (I avoid the term 'phase space', used by most authors, as it is easily confused with unrelated terms, such as *minimum phase*, to characterize filter shapes.) An example of such coordinates is provided by the positions and momenta of particles. Any configuration of the system corresponds to a unique point of this space, and vice versa. Evolution of the system produces a curvilinear path in state space (called an orbit) beginning at the point corresponding to the initial state. This geometrical view, used in dynamics for some time, is an extremely powerful tool for describing chaotic dynamics (Abraham and Shaw, 1983).

Unfortunately, it is rare that an experimentalist can measure all the physical coordinates. Practical considerations often permit measurement of only a few state variables or, even worse, only peripheral quantities that depend indirectly on the state. For example, astronomers have studied the chaotic tumbling of Saturn's satellite Hyperion through space, using observations of its changing brightness (Klavetter, 1989).

Remarkably, extensive measurements of just one variable can reveal the structure of the orbits in the full multivariate state space. Packard *et al.* (1980) and Takens (1981) show that under certain conditions on the dynamical system there is a multidimensional *embedding space*, the coordinates of which can be derived from a single observed variable. The time-series data generate trajectories in this space that are simply related to the system's state-space orbits (see Figure 7.2). (To underscore the difference between state space and embedding space, we use the terms *orbit* and *trajectory* for the trace of the system's evolution in the respective spaces.) This relation is in the form of a smooth map, from the state space to the embedding space, that preserves the topology of the evolutionary paths. Thus the essential features of the unobservable state-space orbits can be understood by studying the accessible trajectories in the embedding space. In this analysis, one does not necessarily need to know the dimension of the state space – or even the identity of its coordinates.

Of the many choices of embedding space coordinates the most common

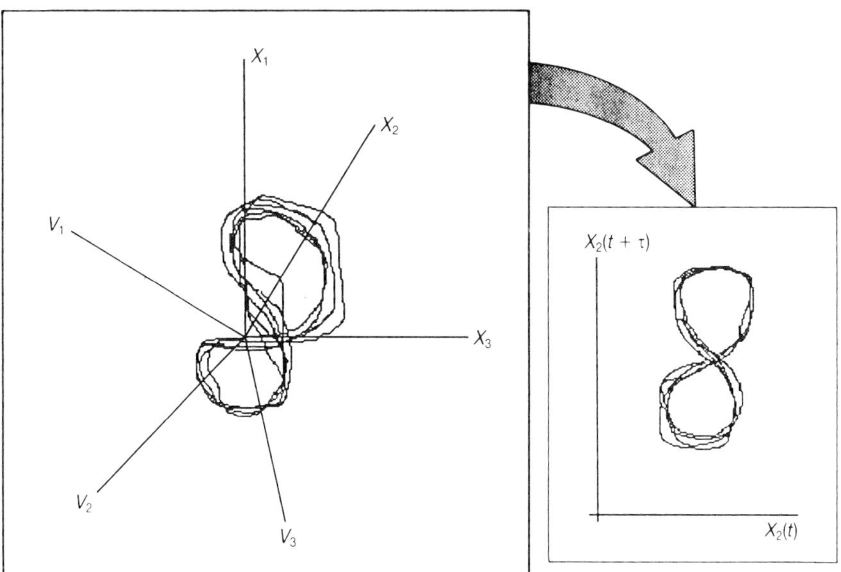

Figure 7.2 Illustration of the relation between the full multidimensional state space (left) of a physical system and the embedding space (right). The arrow represents the topology-preserving transformation between the two spaces.

is the observed variable evaluated at a set of lagged times:

$$X = (X_n, X_{n+k}, X_{n+2k}, \ldots, X_{n+(M-1)k}). \tag{7.3}$$

The lag k and the dimension M are positive integers. In principle the lags need not be equal, but in practice they are almost always chosen to be so. As time goes by, points defined by eq. (7.3) fall on embedding-space trajectories topologically equivalent to the system's state-space orbits. The trajectories can be traced out using time-series data, albeit crudely if the noise level is large. When the system satisfies the requisite mathematical conditions (and M is high enough – see below) we say that we have constructed a *suitable embedding space.*

We thus arrive at the most important tool for the analysis of chaos, the phase portrait. As discussed above, *state-space portrait* might be a better term. However, we follow standard usage and refer to the embedding-space picture of dynamical trajectories depicted by time-series data as a phase-space portrait. A diagram obtained as a surface of section is sometimes called a *return map.*

The phase portrait can be constructed directly from the time-series data. In simplest form it is just a plot of X_{n+1} against X_n. In other cases the process must be described in a space of higher dimensions, and we add coordinates $X_{n+2}, X_{n+3}, \ldots$. Also, the lag k need not be one unit. It may be noted that this way of looking at time-series data is not completely new; it is very similar to 'projection pursuit' (Friedman and Tukey, 1974; Walden, 1992). While most workers have used it only to search for clustering in multidimensional data, projection pursuit may prove to be useful for the sheet-like structures relevant here.

The phase portrait plays a key role in the search for order in chaos, as it is essentially an estimate of the form of the map (Shaw, 1984). For example, the disordered data generated by the logistic equation (Figure 7.1) has a simple parabolic phase portrait (Figure 7.3), as one can see from eq. (7.2).

In practice, with real data, the situation is not so simple. Observational errors always produce noise that somewhat blurs the phase portrait. Yet with a sufficiently large signal-to-noise ratio this plot will reveal the nature of the dynamics in spite of noise.

Another complication is that one does not know *a priori* the values of the lag k and the dimension M of the embedding space. In the theorems justifying the embedding procedure – namely infinite, noise-free data streams – the value of k does not matter. In practice k does matter, and one must figure out a good value from the data. Also, in theory M must be larger than $1 + 2d$ (d is the dimension of the physical state space), but in practice considerably smaller values provide suitable embeddings.

Prediction of a process from observations of its past is as important in the analysis of chaotic time series as it is for random ones. Accordingly a large literature has developed (e.g. Abarbanel *et al.*, 1989; Brock, 1990; Crutchfield, 1989; Crutchfield and McNamara, 1987; Farmer and Sidorowich, 1987; Lewis and Stevens, 1990; Townshend, 1990; and Weigend *et al.*, 1990). Various authors have pursued the related topic *noise reduction* (Farmer and Sidorowich, 1988a, b; Kostelich and Yorke, 1988; Marteau and Abarbanel,

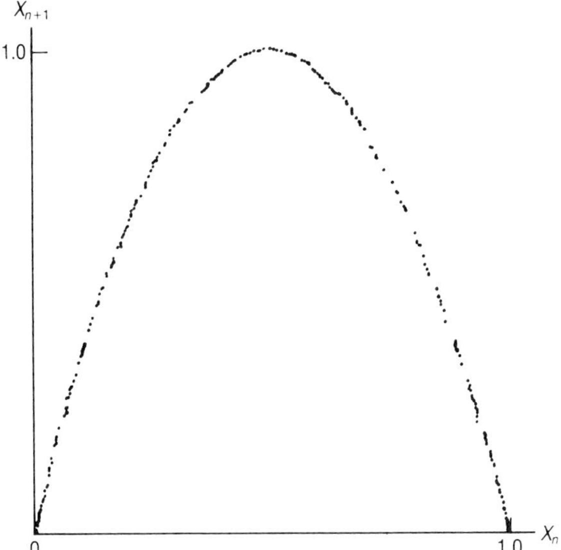

Figure 7.3 Phase portrait constructed from logistic data as in Figure 7.2.

1990; Sugihara and May, 1990). Chaotic processes are deterministic and therefore predictable in principle. In practice, however, SIC greatly limits predictability. Even a tiny uncertainty in the initial state grows rapidly with the passage of time. Below, we make use of prediction to model time series, but only linear prediction. We will see that linear predictors can be surprisingly useful in modeling nonlinear chaotic processes.

This briefest of introductions has discussed only those topics needed to describe the deconvolution procedure in Section 7.4. See references such as Farmer *et al.* (1983), Grassberger and Procaccia (1983a, b), Crutchfield and Packard (1983), Fraser and Swinney (1986), Eckmann and Ruelle (1985), Eckmann *et al.* (1986), Froehling *et al.* (1981), Tong (1990) for general introductions to this rapidly growing field. Gershenfeld (1988) gives a clear overview of embedding and a simplified treatment of the difficult mathematics in Takens (1981).

7.3 Convolutional representation: random and chaotic processes

Physical processes are often stationary, and we assume throughout that all processes have this property. It is not as widely known as it should be that stationarity of a process implies that it has a remarkably explicit, simple, linear form – filtered white noise. Specifically, the Wold (1938) decomposition theorem states: any stationary process can be decomposed into the sum of a purely deterministic and a purely random process; the latter is a *moving average*, i.e. the convolution just described:

The Wold decomposition theorem Any stationary process X can be written as

$$X = R * C + D, \tag{7.4}$$

where D is linearly deterministic, C is a constant (linear) filter, and R is an uncorrelated ('white') process. In addition, C is causal and minimum delay, and R and D are uncorrelated with each other.

It is remarkable that the deterministic aspects (C and D) of any stationary process can be separated from the purely random component (R) in such an explicit and linear way. Practically speaking the Wold theorem is a justification for convolutional models for any physical process known or expected to be stationary. Assume that any linearly deterministic part has been removed, using one of many known methods for identifying and removing trends. Then eq. (7.4) becomes

$$X = R * C \tag{7.5a}$$

or explicitly,

$$X_n = \sum_{k=0}^{\infty} C_k R_{n-k}. \tag{7.5b}$$

This is known as a moving-average (MA) model, because it represents the output process X as a running sum of values of the input process, R. Figure 7.4 shows an example filter used in eq. (7.5) to generate the moving-average data plotted in Figure 7.5.

Another useful time-domain representation is the autoregressive (AR) model,

$$R = X * A \tag{7.6a}$$

or with a minor rearrangement

$$X_n = -\sum_{k} A_k X_{n-k} + R_n, \qquad k = 1, 2, \ldots, \tag{7.6b}$$

representing the current value of X as a linear memory of its previous history plus a random part. Every stationary process has both an MA and an AR representation, as long as noncausal representations are allowed (i.e. k may take on negative values; see Scargle, 1981, hereafter Paper I). As we will see, the AR model is typically easier to estimate, while the MA form has a more direct physical interpretation (as random pulses, or shot noise). It should be kept in mind, however, that the AR and MA models of a given X are not different processes, just different ways to represent the structure of X.

The following comments about the interpretation of the Wold decomposition may clarify its significance. The definition of R_n is the difference between X_n and its projection on the set $\{X_i, i < n\}$ – i.e. R is the residual

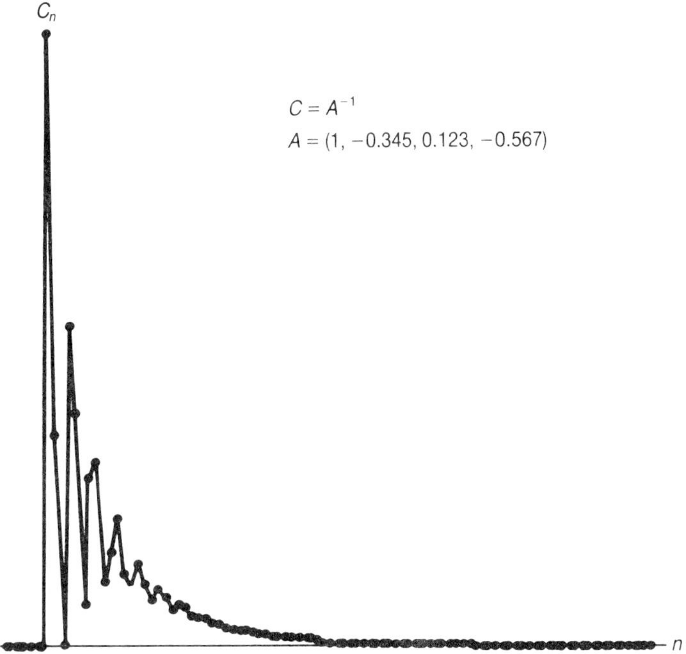

Figure 7.4 A filter, or pulse shape, defined in terms of its autoregressive coefficients A_n but shown here by plotting the corresponding moving average coefficients C_n against n.

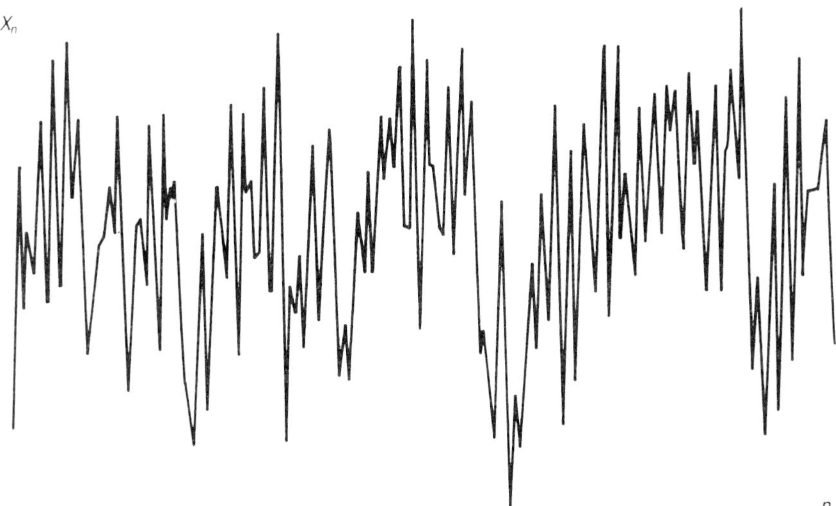

Figure 7.5 Chaotic process generated by convolving the output of the logistic map (cf. Figure 7.1) with the filter shown in Figure 7.4.

in the prediction of the current values of X, based on its past. The summation in (7.5) is therefore the Fourier expansion of X_n in terms of the orthogonal set $\{R_i, i < n\}$. For more details and the proof of the theorem, see Doob (1953) or Brockwell and Davis (1987).

Since chaos is nonlinearly deterministic it resides in the convolutional term, not in the linearly deterministic part, D.

7.4 The deconvolution procedure

Now to the main topic of the paper, a practical modeling procedure for chaotic processes. Here is the chain of reasoning leading from a general problem in data analysis to one in parameter estimation. A nearly universal goal in science is to develop a physically meaningful model of a process from data. Often the measurements are in the form of a time series, $\{X_n, n = 1, 2, \ldots, N\}$. Assume X to be stationary – so it has a moving average representation (Section 7.3). The problem, then, becomes estimation of the elements R, C and D in eq. (7.4). Assuming that one can remove D with detrending methods, the remaining model can be found by estimating a set of MA (or equivalently, as we will see below, AR) coefficients. Because of the generality of the Wold decomposition, this reduction to such an explicit model involves little loss of generality.

I now describe a procedure that starts with conventional predictive deconvolution and incorporates the new concepts introduced in Section 7.2 and the model discussed in Section 7.3. (Throughout this paper *conventional* refers to random-process modeling procedures as described in Robinson (1967), Box and Jenkins (1970) or Paper I. The reader versed in this approach will find the above chain of reasoning and almost all the material below familiar. The only novel item is the cost function.)

Now to the preliminaries. Temporarily ignore measurement errors and other noise sources that inevitably corrupt the observations. Remove any linearly deterministic component from the data, e.g. by detrending. These operations lead to the following model for the process:

$$X = R * C; \tag{7.7}$$

remember that R is an uncorrelated noise process.

From time-series data for X we want to estimate the filter C and the innovation R in this moving average. One does this as follows: Let $\hat{R} = A * X$, with A an arbitrary filter. Next adjust the coefficients in A to make $\hat{R}$ a good estimate of the innovation. If A were C's inverse, $\hat{R}$ would be the actual innovation. But we do not know C. That's what we are solving for! So this seems a hopeless approach.

To understand the chaotic case, consider first the conventional solution to this problem. For random processes one seeks the A that makes $\hat{R}$ uncorrelated. This is done by minimizing, with respect to the parameters in A, a cost function measuring the degree of correlation of $\hat{R}$. The rationale for this choice is as follows. One postulates the best model (say from a given class of models) to be one giving the most accurate predictions. The general linear predictor of the current value, X_n, based on past data X_{n-1}, X_{n-2}, etc., is

$$\hat{X}_n = B_1 X_{n-1} + B_2 X_{n-2} + B_3 X_{n-3} + \cdots. \tag{7.8}$$

The error made at time n is

$$E_n = X_n - \hat{X}_n \tag{7.9}$$

$$= X_n - B_1 X_{n-1} - B_2 X_{n-2} - B_1 X_{n-3} + \cdots \tag{7.10}$$

$$= A_0 X_n + A_1 X_{n-1} + A_2 X_{n-2} + \cdots, \tag{7.11}$$

where we define $A_0 = 1$ and $A_k = -B_k$, $k > 0$. These equations justify the identification of $\hat{R}$ with the prediction error E (cf. eq. (7.6b)) and the terminology prediction-error filter for A. The output of this filter, with the data as input, is just the sequence of errors made by the linear predictor B. One can estimate the model by minimizing the root-mean-square prediction error, or equivalently making the sequence of prediction errors uncorrelated – i.e. white noise (Robinson, 1967; Box and Jenkins, 1970; Brockwell and Davis, 1987, and Paper I).

Random processes are modeled, we have just seen, by finding the linear prediction-error filter A that makes $\hat{R}$ maximally random. We will now see that chaotic processes can be successfully modeled in the same way, only with a different cost function. It seems reasonable to replace the measure of degree of randomness discussed above with some measure of degree of chaos. We could try any of the many measures proposed for chaos (see below), but the following simpler approach works very well.

Since the goal is undoing the effects of a filter, we ask 'How can one differentiate the unfiltered innovation (e.g. Figure 7.1) from the filtered data (Figure 7.5)?' The answer lies in the phase portrait. Figure 7.6 compares the portraits of filtered and unfiltered data. Filtering basically spreads the points. This fact suggests that the effect of a filter C could be removed by these steps: (1) convolve the data with another filter A (meant to be the inverse of C); (2) plot the phase portrait of the resulting data (meant to be the innovation); and (3) minimize with respect to A some measure of spread. A further suggestion is that the area covered by the points in the phase portrait is the quantity to minimize. (In Figure 7.6 note that the area of the unfiltered data is zero – the corresponding points lie along a line – while the filtered data cover a region of nonzero area.) Scargle (1990, hereafter Paper IV) gives more details and an algorithm for evaluating this area.

The following definition generalizes these considerations to embedding spaces of arbitrary dimensions.

Cost function $H(X)$ Given the samples X_n; $n = 1, 2, \ldots, N$:

(a) Construct the M-dimensional embedding space (eq. (7.3));
(b) Define a rectangular grid of (M-dimensional) cells in this space;
(c) Plot the $N - M + 1$ data points in the embedding space;
(d) Then define H as the total area of cells containing data points.

This cost function can be thought of as a measure of the degree to which

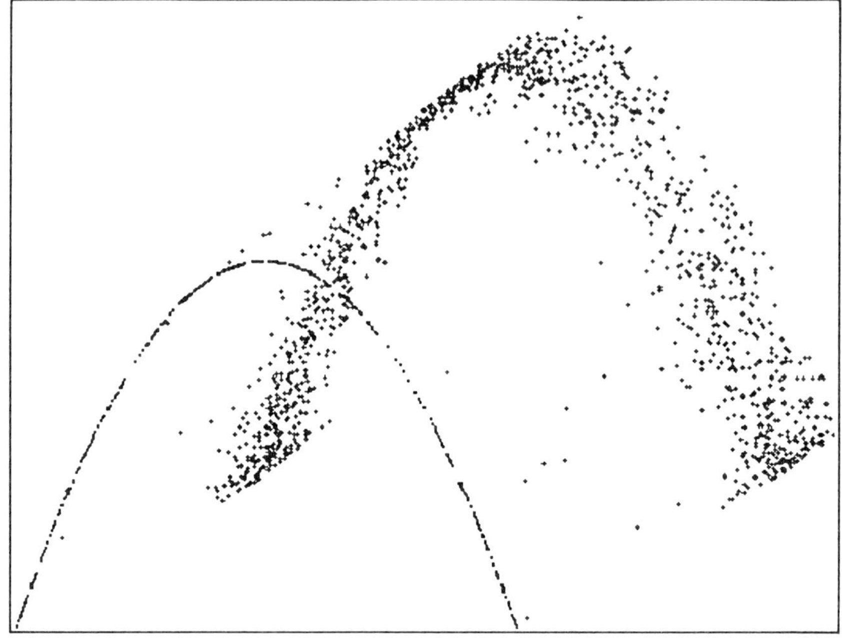

Figure 7.6 Phase portrait, sometimes called return map, for data from the logistic map (points lying on a parabola) and for the same data run through a constant, 'boxcar' filter of length 10 units of time (scattered points).

the points satsify a recurrence relation of the form $X_{n+1} = F(X_n)$, of which eqs. (7.1) and (7.2) are examples. It penalizes widely scattered data and favors distributions that tightly follow a relationship of low dimensionality. It is also closely related to the fractal dimension (sometimes called the capacity) of the corresponding chaotic attractor (Kolmogorov, 1958; Farmer *et al.*, 1983; Froehling *et al.*, 1981), defined as

$$d_C = \lim_{\varepsilon \to 0} \frac{\log N(\varepsilon)}{\log(1/\varepsilon)}, \tag{7.12}$$

where ε represents the width of the state-space cells, and $N(\varepsilon)$ is the number of cells needed to cover the data. In practice ε cannot be arbitrarily small, but (7.12) converges well if there are enough data that the cell populations do not become unreasonably small.

An alternative cost function is

$$H_{\text{info}} = -\sum p_i \log p_i, \tag{7.13}$$

where the sum is over all cells, and p_i, the probability a point will fall in cell i, is estimated with the fraction of points actually in the cell. This measure is related to the information dimension (Farmer *et al.*, 1983)

$$d_I = \lim_{\varepsilon \to 0} \frac{\log I(\varepsilon)}{\log(1/\varepsilon)}, \tag{7.14}$$

where ε is as above and

$$I(\varepsilon) = -\sum_{i=1}^{N(\varepsilon)} p_i \log p_i. \tag{7.15}$$

The *mutual information* between X_{n+1} and X_n (Fraser and Swinney, 1986), closely related to the degree of independence of the two, may itself be useful for modeling, but I have not yet tested it.

These are examples of the many measures arising in chaos theory that may be useful for time-series modeling. Others include Kolmogorov entropy, metric entropy, topological entropy, mutual information and other information measures, and Lyapunov characteristic exponents (Farmer *et al.*, 1983 and references there). Although only indirectly related to degree of chaos, dimensions of chaotic attractors, such as the fractal and Hausdorff dimensions, can be cost functions (small dimension ↔ large degree of chaos).

Elsewhere (Scargle, 1989a, b, Paper IV) I have applied this technique to simulated data – chaotic, random, and mixed data of known characteristics. In these experiments the state space was two-dimensional, and the filter C was the inverse of a low-order autoregressive filter. Figure 7.7 shows the grid used in defining the cost function, superimposed on the phase portrait of the minimum-cost innovation for one test problem. The chaotic process that was filtered to produce these synthetic data was the logistic map, and the estimated innovation closely follows the corresponding parabolic relation.

In the simulations the parameters of low-order autoregressive models were recovered very accurately. Adding noise to the data to simulate observational errors caused moderate degradation of this accuracy. Refining the cost-

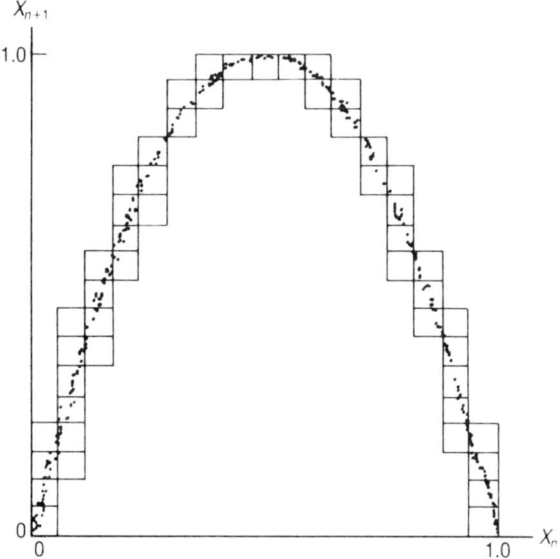

Figure 7.7 Phase portrait from deconvolution of chaotic moving-average process depicted in Figure 7.3. The figure shows only non-empty grid cells.

function grid improved the results up to a point, as expected. In some cases the dimensions of the state space and the order of the model were assumed ahead of time, and therefore did not have to be determined as part of the modeling procedure. In others, a modified 'final prediction error' criterion (see Paper I) determined these numbers.

A surprise came when I applied the technique to MA data with a random innovation. Even though the cost function is tailored for chaos, it is good at deconvolving conventional random processes! It can distinguish chaos from randomness, and is effective at determining the phase properties of models. Since standard deconvolution methods are incapable of determining the phase character (Robinson, 1967) of a random process, special ones have been developed (e.g., Scargle, 1977, Paper I, 1981b; Benveniste *et al.*, 1980; Donoho, 1981; Mendel, 1990; Breidt *et al.*, 1990). Modeling with the cost function defined here readily distinguishes a pulse from its time reverse, and recovers the correct values of the parameters of noncausal, nonsymmetric moving average – normally very difficult problems. Paper IV gives an example of separation of randomness from chaos – in one very simplified case.

Note that none of these cost functions has anything to do with whiteness of the power spectrum or correlation of the time series. The innovations found using these functions therefore need not be the uncorrelated innovation in the Wold representation. In some cases where exact deconvolutions of chaotic processes have been obtained (Scargle, 1991) the innovation is uncorrelated, in others not. In the latter cases the innovation is either exactly or approximately the symbolic dynamics representation of X.

7.5 Discussion

I recommend the method proposed here to model both random and chaotic processes. It should definitely be used instead of L_1 modeling (Scargle, Paper I, 1981b), even for the kinds of problems for which that approach was developed. Simulations show the current method superior in all ways to L_1 (Paper IV).

There are many areas where further work is needed. One stumbling block to some applications is the assumption that the data sampling interval is equal to the recurrence interval – the time interval implicit in the recurrence equation (7.2). While taking data the experimenter rarely knows the recurrence time. In any case it might be impossible or impractical to sample at the corresponding interval.

The cost function discussed here is only one of many possibilities. A systematic study will almost certainly reveal better ones. In unpublished experiments maximizing, not minimizing, the cost function discussed above provided good deconvolutions of some random moving averages.

An alternative to the procedure outlined above, since nonlinearity is essential to chaos, would be to use nonlinear prediction in place of the linear eq. (7.8). Many researchers have studied nonlinear prediction for modeling chaotic time series (references listed in Section 7.2). In addition the projection theorem yields Wold-like representations with nonlinear predictors (Doob, 1953). Nonlinear predictive deconvolution may be fruitful, but the goal of the present work is to find what can be achieved using linear prediction on nonlinear processes.

To recapitulate: Two mathematical theorems set the stage. First, one can study the dynamics of a physical system in its full state space by analyzing a single scalar time series. Second, both stationary random and chaotic processes have representations as filtered uncorrelated processes. Since models can be estimated using embedding-space properties of the data, these two results combine to yield a practical deconvolution method. The result is a powerful tool for modeling chaos and randomness. The convolutional representations that come out of this procedure do not always have an uncorrelated innovation, but are nonetheless useful representations. The deconvolution procedure can detect the presence of chaos or randomness, as well as distinguish and separate them when both are present.

Acknowledgements

I am indebted to many colleagues for useful suggestions, especially Jim Crutchfield, David Donoho, Andrew Walden, Karl Young, John ('Sid') Sidorowich, and the anonymous referee. I am also grateful to Tony Dobrovolskis and Amara Graps for comments on the manuscript, and to Jef Teugels, of the Katholieke Universiteit Leuven, Belgium for his wonderful hospitality, and to the Bernoulli Society, the Institute of Mathematical Statistics, the International Society of Soil Science, and the International Association of Hydrological Sciences for travel support. Funding of various aspects of this research was provided by the NASA Astrophysics Software and Research Aids program (NRA 89-OSSA-8), the NASA-Ames Research Center Director's Discretionary Fund, and a special grant from Mike Smith of the NASA Office of Commercial Programs.

References

Abarbanel, H., Brown, R., and Kadtke, J. (1989). Prediction and system identification in chaotic nonlinear systems: Time series with broadband spectra. *Phys. Lett.* A., **138**, 401–8.

Abraham, R. and Shaw, C. (1983). *Dynamics – The Geometry of Behavior, Part 2: Chaotic Behavior*, Aerial Press, Santa Cruz.

Bai-Lin, H. (1984). *Chaos.* World Scientific Pub. Co., Singapore.

Bai-Lin, H. (1989). *Elementary Symbolic Dynamics and Chaos in Dissipative Systems*, World Scientific Pub. Co., Singapore.

Bellman, R. and Cooke, K.L. (1963). *Differential–Difference Equations*, Academic Press, New York.

Benveniste, A., Goursat, M., and Ruget, G. (1980). Robust identification for nonminimum phase systems: Blind adjustment of a linear equalizer in data communications. *IEEE Trans. Aut. Cont.*, **AC-25**, 385–99.

Berge, P., Pomeau, Y., and Vidal, C. (1984). *Order within Chaos: Towards a Deterministic Approach to Turbulence*, Wiley, New York.

Box, G. and Jenkins, G. (1970). *Time Series Analysis, Forecasting and Control*, Holden-Day, San Francisco.

Breidt, F.J., Davis, R.A., Lii, K.-S., and Rosenblatt, M. (1990). *Maximum Likelihood Estimation for Noncausal Autoregressive Processes.* Preprint.

Brock, W. (1990). Causality, chaos, explanation and prediction in economics and finance. Preprint, to appear in *Beyond Belief: Randomness, Prediction, and Explanation in Science*, Casti, and Karlqvist, (eds), CRC Press, Boca Raton.

Brockwell, P. and Davis, R. (1987). *Time Series: Theory and Methods*, Springer, New York.

Crutchfield, J. (1989). Inferring the dynamic, quantifying physical complexity. In *Measures of Complexity and Chaos*, Abraham, N. *et al.* (eds), Plenum Press, New York.

Crutchfield, J. and McNamara, B. (1987). Equations of motion from a data series. *Complex Systems*, **1**, 417–52.

Crutchfield, J. and Packard, N. (1983). Symbolic dynamics of noisy chaos. *Physica D*, **7**, 201–23.

Donoho, D. (1981). On minimum entropy deconvolution. In *Applied Time Series Analysis II*, Findley, D. (ed.). Academic Press, New York.

Doob, J. (1953). *Stochastic Processes.* Wiley, New York.

Eckmann, J.-P. and Ruelle, D. (1985). Ergodic theory of chaos and strange attractors. *Rev. Mod. Phys.*, **57**, Part I, 617–56.

Eckmann, J.-P., Kamphorst, S., Ruelle, D., and Ciliberto, S. (1986). Liapunov exponents from time series, *Phys. Rev. A.*, **34**, 4971–9.

Farmer, D. and Sidorowich, J. (1987). Predicting chaotic time series. *Phys. Rev. Lett.*, **59**, 845–8.

Farmer, D. and Sidorowich, J. (1988a). Exploiting chaos to predict the future and reduce noise. In *Evolution, Learning and Cognition*, Lee, Y. (ed.), World Scientific Pub. Co., Singapore.

Farmer, D. and Sidorowich, J. (1988b). Optimal shadowing and noise reduction. Preprint submitted to *Physica D.*

Farmer, D., Ott, E., and Yorke, J. (1983). The dimension of chaotic attractors. *Physica D*, **7**, 153.

Feder, J. (1988). *Fractals*, Plenum Press, New York.

Ford, J. (1983). How random is a coin toss? *Physics Today*, **36** (4), 40–7.

Ford, J. (1986). Chaos: Solving the unsolvable, predicting the unpredictable! In *Chaotic Dynamics and Fractals*, Barnsley, M.F. and Demko, S.G. (eds). Academic Press, New York.

Fraser, A. and Swinney, H. (1986). Independent coordinates for strange attractors from mutual information. *Phys. Rev. A.*, **33**, 1134–40.

Friedman, J. and Tukey, J. (1974). A projection pursuit algorithm for explanatory data analysis. *IEEE Trans. Computers*, **C-23**, 881–90.

Froehling, H., Crutchfield, J., Farmer, D., Packard, N., and Shaw, R. (1981). On determining the dimension of chaotic flows. *Physica D*, **3**, 605–17.

Gershenfeld, N. (1988). An experimentalist's introduction to the observation of dynamical systems. In *Directions in Chaos*, Bai-Lin, H. (ed.). World Scientific Pub. Co., Singapore.

Grassberger, P. and Procaccia, I. (1983a). Characterization of strange attractors. *Phys. Rev. Lett.*, **50**, 346–9.

Grassberger, P. and Procaccia, I. (1983b). Estimation of the Kolmogorov entropy from a chaotic signal. *Phys. Rev. A.*, **28**, 2591–3.

Grebogi, C., Ott, E., and Yorke, J.A. (1982). Chaotic attractors in crisis. *Phys. Rev. Lett.*, **48**, 1507–10.

Klavetter, J. (1989). Rotation of Hyperion. II. Dynamics. *Astron. J.*, **98**, 1855–74.

Kolmogorov, A. (1958). *Dokl. Akad. Nauk SSSR*, **119**, 861.

Kostelich, E. and Yorke, J. (1988). Noise reduction in dynamical systems. *Phys. Rev. A.*, **38**, 1649–52.

Lewis, P. and Stevens, J. (1990). Nonlinear modeling of time series using multivariate adaptive regression splines. Preprint.

Lichtenberg, A.J. and Lieberman, M.A. (1983). *Regular and Stochastic Motion*, Springer-Verlag, New York.

Mandelbrot, B. (1983). *The Fractal Geometry of Nature*, Freeman, New York.

Marteau, P. and Abarbanel, H. (1990). Noise reduction in chaotic time series using scaled probabilistic methods. Preprint.

May, R.M. (1976). Simple mathematical models with very complicated dynamics. *Nature*, **261**, 459–67.

Mendel, J. (1990). *Maximum Likelihood Deconvolution*, Springer-Verlag, New York.

Moon, F. (1987). *Chaotic Vibrations: An Introduction for Applied Scientists and Engineers*. Wiley, New York.

Packard, N.H., Crutchfield, J.P., Farmer, J.D., and Shaw, R.S. (1980). Geometry from a time series. *Phys. Rev. Lett.*, **45**, 712–16.

Quinn, T. and Tremaine, S. (1990). Round off error in long-term planetary orbit integrations. Preprint.

Robinson, E. (1967). Recursive decomposition of stochastic processes. In *Econometric Model Building*, Wold, H. (ed.), North-Holland, Amsterdam.

Ruelle, D. (1989). *Chaotic Evolution and Strange Attractors: The Statistical Analysis of Time Series for Deterministic Nonlinear Systems*, Cambridge University Press, Cambridge.

Scargle, J. (1977). Absolute value optimization to estimate phase properties of stochastic time series. *IEEE Trans. Info. Theory*, **IT-23**, 140–3.

Scargle, J. (1981a). Studies in astronomical time series analysis. I: Modeling random processes in the time domain. *The Astrophys. J. Supp.*, **45**, 1–71 (Paper I).

Scargle, J. (1981b). Phase-sensitive deconvolution to model random processes, with special reference to astronomical data. In *Applied Time Series Analysis II*, Findley, D. (ed.), Academic Press, New York.

Scargle, J. (1989a). An introduction to chaotic and random time series analysis. *Int. J. Imag. Syst. and Tech.*, **1**, 243–53.

Scargle, J. (1989b). Random and chaotic time series analysis: Minimum phase-volume deconvolution. In *Nonlinear Structures in Physical Systems, Proceedings of the Second Woodward Conference*, Lam, L. and Morris, H. (eds), Springer-Verlag, New York.

Scargle, J. (1990). Studies in astronomical time series analysis. IV: Modeling chaotic and random processes with linear filters. *The Astrophys. J.*, **343**, 469–82 (Paper IV).

Scargle, J. (1991). Predictive deconvolution of chaotic and random processes. Lecture at the workshop *New Directions in Time Series Analysis* at the Institute for Mathematics and Its Applications, University of Minnesota, 1990. To appear.

Schuster, H. (1988). *Deterministic Chaos*, VCH, New York.

Shaw, R. (1984). *The Dripping Faucet as a Model Chaotic System*, Ariel Press, Santa Cruz.

Sugihara, G. and May, R. (1990). Nonlinear forecasting as a way of distinguishing chaos from measurement error in time series. *Nature*, **344**, 734.

Takens, F. (1981). *Dynamical Systems and Turbulence, Warwick, 1980*. Lecture Notes in Mathematics 898, Springer-Verlag, Berlin.

Thompson, J. and Stewart, H. (1986). *Nonlinear Dynamics and Chaos*, Wiley, New York.

Tong, H. (1990). *Non-Linear Time Series, A Dynamical System Approach*, Oxford University Press, Oxford.

Townshend, B. (1990). Nonlinear prediction of speech signals. Preprint, submitted to *IEEE Trans. Signal Processing*.

Walden, A.T. (1992). Clustering of attributes by projection pursuit for reservoir characterization. In *Automated Pattern Analysis in Petroleum Exploration*, Palaz, I. and Sengupta, S., (eds). Springer, New York.

Weigend, A., Huberman, B., and Rumelhart, D. (1990). Predicting the future: A connectionist approach. Preprint submitted to *Int. J. Neur. Sys.*

Wold, H. (1938). *A Study in the Analysis of Stationary Time Series*, Almqvst and Wiksell, Uppsala.

Wolfram, S. (1985). Origins of randomness in physical systems. *Phys. Rev. Lett.*, **55**, 449–52.

Chapter 8

Deconvolving nonGaussian time series: the seismic experience

A. T. Walden

An important problem in reflection seismology is the deblurring (or deconvolution) of seismic traces. Deconvolution may be carried out in two steps: spectral whitening followed by phase correction. Recently, the maximizing of a chosen statistic (or projection index), has been used for the estimation of a phase-lag correction for seismic data. Unfortunately, spectral whitening cannot be accomplished across the full frequency range (zero to Nyquist) because of a rapid decay in energy at low and high frequencies, and the ubiquitous seismic noise (i.e. the data is band-limited). A statistic which is superior to another in the ideal full-bandwidth case can be substantially inferior in the band-limited case. For the latter type of data, including real seismic data, kurtosis – not attractive in the ideal case – proves a remarkably satisfactory choice of statistic. Assuming additive noise, the relative performance of the competing objective functions is also shown to depend on the signal-to-noise ratio.

8.1 Introduction

Each year in the search for oil and gas the seismic industry records about one billion (10^9) time series. These time series, plotted vertically, are used to build up 'cross-sections' of the earth – seismic sections – which are processed to enhance their resolution. A noise-free seismic trace is often written in the linear process form

$$s_t = \sum_j a_j e_{t-j} = a_t * e_t, \tag{8.1}$$

i.e. the convolution of a 'blurring' filter $\{a_j\}$ with an independent and identically distributed (i.i.d.) zero-mean innovations sequence $\{e_t\}$. The filter values $\{a_j\}$ are the values of the seismic pulse which passes through the earth, while the innovations sequence $\{e_t\}$ corresponds to the reflection coefficients at the interfaces of different materials making up the layers of the earth. (We shall sometimes refer to $\{a_j\}$ as the (seismic) wavelet filter to distinguish it from other filters we need to introduce.) The purpose of reflection seismics is to clearly delineate these 'layers', but this will not be possible if the filter $\{a_j\}$ is long and asymmetric, since a mix of interference effects will be seen instead.

The technique of deconvolution attempts to remove or reduce the blurring effect of $\{a_j\}$.

8.1.1 NonGaussian innovations

After a well has been drilled the innovations (reflectivity) sequence $\{e_t\}$ can in fact be computed from data obtained from instruments placed in the borehole. Examination of such sequences computed on a world-wide basis shows that their distributions are generally symmetric with mean zero and can be modeled by the generalized Gaussian family of distributions (Gray, 1979; Walden and Hosken, 1986) having p.d.f.

$$f(e;\alpha,\beta)=(\alpha/\{2\beta\Gamma(1/\alpha)\})\exp(-\{|e|/\beta\}^{\alpha}),\qquad -\infty<e<\infty,$$

where Γ is the gamma function, $\beta>0$ is a scale parameter, and $\alpha>0$ is a shape parameter typically less than one (i.e. the distribution is more spiky and long-tailed than the Laplace distribution). The linear process (8.1) is thus driven by a *nonGaussian* sequence, and the convolution makes the distribution of $\{s_t\}$ closer to the Gaussian (Mallows, 1967). Since any drilling occurs after the processing of the seismic data, the appropriate shape parameter α of the innovations distribution is unknown at the deconvolution stage.

8.1.2 NonGaussianity and phase

The frequency response function or transfer function corresponding to the filter $\{a_j\}$ is

$$A(\omega)=\sum_j a_j e^{-i\omega j}=|A(\omega)|\exp\{i\theta_A(\omega)\}$$

where $|A(\omega)|$ is the amplitude and $\theta_A(\omega)$ the phase responses of the filter. If $\theta_A(\omega)=0$ for all ω the wavelet filter is said to be *zero phase*; generally $\theta_A(\omega)\neq 0$ for all ω.

Deblurring by filtering seeks to remove the effects of $\{a_j\}$ by finding an operator which both corrects for the shaping of the amplitude response by $|A(\omega)|$ – 'spectral whitening' – and corrects for any non-zero phase response – 'zero phasing'.

Consider a (moving average) filter $\{a_j\}$ having $q+1$ terms. With real distinct roots there are 2^q ways of specifying the roots of the corresponding z-polynomial $A(z)=\Sigma a_j z^j$ which give the same amplitude response $|A(\omega)|$ but *different* phase responses $\theta(\omega)$. If the innovation sequence $\{e_t\}$ were Gaussian, the probability structure of the process $\{s_t\}$ would be the same for all choices of the roots, and the phase response could not be uniquely determined. However, if, as here, the innovations sequence is nonGaussian, then the probability structure of $\{s_t\}$ will be different according to the set of roots present (see Lii and Rosenblatt, 1982). This then provides a route into estimating the phase response of the filter, enabling conversion to zero phase. Note that the standard time-series approach of Box and Jenkins (1970) assumes Gaussian innovations, and takes the roots to be the unique set which are all outside the unit circle (the 'invertibility' condition to statisticians, and 'minimum-phase' condition to engineers).

Figure 8.1 shows two filters having the identical 'power' (or squared amplitude) responses, but very different phase responses. The acausal symmetric filter has a phase response which is (approximately, due to truncation effects and rounding errors) zero throughout the pass-band of the signal (10–65 Hz, say), while the causal, asymmetric filter has a (detrended) phase response throughout the pass-band, centered about 90°. The causal asymmetric filter is typical of the blurring filter $\{a_j\}$ found on seismic data, and the deblurring (deconvolution) filtering is aimed at changing it into the acausal symmetric type of filter (zero-phasing) and additionally flattening the amplitude response to produce a narrower filter (thus increasing resolution).

To illustrate the advantage of having a symmetric acausal filter after deconvolution, consider Figure 8.2. The first time series is an actual earth reflection coefficient series $\{e_t\}$, calculated from measurements down an oil well. (The points are joined together and positive runs coloured black and negative runs white, a traditional display for seismic measurements.) Time series 2–6 are five repeats, for clarity, of the result of the convolution (8.1) where the blurring filter is the acausal symmetric filter of Figure 8.1. Time series 7–11 are five repeats of the convolution in (8.1) where the blurring filter is now the causal asymmetric filter of Figure 8.1. In practice, before drilling, we would only have the observed convolution from which to draw inferences about the reflection coefficient series. From Figure 8.2 it is easy to see that time series 2–6 are much more like the series of interest, series 1, than are series 7–11. The latter has peaks at the incorrect (shifted) times and interference effects from the lobes of the filter blurs details of interest.

8.2 Deconvolution

8.2.1 Deconvolution in one step

A deconvolution filter $\{f_j\}$ of length L applied to $\{s_t\}$ gives

$$\hat{e}_t = f_t * s_t = f_t * a_t * e_t,$$

an approximation to $\{e_t\}$. So-called MED-type deconvolution (minimum entropy-type deconvolution) methods find as the deconvolution filter the one which maximizes some statistic T formed from the $\{\hat{e}_t\}$ values. This optimization must thus be carried out over the L parameters of the filter. The idea is that the distribution of the $\{\hat{e}_t\}$ should be maximally nonGaussian. As Huber (1985, pp. 467–8) points out, MED-type deconvolution is thus a projection pursuit method, and choosing a statistic corresponds to choosing a projection index to define 'nonGaussianity'. Various statistics have been utilized (see. e.g., Walden, 1985). Gray (1979) favored

$$T = O_1^2 = \frac{\Sigma \hat{e}_t^2 / N}{(\Sigma |\hat{e}_t| / N)^2}. \tag{8.2}$$

Wiggins (1978) chose

$$T = O_2^4 = \frac{\Sigma \hat{e}_t^4 / N}{(\Sigma \hat{e}_t^2 / N)^2}, \tag{8.3}$$

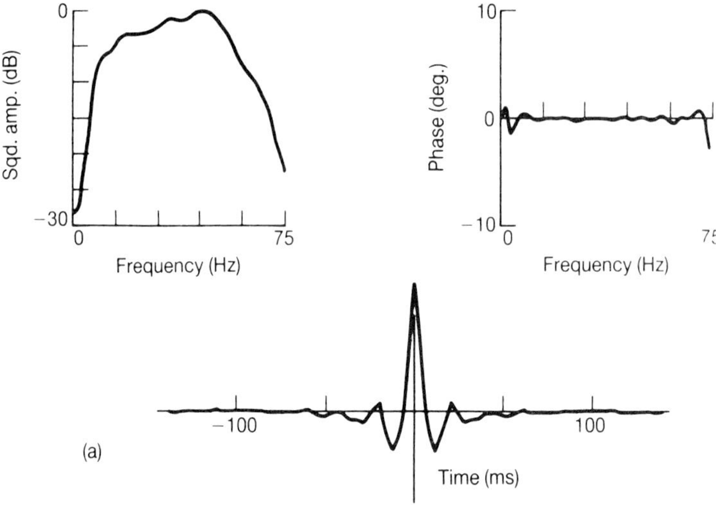

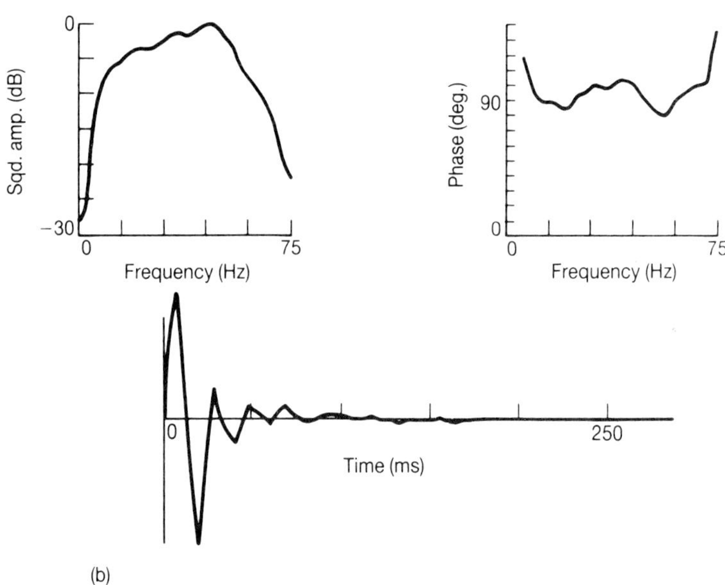

Figure 8.1 Two filters, and their squared amplitude and phase responses over the frequency pass-band for (a) an acausal symmetric filter, and (b) a causal asymmetric filter (with detrended phase response). Note that the squared amplitude responses are identical, but the phase responses are very different.

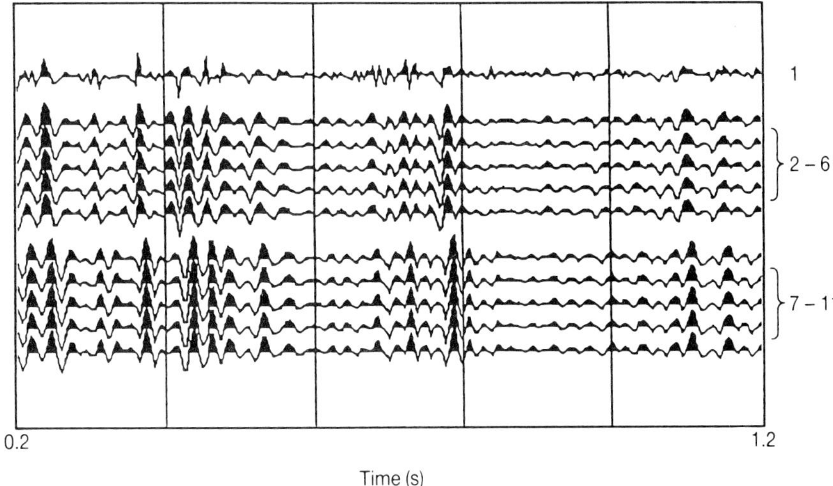

Figure 8.2 Series 1 is an actual reflection coefficient series, series 2–6 are repeats of series 1 convolved with the filter in Figure 8.1(a), while series 7–11 are repeats of series 1 convolved with the filter in Figure 8.1(b). The series 2–6 reproduce the important characteristics of series 1 much better than do series 7–11.

where N is the sample size. O_1^2 is the square of the statistic, $\sqrt{(N\Sigma\hat{e}_t^2)}/\Sigma|\hat{e}_t|$ which provides the most powerful scale invariant test of the hypothesis that $\{\hat{e}_t\}$ has the Laplace rather than Gaussian distribution (large values indicating the former); see Hogg (1972). Hence by maximizing O_1^2 it is expected that the deconvolution filter will act to change the distribution of $\{s_t\}$ from near Gaussian to near Laplacian (i.e. to nearer that of $\{e_t\}$). The objective function O_2^4 is (apart from non-subtraction of the sample mean) a common estimator of the kurtosis of the sample. Kurtosis has been used as a test statistic for detecting statistical *outliers* from a Gaussian distribution (Ferguson, 1961) and Wiggins' intention was to change $\{s_t\}$ towards a small number of strong amplitude events, assuming this form for $\{e_t\}$. This was the original sense of 'entropy' minimization, i.e. a move towards a few large events.

An approximation to the asymptotic variance–covariance matrix of MED-type deconvolution filters was derived by Donoho (1981), using the theory of M-estimators. The variance–covariance matrix $\mathbf{C}$ was expressed as the product of a scalar figure of merit, dependent on the chosen statistic T and the probability density function f of the innovations, and a matrix $\mathbf{F}$ dependent only on the wavelet filter $\{a_j\}$, i.e.

$$\mathbf{C} = M(T, f)\mathbf{F}(\{a_j\}).$$

Hence the relative performance of various statistics could be analyzed by computing the figure of merit $M(T, f)$ for various innovations distributions (see also Walden, 1985) with improving estimation corresponding to a *decreasing* figure of merit. A consequence of the form for the distributions of the innovations derived from well data is that, using Donoho's results,

Gray's favored statistic (8.2) would be expected to substantially outperform Wiggins' kurtosis (8.3). For example, for a Laplace innovations distribution, the figure of merit is six times smaller (i.e. better) with O_1^2 than with O_2^4. Other statistics closely related to O_1^2, and certain of other families of statistics, would similarly be expected to outperform kurtosis (Walden, 1985).

Two necessary assumptions for Donoho's theory are that the data $\{s_t\}$ are noise-free and the wavelet filter can be inverted back to a 'spike'. The latter assumption means that although the amplitude response of the wavelet filter need not be flat, it must not be totally deficient of amplitude in any particular frequency range. Both these assumptions are unrealistic for reflection seismics. Why? There is always a substantial amount of noise on seismic data; if we replace (8.1) by the more realistic model

$$s_t = a_t * e_t + n_t \tag{8.4}$$

where $\{n_t\}$ is the noise, then the ratio of the variance of the signal $\{a_t * e_t\}$ to the variance of the noise $\{n_t\}$ (i.e. the signal-to-noise ratio or SNR) can vary between about 6 and 30 depending on, e.g., environmental factors during acquisition, and geological effects. Secondly, the amplitude spectrum of the wavelet filter is deficient in low and high frequencies due to the inability of seismic sources to generate low-frequency energy, and preferential conversion of high-frequency energy into heat; see, e.g., Figure 8.1. Other factors, including recording filters, also play their part. Typically, then, the amplitude response of $\{a_j\}$ decays very rapidly at low and high frequencies, and with noise level fairly constant with frequency (i.e. fairly white) it follows that below some low frequency (say 4 Hz) and above some high frequency (say 80 Hz) the noise dominates the signal. Full inversion of the amplitude spectrum of the wavelet filter in the range 0 to Nyquist (typically 125 Hz) is thus impossible because the noise at low and high frequencies would be drastically inflated. The wavelet filter is said to be *band-limited*.

8.2.2 A two-step approach to deconvolution

Recently, attention has turned from the estimation of a complete MED-type deconvolution filter, which has proved very disappointing in practice – not least due to the multi-parameter optimization problem (Wiggins, 1985 and Longbottom *et al*., 1988) – to deconvolution using the two parts of the frequency-domain representation, namely amplitude and phase (Levy and Oldenburg, 1987 and Longbottom *et al*., 1988). First, the amplitude spectrum of the process is whitened (flattened) using conventional methods. Basically this involves fitting an autoregressive model, from which the residuals form the whitened process. In estimating the autoregressive parameters, the diagonal of the correlation matrix is increased by a few percent – a form of ridge regression – which stabilizes the amplitude spectrum inversion by not inflating the noise at low and high frequencies. The appropriate filter is the prediction error filter of length p, say,

$$(1, -g_0, \ldots, -g_{p-1})$$

where $g_0, \ldots, g_{p-1}$ are the estimated autoregressive parameters. Denote this filter, suitably scaled, by $\{g_j\}$. Applying it to (8.4) gives

$$g_t * s_t = g_t * a_t * e_t + g_t * n_t.$$

Taking Fourier transforms throughout

$$G(\omega)S(\omega) = G(\omega)A(\omega)E(\omega) + G(\omega)N(\omega).$$

The signal component can be written

$$\begin{aligned} G(\omega)A(\omega)E(\omega) &= |G(\omega)| \exp\{i\theta_G(\omega)\} |A(\omega)| \exp\{i\theta_A(\omega)\} E(\omega) \\ &= |B(\omega)| \exp\{i\theta_B(\omega)\} E(\omega), \end{aligned} \tag{8.5}$$

where $\theta_B(\omega) = \theta_G(\omega) + \theta_A(\omega)$ and $|B(\omega)| = |G(\omega)|\,|A(\omega)|$. Since $|G(\omega)|$ is the inverse of $|A(\omega)|$ in the signal dominated frequency range, $|B(\omega)|$ is flat in this range, and falls off at low and high frequencies.

Next, it is assumed that the phase response of the effective filter is linear, i.e.

$$\theta_B(\omega) = \theta_0 + \ell\omega.$$

Hampson and Galbraith (1981, p. 15) considered that filters which are observed in practice (instrument responses, source signatures, etc.) often have phase spectra which are essentially linear *over a limited bandwidth*, and gave examples. Longbottom *et al.* (1988) also gave an example, and others may readily be found. The slope ℓ corresponds to a time shift – this can never be determined from a convolutional model such as (8.1) or (8.4) since $\{s_t\}$ can be kept the same if $\{a_j\}$ is advanced in time and $\{e_t\}$ delayed in time. A bulk time shift is in this sense of no interest. The intercept θ_0 corresponds to a phase-lead. As stated above, we wish to reduce $\{a_j\}$ to zero-phase, hence we need to apply a phase-lag filter $\{c_j(\phi_0)\}$ with frequency-response function

$$C(\phi_0) = \exp\{-i\phi_0\}$$

with $\phi_0 = \theta_0$ since then the signal component is

$$\exp\{-i\theta_0\} \exp\{i(\theta_0 + \ell\omega)\} |B(\omega)|\, E(\omega) = \exp\{i\ell\omega\} |B(\omega)|\, E(\omega),$$

giving an effective filter acting on the innovations which is zero phase (apart from a time shift), with a flat amplitude spectrum over signal-dominated frequencies. The phase correction is determined by finding the ϕ_0 which maximizes a statistic calculated from

$$\hat{e}_t = c_t(\phi_0) * g_t * a_t * e_t + c_t(\phi_0) * g_t * n_t. \tag{8.6}$$

The multi-parameter optimization of MED-type deconvolution, with all its attendant numerical problems, is thus reduced to a straightforward single-parameter optimization.

Our interest in this work is to choose an appropriate statistic to accomplish this. Do the broad conclusions from Donoho's theory regarding the inefficiency of kurtosis also apply in the problem of phase-lag estimation? What about the effect of band limitation and noise? How do band limitation and noise interact with the amplitude distribution of the innovations, $\{e_t\}$, in determining the relative performance of different statistics? These important practical questions are addressed in this chapter.

8.3 Investigation of phase-lag estimation by simulation

To begin to understand phase-lag estimation by maximization of a chosen statistic, it is useful to carry out a simulation experiment making the following assumptions (unrealistic for seismic data): (1) the effective filter has *constant* phase throughout the frequency band zero to Nyquist; (2) its amplitude spectrum has been perfectly spectrally whitened (flattened) throughout this range; (3) the innovations sequence is independent. In this model $\{s_t\}$ is thus represented as an i.i.d. sequence (with as yet unspecified amplitude distribution) convolved with a phase-lead filter, i.e. (8.5) becomes simply

$$G(\omega)A(\omega)E(\omega) = \exp\{i\theta_0\}E(\omega)$$

It is convenient to choose $\theta_0 = 0$, and then vary the phase lag ϕ_0 of (8.6) between 0° and 180°. As the phase-lag filtering applied to the artificial innovations (reflection) sequence is changed, so is the amplitude distribution of the resulting values. If a chosen statistic T is to be at all useful, it should take an overall maximum for ϕ_0 close to zero (i.e. no phase-lag correction needed, since $\theta_0 = 0$), thus closely identifying the innovations sequence.

Given a simulated i.i.d. sequence $\{e_t\}$ (see the Appendix for details), it is easy enough to introduce a phase lag ϕ_0. The sequence $\{x_t\}$ corresponding to a phase-lag ϕ_0 relative to $\{e_t\}$ is

$$x_{t;\phi_0} = e_t \cos\phi_0 + \mathscr{H}(e_t)\sin\phi_0 \qquad (8.7)$$

where $\mathscr{H}(e_t)$ is the discrete Hilbert transform of $\{e_t\}$, i.e. a 90° phase lag has been applied to $\{e_t\}$. If it is desired to filter $\{e_t\}$ by a whole series of phase-lag filters, this can be done by computing $\mathscr{H}(e_t)$ once, and then applying (8.7) for each required phase-lag.

8.3.1 Large-scale simulation: full-bandwidth case, no noise

Under conditions (1)–(3) above, a large-scale simulation experiment was carried out using the following steps:

(a) Generate an independent sequence from a generalized Gaussian distribution (see the Appendix) with shape parameter α. For convenience give the sequence a phase-lead of $\theta_0 = 0$. The sequence length is specified in seconds, with sample interval $\Delta t = 4$ milliseconds (ms), as usual in exploration seismology. The default length here is 1 s (251 points).

(b) Apply a phase-lag for $\phi_0 = 0, 1, 2, \ldots, 180°$. This can be done simply using (8.7): it is not necessary to calculate and apply a different filter each time.

(c) Record the value of ϕ_0 giving a maximum to O_1^2 and O_2^4, and the errors E_1^2 and E_2^4 say. An error of 180° corresponds to a polarity inversion for $\{\hat{e}_t\}$. This does not matter because the polarity of $\{e_t\}$ cannot be ascertained from the convolutional model, since both $\{a_j\}$ and $\{e_t\}$ can be multiplied by -1 without changing $\{s_t\}$. Hence, if the errors exceed 90°, then it is the error minus 180° that is recorded.

(d) Carry out steps (a)–(c) 500 times.

Independent simulation experiments were performed for time series of length 1 s (251 points) for $\alpha = 0{\cdot}6$ (more long-tailed than the Laplace, kurtosis $= 15{\cdot}6$) and 1 (Laplace, kurtosis $= 6$). For brevity we report only the case $\alpha = 0{\cdot}6$ in detail in this paper. Cumulative plots were produced, giving the percentage of the 500 cases with error magnitudes less than E for $5^\circ \leqslant E \leqslant 40^\circ$ in steps of 5°. These cumulative plots are given in Figure 8.3, where it is seen that using O_1^2 and $\alpha = 0{\cdot}6$, the error is less than 5° in 97 per cent of cases, and less than 10° in 100 per cent of cases! O_2^4 performs less well for $\alpha = 0{\cdot}6$, but nevertheless, the error is less than 15° in 97 per cent of outcomes. O_1^2 also outperforms O_2^4 for a Laplace innovations sequence ($\alpha = 1{\cdot}0$). Additional replications indicate that the lines in Figure 8.3 are accurate to within a few per cent. Note that in Figure 8.3, and in all subsequent similar plots, the curve for O_2^4 is denoted by a dashed line, while for O_1^2 it is denoted by a solid line. Generally speaking both O_1^2 and O_2^4 perform well

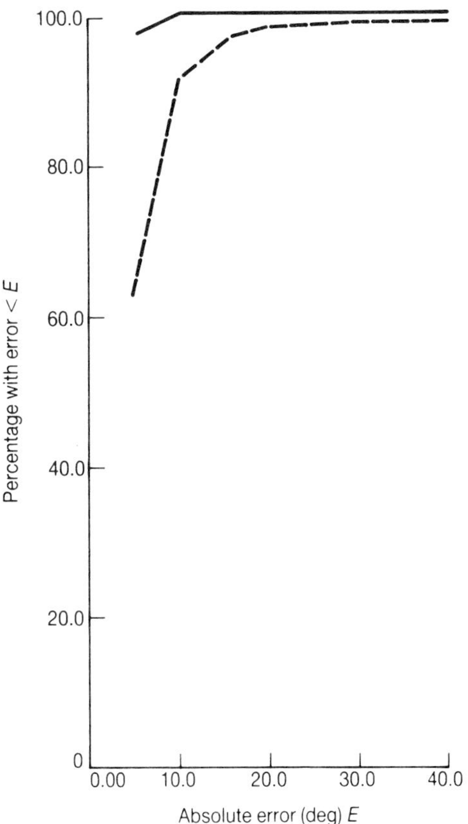

Figure 8.3 Phase-lag estimation error curves. The generalized Gaussian shape parameter α for the distribution of innovations takes the value 0·6. Full-bandwidth time series of length 1 (251 points) were used. Signal-to-noise ratio $= \infty$ (no noise). The solid line is for O_1^2, the dashed line is for O_2^4.

here. The fact that O_1^2 outperforms O_2^4 means that the same preference for O_1^2 is exhibited here as it is with the figure of merit results for the deconvolution filter in the MED-type methods (Donoho, 1981; Walden, 1985), under the no-noise, full-band assumption.

It is worth pointing out that the observed error distributions, both here and in forthcoming sections, are all approximately symmetric about zero; hence bimodality is implied when the absolute errors cluster at large values. When present, such a tendency can be readily discerned from the cumulative absolute error plots (not present in Figure 8.3).

8.3.2 Full bandwidth plus additive noise

Suppose now an additional step is inserted into the sequence (a)–(d), by adding white Gaussian noise after step (b). This was done to give a range of signal-to-noise ratios (SNRs). For each ratio an independent 500-size simulation was performed using generalized Gaussian-distributed $\{e_t\}$ with both $\alpha = 0{\cdot}6$ and $\alpha = 1{\cdot}0$. (Note the sequence used to form the noise is also independent of the sequence used in step (a).) We recall that in Figure 8.3 (for $\alpha = 0{\cdot}6$) the SNR was effectively ∞; Figure 8.4 is the equivalent of Figure 8.3 but now with an SNR of 3. Both statistics still perform well with an SNR of 3, with O_2^4 now much closer to O_1^2 for small absolute errors.

For Laplace-distributed $\{e_t\}$ it appears that O_2^4 can outperform O_1^2 at such low SNRs, so that whether O_2^4 outperforms O_1^2 at a certain SNR depends on the innovations distribution.

8.3.3 Band-limited, no noise

In the full-band case and high SNRs both O_1^2 and O_2^4 perform well. Both statistics successfully detect filtering of the initial i.i.d. sequence $\{e_t\}$. Now relax assumption (2) concerning the perfect whitening of the amplitude spectrum. Typically the amplitude spectrum of the effective filter after standard whitening deconvolution would be quite flat within the seismic bandwidth, but not outside. Such an amplitude spectrum should be closely approximated by that of a Butterworth filter with corner frequencies of 8 and 64 Hz, with a decay rate at low frequencies of 18 dB/octave (i.e. 18 dB every time the frequency halves) and a decay rate at high frequencies of 72 dB/octave (i.e. 72 dB every time the frequency doubles), denoted 8/18–64/72.

Another set of independent simulations incorporating steps (a)–(d) was carried out, including an additional step, (between (b) and (c)), that of (zero-phase) Butterworth filtering to shape the amplitude spectrum as defined above. Independent innovations sequences were again generated for $\alpha = 0{\cdot}6$ and 1. In Figure 8.5 the cumulative displays of phase-lag error magnitudes derived from segments of length 1 s (251 points) are given for $\alpha = 0{\cdot}6$. O_2^4 performs very much better than O_1^2! As the data length increases, the performance of O_1^2 increases only very gradually, while that of O_2^4 shows large increases.

The picture for Laplace-distributed innovations sequences is essentially

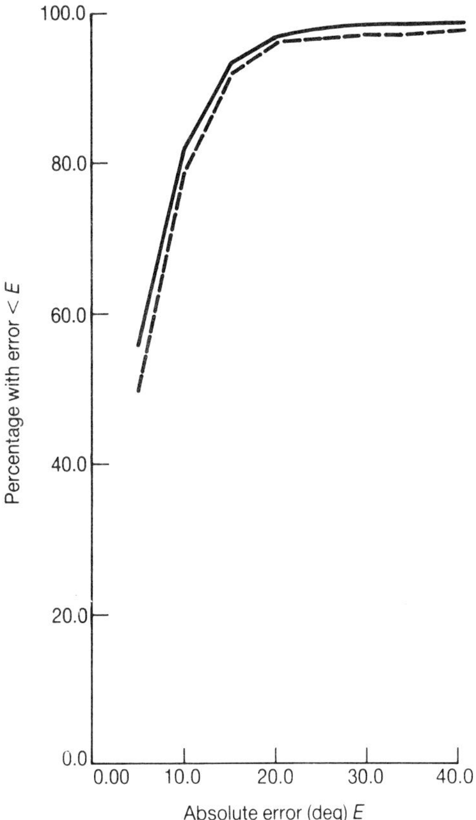

Figure 8.4 Phase-lag estimation error curves. The generalized Gaussian shape parameter α for the distribution of innovations takes the value 0·6. Full-bandwidth time series of length 1 (251 points) were used. Signal-to-noise ratio = 3. The solid line is for O_1^2, the dashed line is for O_2^4.

similar, only this time the percentage of cases with low errors decreases for both O_1^2 and O_2^4.

Making a comparison with the full-band case (compare Figures 8.3 and 8.5) it can be seen that the effect of band-limitation on the performance of O_1^2 for phase-lag estimation is disastrous! Of course the expected effect of filtering an innovations sequence with the Butterworth filter – or any band-limiting filter – would be to make its distribution closer to Gaussian and no longer i.i.d. (In addition the band-limiting filter is special in that it cannot effectively be inverted back to a 'spike'.) Clearly O_2^4 works better under these circumstances than does O_1^2.

8.3.4 Effect of low-cut frequency

It has been demonstrated that band-limiting a synthetic time series using an 8/18–64/72 Hz Butterworth filter results in O_2^4 outperforming O_1^2, unlike the

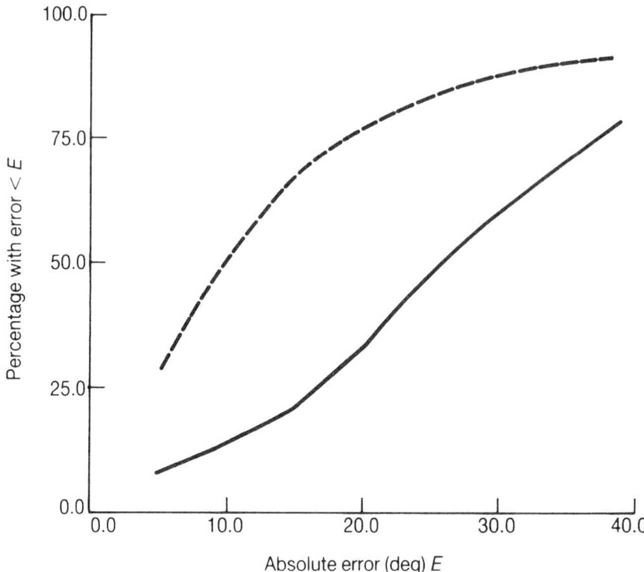

Figure 8.5 Phase-lag estimation error curves. The innovations distribution is generalized Gaussian with $\alpha = 0{\cdot}6$. There is no noise, but the data has been band-limited using an 8/18–64/72 Hz zero-phase Butterworth filter. Time-series length used was 1 s (251 points). The solid line is for O_1^2, the dashed line is for O_2^4.

full-band case. But how sensitive are the characteristics of the behavior of these two objectives to the particular low-cut frequency (i.e. the frequency at which the rapid low frequency decay begins)? To investigate this for the no-noise case, a new set of simulations was carried out using a low-cut of 2 Hz, with the other parameters unchanged (i.e. a 2/18–64/72 Hz Butterworth). The cumulative displays of phase-lag error magnitudes derived from segments of length one second are given for $\alpha = 0{\cdot}6$ in Figure 8.6. O_2^4 still outperforms O_1^2 (until the error exceeds 30°) but the difference is considerably less than when the low-cut was 8 Hz (Figure 8.5). In fact the performance of O_2^4 is little changed, while O_1^2 has responded to the lower frequencies available by increasing performance considerably (e.g. an error $\leqslant 15°$ is achieved in 21 per cent of cases when the low-cut is 8 Hz, but this jumps to 49 per cent for the 2 Hz low-cut).

For Laplace-distributed innovations ($\alpha = 1{\cdot}0$) O_1^2 again performs much better than in the 8 Hz low-cut case, while O_2^4 is little changed.

It should be emphasized that for understanding likely behavior in seismic applications, the previous Butterworth filter (8/18–64/72) is much more realistic than that with the 2 Hz low-cut; hence the results presented in this section are for general understanding rather than for making inferences for seismic analyses. The discussion which follows applies to the 8 Hz low-cut Butterworth band-limiting.

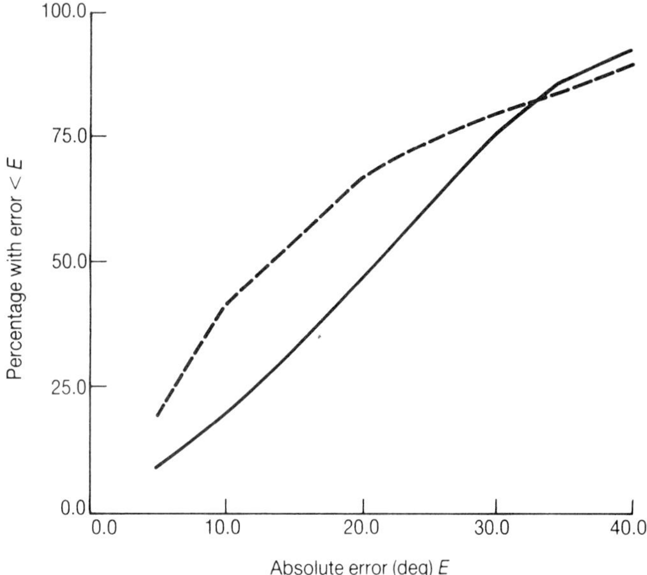

Figure 8.6 Phase-lag estimation error curves. The innovations distribution is generalized Gaussian with $\alpha = 0{\cdot}6$. There is no noise, but the data has been band-limited using an 2/18–64/72 Hz zero-phase Butterworth filter. Times-series length used was 1 s (251 points). The solid line is for O_1^2, the dashed line is for O_2^4.

8.3.5 Band-limited plus additive white noise

Consider now the effect of additionally including additive white noise in the synthetic time series from which the phase lag is to be estimated. After step (b) in the simulation, two steps are inserted, viz. zero-phase Butterworth filtering followed by the addition of white Gaussian noise at a specified SNR. This was again done to give a range of signal-to-noise ratios (SNRs); for each ratio an independent 500-size simulation was performed using generalized Gaussian-distributed $\{e_t\}$ with $\alpha = 0{\cdot}6$. By way of example, the cumulative displays of errors are compared in Figure 8.7 for an SNR of 3. O_2^4 is clearly superior to O_1^2, but both methods perform poorly. For Laplace-distributed innovations, the performance difference is slightly less pronounced, and the difference also narrows with decreasing SNR.

8.4 Application to seismic data

A recently developed deconvolution procedure in oil and gas exploration consists of first spectrally whitening the seismic traces (8.5), and then finding the phase lag to maximize a chosen statistic (8.6). From the simulations reported herein based on band limitation and noise typical of seismic data, kurtosis should perform better than the competing statistic O_1^2 given in (8.2). 100 traces of good quality seismic data were used; these data are plotted in

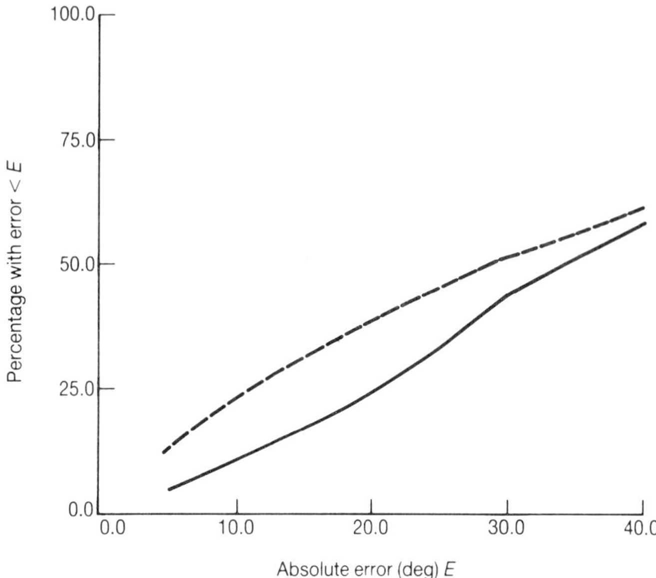

Figure 8.7 Phase-lag estimation error curves. The generalized Gaussian shape parameter α for the distribution of innovations takes the value 0·6. Full-bandwidth time series of length 1 (251 points) were used. Signal-to-noise ratio = 3. The data has been band-limited using an 8/18–64/72 Hz zero phase Butterworth filter prior to the addition of white noise. The solid line is for O_1^2, the dashed line is for O_2^4.

Figure 8.8. The 'traces' are simply time series of the reflection response (8.4) plotted vertically below the coincident source and receiver positions on the Earth's surface. They give rise to a pseudo-cross-section through the Earth. The whitening of the amplitude spectra of the traces using the filter $\{g_j\}$ has already been carried out. The form of the traces varies only slowly laterally (because the lithology does likewise) and several strong events ('layers in the earth') persist right across the section, so that clearly the vertical time series are quite heavily correlated. The type and detail of the spatial correlation of traces is too detailed and variable to have attempted to include it in the simulations.

We are trying to correct for the phase-*lead* filter acting on the innovation sequence $\{e_t\}$ by designing a phase-*lag* correction filter to cancel it out. With a drilled well it is possible to estimate the sequence $\{e_t\}$ and, by using frequency-response function estimation methods (Jenkins and Watts, 1968; White, 1984) to estimate the phase response of the effective filter and its pointwise 90% confidence intervals (e.g. Walden, 1986). This phase response is plotted over the signal-dominated frequency range in Figure 8.9; note how linear it is. The intercept of the best-fit straight line with the axis at frequency zero gives a phase lead of about 58°. This represents our best estimate of the phase lead, but is only available after drilling a well. The deconvolution task considered in this paper is carried out prior to drilling a well.

Each time series (of duration 2 s, hence 501 points at 4 ms sampling) was

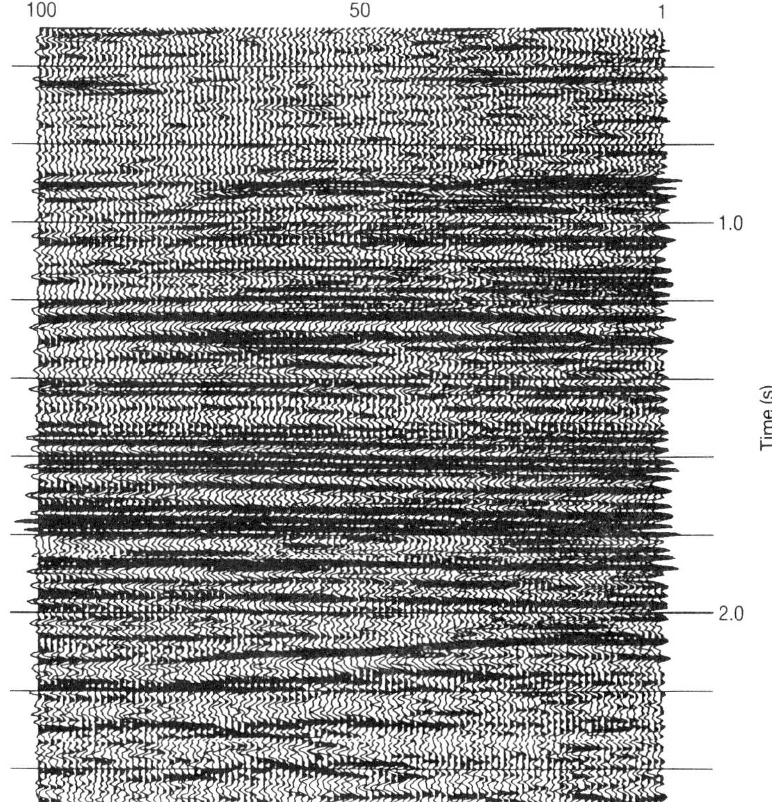

Figure 8.8 One hundred seismic traces, plotted vertically. Here 'time' is the two-way reflection time.

phase-lag corrected using (8.7) in 1° intervals from 0 to 180°, and the statistics O_1^2 and O_2^4 calculated for each angle. For each trace, the phase-lag ϕ_0 maximizing each of the statistics was found and recorded. Suppose we accept that the phase lead, with point estimate 58°, simply lies in the interval [55, 60°). Then, for example, ϕ_0 values falling in [50, 65°) are taken to have an absolute error less than or equal to 5°. The familiar plot of percentage of outcomes with an error less than a specified quantity is shown in Figure 8.10, and has a familiar look to it (compare with Figure 8.7, for example) despite the fact that here the traces are notably correlated. The SNR of the seismic data can be estimated using multiple coherence analysis (Jenkins and Watts, 1968; White, 1984) and turns out to be about 6·5 (between the cases 3 and ∞ discussed in detail in this paper).

Another way of demonstrating the superiority of O_2^4 over O_1^2 involves using a large number of traces at once in computing the statistics, i.e. the sums involved in (8.2) and (8.3) are accumulated over m traces rather than just 1. Using the first fifty traces gives a maximum for O_1^2 at $\phi_0 = 24°$ but a

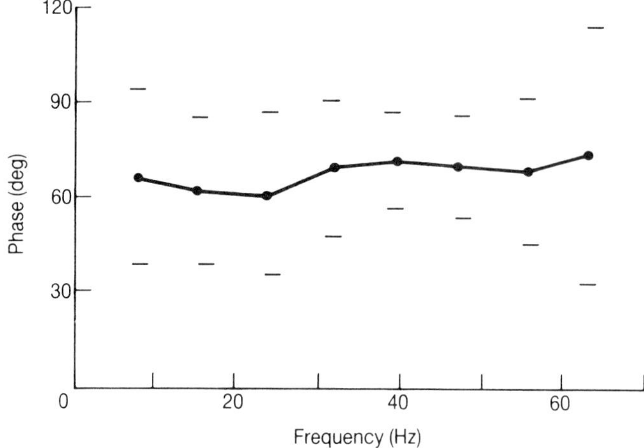

Figure 8.9 Estimated phase response, with 90 per cent confidence intervals, for effective wavelet. Only the response over the signal dominated frequency range is given.

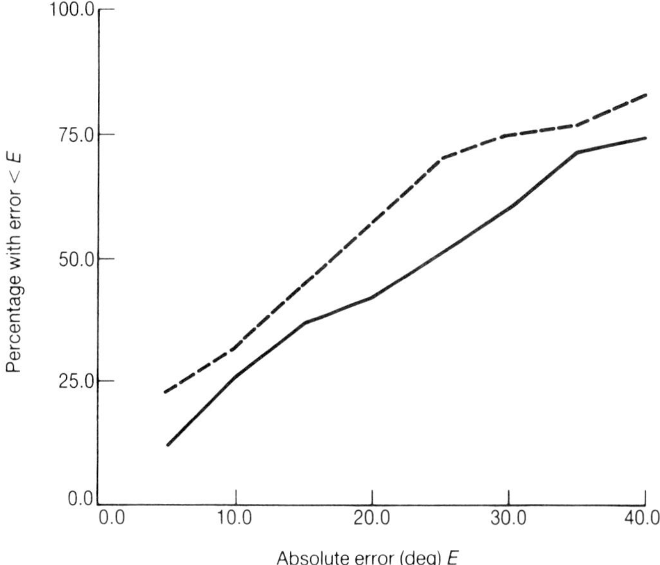

Figure 8.10 Phase-lag estimation error curves compiled from the 100 seismic traces assuming the optimum phase-lag to be in the interval [55, 60°).

maximum for O_2^4 at $\phi_0 = 49°$. Using the second group of 50 traces gives 46 and 54° respectively. Hence O_2^4 is the less variable, and apparently more accurate (taking 58° as correct). Using all 100 traces, O_1^2 gives 45° while O_2^4 gives 50°; again the latter is more accurate. Practical estimation uses as many consecutive traces as are available/suitable for estimation.

The result of applying the phase-lag correction to the 100 traces is hard to appreciate unless a large scale is used. Figure 8.11(a) shows a detail – traces 1–50 in the interval 2·0–2·2 s of the original data. In Figure 8.11(b) the phase-lag correction estimated from the 100 traces using O_2^4, namely 50°, has been applied. Reflection events are clearer in Figure 8.11(b), making interpretation easier.

It is remarkable that the technique discussed herein is capable of estimating a phase-lag using only the convolution result in (8.4) to within a few degrees of the phase-lag estimated when $\{e_t\}$ is 'known' after drilling (especially as $\{e_t\}$ is usually, in practice, slightly correlated, rather than i.i.d.). See Hosken *et al.* (1987), and Longbottom *et al.* (1988) for some cautionary details.

8.5 Summary and conclusions

In this paper it has been shown by simulation using innovations distributions with high kurtosis that maximization of Wiggins' statistic O_2^4 provides a poorer phase-lag estimate than does Gray's O_1^2, *when the data is noise-free and full-band.* With additive white noise the performance of these statistics depends on SNR and the innovations distribution.

If the time series, with or without noise, is band-limited to a frequency content typical of seismic data (low-cut 8 Hz) O_2^4 substantially outperforms O_1^2. This is true for innovations with kurtosis of 6 or 15·6, suggesting that in the band-limited case the potential effect of the innovations amplitude distribution is small with respect to the relative performances of competing statistics. The performance difference between O_2^4 and O_1^2 in the band-limited case can be reduced by lowering the low-cut frequency.

The results of the tests reported here strongly support the use of Wiggins'

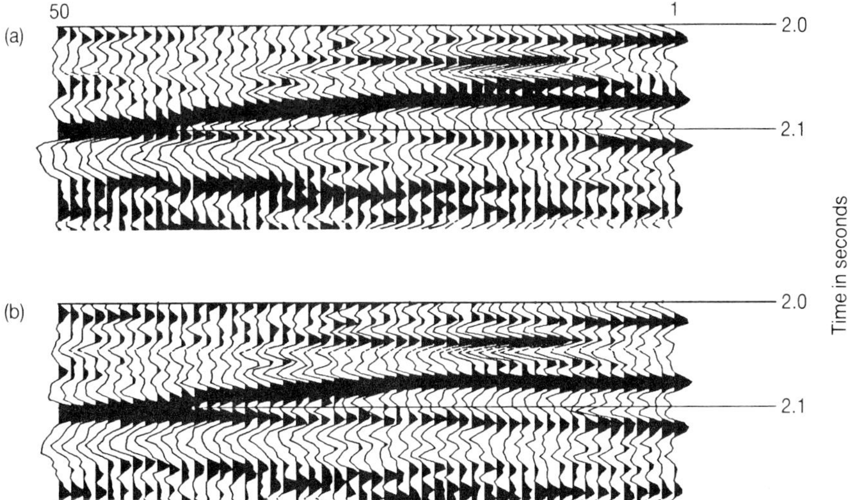

Figure 8.11 Part of the seismic section at an increased scale, i.e. traces 1–50 and time interval 2·0–2·2 s; (a) before, and (b) after the 50° phase-lag correction.

objective for practical phase-lag estimation. Indeed this objective has performed well in practice (Hosken *et al.*, 1987). It is also hoped that the results will stimulate the statistical community – and in particular time-series analysts – to pay more attention to the practical effects of band limitation.

8.6 Appendix: Generation of random sequences with generalized Gaussian distributions

In order that the amplitude distribution of simulated innovations sequences should match those found in practice, it was necessary to generate sequences from the generalized Gaussian distribution. A sequence of pseudo-random uniform variates on (0, 1) was generated, and for each outcome u a generalized Gaussian sample value was found by solving for x in the probability integral transform

$$u = F(x) = (\alpha/\{2\beta\Gamma(1/\alpha)\} \int_{-\infty}^{x} \exp(-\{|t|/\beta\}^{\alpha})\,dt,$$

where $F(x)$ is the distribution function. Walden and Hosken (1986) showed that this distribution function can be written

$$F(x) = 0{\cdot}5\{1 \pm \{\Gamma([1/\alpha]; [x/\beta]^{\alpha})/\Gamma(1/\alpha)\}\}$$

with a positive sign if $x > 0$ and negative if $x < 0$, where $\Gamma(z; \omega)$ is the incomplete gamma function defined by

$$\Gamma(z; \omega) = \int_{0}^{\omega} t^{z-1} \exp(-t)\,dt,$$

and the incomplete gamma function ratio $\Gamma([1/\alpha]; [x/\beta]^{\alpha})/\Gamma(1/\alpha)$ may be computed using the algorithm of Lau (1980). The required value is then found as the solution of $u - F(x) = 0$. To generate Laplace-distributed sequences ($\alpha = 1$) it is quicker to generate pseudo-random variates from the negative exponential distribution

$$f(y; \beta) = (1/\beta) \exp(-y/\beta), \qquad 0 \leqslant y < \infty,$$

and negate the value with a probability of $\frac{1}{2}$ to give a sample from the Laplace distribution

$$f(x; \beta) = (1/\{2\beta\}) \exp(-|x|/\beta), \qquad -\infty < x < \infty.$$

In either case the sample can be rescaled to variance unity by noting that the unscaled variance is $\{\Gamma(3/\alpha)/\Gamma(1/\alpha)\}\beta^2$.

Acknowledgements

This work was carried out while the author was with The British Petroleum Company plc. The author would like to thank the Chairman and Board of

Directors of The British Petroleum Company plc for permission to publish the paper, and the referee for suggestions leading to a much more concise exposition.

References

Box, G.E.P. and Jenkins, G.M. (1970). *Time Series Analysis, Forecasting and Control.* Holden-Day, San Francisco.

Donoho, D.L. (1981). On minimum entropy deconvolution. In *Applied Time Series Analysis II*, Findley, D.F. (ed.), Academic Press, New York.

Ferguson, T.S. (1961). On the rejection of outliers. In *Proc. 4th Berkeley Symp. Math. Statist. and Probability*, Vol. 1, Neyman, J. (ed.), University of California Press, California.

Gray, W.C. (1979). *Variable Norm Deconvolution.* Unpublished Ph.D. dissertation, Stanford University, Dept. of Geophysics.

Hampson, D. and Galbraith, M. (1981). Wavelet extraction by sonic log correlation. Presented at the 1981 *Canadian Society of Exploration Geophysicists National Convention*, 12–14 May, Calgary, Canada.

Hogg, R.V. (1972). More light on the kurtosis and related statistics. *J. Amer. Statist. Assoc.*, **67**, 422–24.

Hosken, J.W.J., Longbottom, J., Walden, A.T., and White, R.E. (1987). Maximum kurtosis phase and the phase of the seismic wavelet. In *Proc. 1986 Research Workshop on Deconvolution and Inversion*, M. Bernabini *et al.* (eds), Blackwell, Oxford.

Huber, P.J. (1985). Projection pursuit. *Ann. Statist.*, **13**, 435–75.

Jenkins, G.M. and Watts, D.G. (1968). *Spectral Analysis and its Applications*, Holden-Day, San Francisco.

Lau, C. (1980). A simple series for the incomplete gamma integral. *Appl. Statist.*, **29**, 113–14.

Levy, S. and Oldenberg, D. (1987). Automatic phase correction of CMP stacked data. *Geophysics*, **52**, 51–9.

Lii, K.S. and Rosenblatt, M. (1982). Deconvolution and estimation of transfer function phase and coefficients for nonGaussian linear processes. *Ann. Statist.*, **10**, 1195–1208.

Longbottom, J., Walden, A.T., and White, R.E. (1988). Principles and application of maximum kurtosis phase estimation. *Geophys. Prospecting*, **36**, 115–38.

Mallows, C. (1967). Linear processes are nearly Gaussian. *J. Appl. Prob.*, **4**, 313–29.

Walden, A.T. (1985). NonGaussian reflectivity, entropy, and deconvolution. *Geophys.*, **50**, 2862–88.

Walden, A.T. (1986). Estimating confidence intervals for the gain and phase of frequency response functions. *Geophys. J. Roy. Astron. Soc.*, **87**, 519–37.

Walden, A.T. and Hosken, J.W.J. (1986). The nature of the nonGaussianity of primary reflection coefficients and its significance for deconvolution. *Geophys. Prospecting*, **34**, 1038–66.

White, R.E. (1984). Signal and noise estimation from seismic reflection data using spectral coherence methods. *Proc. Inst. Electrical and Electronics Engrs.*, **72**, 1340–56.

Wiggins, R.A. (1978). Minimum entropy deconvolution. *Geoexploration*, **16**, 21–35.

Wiggins, R.A. (1985). Entropy guided deconvolution. *Geophys.*, **50**, 2720–6.

Chapter 9

Deconvolution of short period teleseismic and regional time series

Z. A. Der, A. C. Lees, K. L. McLaughlin, and R. H. Shumway

Deconvolution methods can be useful for analyzing seismic recordings of earthquakes and explosions located in source areas of limited extent and recorded at arrays of multiple sensors. Seismic data have been shown to consist of convolutions of source and site (sensor)-related time series. The site effect is the result of the distortion of waveforms by geological structures near the receiving sensors. In the frequency domain, the Fourier transforms of individual recordings can be 'factored' into site and source frequency-response functions.

The problem of estimating simultaneously the site and source functions is treated here as the statistical problem of estimating two convolved vector time series corrupted by additive noise. By regarding one of the series as being stochastic and the other fixed, it is shown that the fixed component can be estimated by maximizing a frequency-domain approximation to the likelihood. The stochastic series is estimated as the minimum mean square error or Wiener filter. Relations to random coefficient and ridge regression estimators for linear time series models are explored. Application of the EM algorithm yields simultaneous estimators for receiver and source functions as outcomes of alternate weighted receiver and source stacking operations. Tests of hypotheses for receiver and source functions are given.

Array *P* wave recordings at conventional and three-component arrays, processed using the multichannel deconvolution techniques, show that, by reducing site effects, the technique facilitates the recognition of secondary arrivals, the surface reflection *pP*, spall and scattered *P* waves from the source region. The effects of interferences between the direct *P* waves and associated surface reflections may increase the magnitude bias between the Soviet (Kazakh) and US Nevada test sites beyond that due to differences in mantle attenuation. Analyses of regional arrivals at small arrays show that spectral factorability can be useful for detecting small differences in location or source mechanisms of small regional events.

9.1 The structure of seismic signals

The practice of including short-period teleseismic *P* wave data in studies of seismic sources offers the advantages of increased time–space resolution and bandwidth. It has been shown that significant energy exists in teleseismic *P* waves to frequencies up to 10 Hz, along low-attenuation paths, and up to 4 Hz along high-attenuation paths. Utilizing higher frequencies than those being matched in studies of long-period waveforms is imperative in nuclear monitoring where small changes in source depth, of the order of a few hundred meters, can have a profound effect on the wave characteristics and amplitudes. Unfortunately the utilization of high-frequency waves for extracting source information is hindered by the distortions of the waveforms due to the geological complexities near the recording sites and by similar effects near the sources. The multi-channel deconvolution method, introduced by Shumway and Der (1985) which estimates the source and site effects iteratively from a large matrix of seismograms, consisting of *P* waves from multiple events recorded at large seismic arrays, can remedy this problem by approximately correcting the data for site distortion.

The need for such techniques arises because we simply do not know what *P* waves emerging from an actual source region would look like if we could observe them on some hypothetical sensor located in the Earth's mantle. Our early ideas about the structure of teleseismic body waves were formed more by theoretical preconceptions than facts. Early models of teleseismic *P* waves were constructed by embedding a point source in an elastic half-space. In the case of nuclear explosions this will result in a *P* wave followed by a *pP* arrival, a reflection from the free surface (Figure 9.1). Besides the direct *P* and *pP* phases, indications of numerous other arrivals such as spall and Rayleigh-to-*P* conversions were found in *P* waves when data were analyzed by various methods (Bakun and Johnson, 1973; and Stewart and Douglas, 1983). Moreover, theoretical numerical simulations of *P* wave radiation from nuclear explosions, including near source nonlinear behavior, indicated that the actual *P* waves may be considerably more complex than those predicted by linear elastic models (McLaughlin *et al.*, 1988). Therefore, methods for estimating the source functions should not be based on very restrictive theoretical models such as the assumption that a direct *P* and a *pP* pulse convolved with some source-time function and possibly a site-response function make up the *P* wave seismogram (we shall call this the $P + pP$ model).

A major advance in the understanding of the structure of teleseismic *P* waves was the finding by Filson and Frasier (1972) that *P* wave amplitude spectra from a set of events observed at a narrow range of azimuths at multiple array sites may be factored as (ν is frequency in hertz)

$$P_{ij}(\nu) = R_i(\nu)S_j(\nu) \tag{9.1}$$

where $P_{ij}(\nu)$ are the *P* wave spectra, $S_j(\nu)$ are source factors, $R_i(\nu)$ are factors appropriate to the receiver sites, and $i = 1, \ldots, m, j = 1, \ldots, n$ correspond to the individual receivers and sources, respectively. The receiver factors seem to be functions of the slowness vector of the particular arrival. They reflect the distorting effects of the near-receiver geological structures which cause the waveforms observed at large seismic arrays to vary strongly from sensor to sensor. For a given array sensor they can be treated as invariant, but frequency-dependent, transfer functions for a group of events, as long as the slowness vectors at the given sensor remain approximately the same. Since

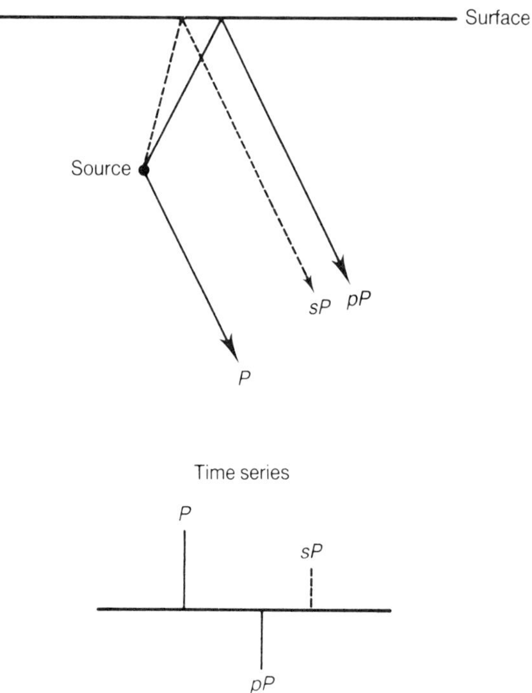

Figure 9.1 Generation of the *pP* phase. The surface reflected *P*, called the *pP* phase, follows the direct *P* wave. The polarity of the pulse is inverted by the reflection from the free surface.

short-period waveforms at individual seismic sensors depend on these unknown transfer functions, they cannot be reliably inverted for source properties without corrections. The validity of the factorability condition described by eq. (9.1) has been verified for a number of seismic arrays as well as more widely separated receivers where the idea was also exploited for deriving 'relative receiver functions', which are time-domain representations of intersite transfer functions (Mellman and Kaufman, 1981) including attenuation effects.

Deconvolution methods appear to be eminently suitable for solving the problem of estimating the source-radiated pulse sequences from *P* wave data since they utilize both the amplitude and phase information, require no assumptions about the signal model, and can be adapted to reducing the site effects (Shumway and Der, 1985). Moreover, in deconvolving approximately the effects of the instrument and the path attenuation in the first step, we may reduce the ambiguity in the results relative to elaborate approaches such as complete waveform modeling which involves specific source models. More work needs to be done to interpret such results in the future and to understand the physical nature of the source time functions derived. We present the results of deconvolutions of data from explosions at most of the major test sites: the Soviet sites Degelen, Shagan, the French sites Sahara and the Chinese test site Sinkiang, Tuamotu, and the US site, Pahute Mesa. Deconvolutions have also been performed on three-component regional teleseismic data.

An additional application of such methods is the analysis of particle motion of body waves. The particle motion of a P wave arriving at a three-component station will be characteristic of the source region, but will not necessarily be in the vertical plane that includes the source. This would be the case only if the geology near the station were laterally homogeneous. Instead, the actual particle motion can be shown to be three-dimensional and very complicated. The complex transfer functions among the various components of motion could be used to help identify the source region.

The same concepts also apply to regional arrivals such as the P_n and L_g phases. It can be shown that the spectra of these arrivals can also be decomposed into source and site factors. In the case of regional arrivals which consist of multiple reverberations of body waves in the crust, the source factor cannot be estimated uniquely. Nevertheless, the site factors, even though they are non-unique, are characteristic of a given source location and mechanism. Thus, spectral factoring can be used to group events with respect to these non-unique components.

9.2 Statistical considerations

In this section, we examine the statistical questions of interest for linear models that are consequences of the spectral factorization given in (9.1). The model (9.1) implies, for example, that the observed data are generated in the time domain as a convolution of the receiver and source functions for a number of events (sources) at a number of sites (receivers). If additive noise is assumed, we may write a time domain version of (9.1) in the form

$$y_{ijt} = \sum_{u=-\infty}^{\infty} r_{iu} s_{j,t-u} + v_{ijt}, \tag{9.2}$$

$t = 0, \pm 1, \pm 2, \ldots,$ where y_{ijt} denotes the received data at receiver $i, i = 1, \ldots, m$ for event $j, j = 1, \ldots, n$. We will assume generally that each event is observed at all receivers although this is not necessary for the theoretical development and simply leads to more tractable equations. Generally, it is the case that neither the receiver functions, r_{it}, nor the source functions s_{jt} are known in advance, and ultimately both must be estimated using the available data.

Particularly tractable statistical assumptions for the components of the linear model defined by (9.2) are that (a) the noise terms v_{ijt} are independent over receivers i and events j but are stationarily correlated over time t with a common autocovariance function, (b) the signals are independent stationary series with identical autocovariance functions and (c) the receiver functions are deterministic unknown functions. It is easy enough to interchange the assumptions made about receivers and sources so that the sources are deterministic and the receivers are random. Under certain conditions, it is even convenient to regard both series as deterministic but unknown.

We discuss first the problem of estimating the stochastic source function, called 'deconvolution', for general linear models in which the receiver functions are known. This is the multivariate generalization of Wiener filtering and reduces to the time-series analog of the random coefficient regression problem as discussed in Shumway (1988, pp. 245–62). If the autocovariance functions (or equivalently, the spectra) of the signal and noise processes are assumed to be known, the single-receiver, single-source version of this model is applied in Oldenburg (1981) to the deconvolution of seismic data. A robust (over possible receiver functions) solution is proposed in the univariate case by Walden (1988). Deregowski (1971) also works with deconvolution in the frequency domain as does Hunt (1972).

It is important to consider the effect on these deconvolution procedures of changing the spectra of the signal and noise processes. Estimating the signal and noise spectra is possible using maximum likelihood or by spectral analysis of variance arguments given in the sections to follow. Measures of how adequately the factorization model fits the data are derived from tests of hypotheses for presence or absence of the signal. These measures turn out to be expressible in terms of the multiple coherence between the observed and predicted series under the model defined by (9.2). The bias and resolution capabilities of the Wiener filters for resolving the $P + pP$ signal source are illustrated.

Finally, we extend the procedure to the case where the receiver functions are unknown by showing how they can be estimated by maximum likelihood using the EM (expectation maximization) algorithm (Shumway and Der, 1985). The simultaneous estimation of receiver and source functions is exhibited as an iterative process of stacking alternately over receivers and sites.

It should be noted that there is a substantial literature on deconvolution of single time series based on entropy arguments or on the assumption that the underlying signals are nonGaussian. These methods advocate using, as objective functions, various ratios of higher-order moments and often work quite well on strictly positive nonGaussian signals although this assumption is not always explicit in the derivations. Examples of this kind of approach are Wiggins (1978), Donoho (1981), Walden (1985) and Lii and Rosenblatt (1982). However, seismic signals from P-waves are not, in general, strictly positive or simple in structure and may, in our opinion, be conceptually closer to Gaussian time series.

One might also mention the use of various parametric models which attempt to model the known delay structure of the $P + pP$ specifically. For example, the delay structure induces a known rippling behavior in the spectrum if the spectrum can be reliably observed over broad bands. Hence, techniques based on the cepstrum (Bogert *et al.*, 1962) or on maximizing the Whittle likelihood (Hannan and Thomson, 1974) have been applied. Our feeling is that such approaches have not been particularly successful in the seismic case because of complexities in the source function and the fact that we are observing this complex behavior over a fairly narrow frequency band. A time-domain approach using the state-space model is in Kormylo and Mendel (1983).

The approach in the following sections is different from previous approaches in the way it uses the multivariate structure of the data to estimate simultaneously the receiver and source functions. We begin by reviewing the appropriate results for a general linear model with random signal coefficients.

9.2.1 The general linear model

If the observed values y_{ijt}, $i = 1, \ldots, m$, $j = 1, \ldots, n$ in the linear time series model (9.2) are stacked in an $mn \times 1$ vector, first over the m receivers within each of the n events, say

$$\mathbf{y}_t = (y_{11t}, \ldots, y_{m1t}, \ldots, y_{1nt}, \ldots, y_{mnt})'$$

we may write the equations in vector form as

$$\mathbf{y}_t = \sum_{u=-\infty}^{\infty} \mathbf{z}_u s_{t-u} + \mathbf{v}_t \tag{9.3}$$

where $\mathbf{s}_t = (s_{1t}, \ldots, s_{nt})'$ is the $n \times 1$ signal vector and $\mathbf{v}_t$ is the $mn \times 1$ vector of noises, stacked in the same order as are the observed series in $\mathbf{y}_t$. The $mn \times n$ design matrix $\mathbf{z}_t$ has the special form

$$\mathbf{z}_t = \mathbf{I}_n \otimes \mathbf{r}_t \tag{9.4}$$

where $\mathbf{r}_t = (r_{1t}, r_{2t}, \ldots, r_{mt})'$ is the $m \times 1$ vector of receiver functions, $\mathbf{I}_n$ is the $n \times n$ identity matrix and $\otimes$ denotes the Kronecker or direct product. The signal and noise vectors are assumed to be weakly stationary Gaussian zero-mean time series with spectral matrices $f_s(\nu)\mathbf{I}_n$ and $f_v(\nu)\mathbf{I}_{mn}$, where $f_s(\nu)$ are the spectral densities of the signals and $f_v(\nu)$ are the common spectral densities of the noise processes. The argument is frequency ν, $-5 \leqslant \nu \leqslant 0{\cdot}5$, measured in cycles per unit time. This primitive set of assumptions can be easily generalized to more complicated cases; we adopt the assumption that the spectral matrixes are in this form purely for ease in exposition. The $mn \times mn$ spectral matrix, of the observed vector $\mathbf{y}_t$ then becomes (* denotes the complex conjugate transpose)

$$\mathbf{f}_\mathbf{y} = f_s \mathbf{Z}\mathbf{Z}^* + f_v \mathbf{I}_{mn} \tag{9.5}$$

where we introduce, for simplicity, a notation that suppresses the frequency argument in the notation for the spectra. The elements of the $mn \times n$ matrix $\mathbf{Z}(\nu)$ are the infinite Fourier transforms

$$\mathbf{Z}(\nu) = \sum_{t=-\infty}^{\infty} \mathbf{z}_t e^{-2\pi i \nu t} \tag{9.6}$$

where we make the regularity assumption

$$\sum_{u=-\infty}^{\infty} |t| \, \|\mathbf{z}_t\| < \infty$$

with $\|\mathbf{A}\|^2 = \text{tr}\, \mathbf{A}\mathbf{A}'$ the norm of the matrix $\mathbf{A}$ and tr denoting the trace. The signal and noise processes are assumed to be cumulant mixing in the sense of Brillinger (1969); the reader is referred to Pawitan and Shumway (1989) for theoretical justification of the statements made during this development.

9.2.2 Deconvolution of the source function

Under the assumptions given in Section 9.2.1, it is easy to verify that the minimum mean-square estimator for the signal vector is (Shumway, 1988, p. 247)

$$\hat{\mathbf{s}}_t = \sum_{u=-\infty}^{\infty} \mathbf{h}_u \mathbf{y}_{t-u} \tag{9.7}$$

where the $n \times mn$ matrix function $\mathbf{h}_t$ is the inverse Fourier transform of the $n \times mn$ matrix $\mathbf{H}(\nu)$, defined as

$$\mathbf{H} = (\mathbf{Z}^*\mathbf{Z} + \theta \mathbf{I}_n)^{-1}\mathbf{Z}^* \tag{9.8}$$

with

$$\theta(\nu) = \frac{f_v(\nu)}{f_s(\nu)} \tag{9.9}$$

the inverse of the signal-to-noise ratio. The estimator above reduces to the usual Wiener filter when $m = n = 1$ as used, for example, by Deregowski (1971), Oldenburg (1981), Hunt (1972) or Walden (1988). The Wiener filter is equivalent to statistical ridge-regression approaches where the noise-to-signal parameter takes the role of the ridge parameter for each frequency. Taking the ridge parameter as 0 leads to the usual regression estimator, computed under the assumption that the signal vector is deterministic. The usual estimator with $\theta = 0$ leads to instability in the presence of noise (see Shumway and Der, 1985), although θ is made quite small in the examples given here. If one chooses to behave as if the signal-to-noise ratio is known, then the estimator for a finite time sample is computed by convolving a finite-length, say M, filter approximation to $\mathbf{h}_t$, calculated as the discrete Fourier transform of $\mathbf{H}(\nu)$, with the data using (9.8). The mean-square error due to the truncation of this approximation is shown to be $o(M^{-2})$ by Pawitan and Shumway (1989).

The source signal estimators for the particular model assumed in (9.2) fall easily out of the general form (9.8) with

$$\mathbf{Z} = \mathbf{I}_n \otimes \mathbf{R} \tag{9.10}$$

where $\mathbf{R}(\nu)$ is the $m \times 1$ vector of Fourier transforms of $\mathbf{r}_t$. This yields the solution

$$H_i = \frac{R_i^*}{\Sigma_{k=1}^m |R_k|^2 + \theta}, \tag{9.11}$$

which exhibits the time domain estimator for the signal (9.7) as the result of the simple averaging operation

$$\hat{s}_{jt} = \sum_{i=1}^{m} \sum_{u=-\infty}^{\infty} h_{iu} y_{ij,t-u}, \tag{9.12}$$

which is similar to the procedure of 'stacking over receivers' in the seismic literature.

9.2.3 Estimation of signal and noise spectra; multiple coherence

When the signal and noise spectra are not assumed to be known, then they must be estimated somehow from the data. Two reasonable alternatives are (1) an analysis of power approach analogous to what is usually done in random effects analysis of variance and (2) maximum likelihood estimation using the Whittle approximation to the likelihood function. The analysis of power approach given in this section depends on knowing the receiver functions exactly. It is reasonable that the receiver functions as estimated in Section 3.4 can be used in place of the true values. An approach to estimating the spectra by maximum likelihood using the EM algorithm is given in Section 9.2.4 and in Shumway and Der (1985).

Both approaches depend on using the asymptotic properties of the discrete Fourier transform of a finite data sample $\mathbf{y}_t$, $t = 0, 1, \ldots, T-1$, say,

$$\mathbf{Y}(k) = \frac{1}{\sqrt{T}} \sum_{t=0}^{T-1} \mathbf{y}_t \mathrm{e}^{-2\pi i \nu_k t}$$

of the data, evaluated at some frequency of the form $\nu_k = k/T$ near to the target frequency ν. Using the asymptotics implied by the strict stationarity and cumulant mixing conditions, it follows (see Pawitan and Shumway, 1989) that the $mn \times 1$ complex vector $\mathbf{Y}(k)$ has a complex Gaussian distribution with mean zero and covariance matrix given by (9.5) evaluated at $\nu = \nu_k$. The same holds for a collection of vectors sampled in a band of frequencies in the neighborhood of ν.

It is useful to write the consequences of the preceding statements in terms of a linear model (the approximation is $o_p(1)$)

$$\mathbf{Y} = \mathbf{ZS} + \mathbf{V} \tag{9.13}$$

for the discrete Fourier transforms of $\mathbf{y}_t$, $\mathbf{s}_t$ and $\mathbf{v}_t$, with $\mathbf{Z}$ still defined as the Fourier transform (9.6). In terms of the component receiver functions, we may also write (9.13) as

$$Y_{ij} = R_i S_j + V_{ij} \tag{9.14}$$

which corresponds directly to the time-domain representation (9.2).

The usual procedure in analysis of variance arguments for the random effects case is to consider the expectations of the power components that would result from a test for no signal in the above model when the signal vector is assumed to be fixed (see Shumway, 1988, p. 260). This leads to Table 9.1, where we can get analysis of power estimators by equating the power components to their expectations. From Table 9.1, an asymptotically

Table 9.1 Analysis of power

Source	Power	df	Expected mean power
Signal (SP)	$\mathbf{Y^*Z(Z^*Z)^{-1}Z^*Y}$	n	$f_v + f_s \Sigma_{i=1}^m \lvert R_i \rvert^2$
Error (EP)	Difference	$n(m-1)$	f_v
Total (TP)	$\mathbf{Y^*Y}$	nm	

unbiased estimator for the noise power is the mean error power used directly, whereas the signal power can be estimated by subtracting the noise power estimate from the mean signal power with the scaling as indicated in the last column of Table 9.1.

One can also use Table 9.1 to develop a test that the signal spectrum is zero. For example, under the assumption that $f_s = 0$, the large-sample distribution of

$$F = \frac{SP}{EP}(m-1) \tag{9.15}$$

is F with $2n$ and $2n(m-1)$ degrees of freedom. If L adjacent spectral components are smoothed in the power components, the degrees of freedom are multiplied by $L = BT$, where B is the bandwidth in hertz and T is the sample length in seconds. The multiple coherence (see Shumway, 1988, p. 227) between the observed vector and the fixed regression ($\theta = 0$) predicted values, say $Y_{ij} = R_i \hat{S}_j$, is given by

$$R^2 = \frac{SP}{TP} = \frac{|\Sigma_{i,j} Y_{ij} \hat{Y}_{ij}^*|^2}{\Sigma_{i,j}|Y_{ij}|^2 \Sigma_{i,j}|\hat{Y}_{ij}|^2} \tag{9.16}$$

and we have the relation

$$F = \frac{R^2}{1-R^2}(m-1), \tag{9.17}$$

showing that the multiple coherence is a direct measure of model adequacy in the above case that can be compared with critical values of the F-distribution to determine statistical significance. The sums in the numerator and denominator can also be taken over a frequency band of width B and we obtain an **F**-distribution with approximately $2BTn$ and $2BTn(m-1)$ degrees of freedom.

9.2.4 Maximum likelihood estimation of receiver effects

In the case where the receiver functions are unknown, one can consider maximizing the approximate Whittle likelihood

$$\ln L = -\ln |\mathbf{f}_y| - \mathbf{Y}^* \mathbf{f}_y^{-1} \mathbf{Y} \tag{9.18}$$

where $\mathbf{f}_y$ is the complex covariance matrix (9.5). This exhibits the likelihood function as a rather involved function of the parameters f_s, f_v, and $R_1, R_2, \ldots, R_m$ at each center frequency ν_k. Since this is somewhat intractable, we consider using the EM algorithm as in Shumway and Der (1985).

Consider writing the likelihood of the complete-data in terms of **S** and **V** of (9.13), which gives

$$\ln L' = -n \ln f_s - \frac{\|\mathbf{S}\|^2}{f_s} - mn \ln f_v - \frac{\|\mathbf{V}\|^2}{f_v}. \tag{9.19}$$

The last term can be rewritten directly in terms of the unknown receiver functions using the equivalent version of the linear model (9.13). This gives

$$\mathbf{Y} = \mathbf{W}\mathbf{R} + \mathbf{V}, \tag{9.20}$$

where

$$\mathbf{W} = \mathbf{S} \otimes \mathbf{I}_m \tag{9.21}$$

is the Kronecker product of the $n \times 1$ signal vector and the $m \times m$ identity matrix $\mathbf{I}_m$.

The EM algorithm in this case defines successive iterates for the estimated parameters as maximizers of the expectation

$$Q(\alpha_0, \alpha) = E_0(\ln L' \mid \mathbf{Y}) \tag{9.22}$$

where $\alpha = (f_s, f_v, R_1, \ldots, R_m)'$ denotes the vector of parameters and α_0 is the vector of current values for these parameters. The expectation E_0 in (9.22) is taken under the parameter values α_0 and we maximize (9.22) at each step with respect to α.

Now, using the form of the log-likelihood (9.19), in view of (9.20) and (9.22), it is clear that

$$\hat{\mathbf{R}} = [E_0(\mathbf{W}^*\mathbf{W} \mid \mathbf{Y})]^{-1} E_0(\mathbf{W}^* \mid \mathbf{Y})\mathbf{Y} \tag{9.23}$$

Each step then requires the conditional expectations under the complex Gaussian assumption. Then, using the specification (9.13), it is easy to show that the conditional expectation $\hat{S}_j = E_0(S_j \mid \mathbf{Y})$ is given by

$$\hat{S}_j = \frac{\Sigma_{i=1}^m R_i^* Y_{ij}}{\Sigma_{i=1}^m |R_i|^2 + \theta} \tag{9.24}$$

and the mean squared error is

$$\sigma_j^2 = \frac{f_v}{\Sigma_{i=1}^m |R_i|^2 + \theta}, \tag{9.25}$$

where $\alpha_0 = (f_s, f_v, R_1, \ldots, R_m)'$ in this case. The form for (9.24) shows that this deconvolution involves the same stacking procedure as (9.12), only applied in the frequency domain.

Then, substituting (9.24) and (9.25) into (9.23) leads to the iterative updating equation

$$\hat{R}_i = \frac{\Sigma_{j=1}^n \hat{S}_j^* Y_{ij}}{\Sigma_{j=1}^n (|\hat{S}_j|^2 + \sigma_j^2)} \tag{9.26}$$

for the receiver functions, expressed as a weighted stack, with weights proportional to the source signal estimators.

As an alternative to the analysis-of-power approach to estimating the signal and noise spectra, f_s and f_v, note that applying the EM algorithm (see Shumway, 1988, p. 250) gives the updates

$$\hat{f}_s = \frac{1}{n} \sum_{j=1}^{n} (|\hat{S}_j|^2 + \sigma_j^2) \tag{9.27}$$

and

$$\hat{f}_v = \frac{1}{mn} \sum_{i,j} (|Y_{ij} - \hat{R}_i \hat{S}_j|^2 + |\hat{R}_i|^2 \sigma_j^2). \tag{9.28}$$

As an alternative to using the sample multiple coherence as a measure of model adequacy, one might use the likelihood ratio test of $R_1 = R_2 = \cdots = R_m = 1$ as a measure of factorability. This would involve comparing the difference between twice the maximized log-likelihood (9.18) with $\hat{\mathbf{R}}$ and with $R_i = 1$ to a chi-squared random variable with $2m$ degrees of freedom.

9.2.5 Practical application of the deconvolution algorithms

The deconvolution is performed in the frequency domain. The first step is to factor out the 'known' effects due to attenuation, the approximate instrument response, and, optionally, a von Seggern and Blandford (1972) source pulse, henceforth designated by VSB in this paper. A good initial model for site functions is a delta function with some random noise; this would explain the fact that, aside from minor variations in detail, most teleseismic P vertical component waveforms look similar at most sensors. Hence, at the start of the iteration process the source and site functions are each initialized to a delta function in the time domain. The mathematical details and derivation of the algorithm are given in Section 9.2.3; the essential procedure is described below in an intuitive manner. Most of the complexities of the algorithm are dictated by the need to minimize the amplification of background noise in the results using the noise-to-signal parameter θ.

If we have m receivers (sites) and n sources, we may describe $(m \times n)$ P wave spectra (or time series) in terms of $m + n$ factors; there is thus a considerable degree of redundancy with m and n sufficiently large. The sizes of data sets varied due to data availability; typical data sets consisted of a dozen events recorded at a dozen sensors. This helps us to separate the $R_i(\nu)$ from the $S_j(\nu)$. This cannot be achieved uniquely, however, unless something is known *a priori* about the properties of each. Otherwise, common spectral factors can be assigned to either the source of receiver factors, or canceling reciprocal factors may be assigned to both the receiver and source factors. This is the most dangerous aspect in any simultaneous determination of source and site factors. Fortunately, if we estimate the source factors first by averaging over receiver sites we may expect that, in the first step, the site effects will be greatly reduced for teleseismic P waves. This expectation is based on the assumption that the site factors will tend to be quite random and dissimilar at the various array sensors, which appears to be the case at most arrays. In the algorithm the factors are refined without too much change and thus the chances of introducing major distortion into the estimated source and site spike sequences is minimized.

The deconvolution process starts with the estimation of the source factors by deconvolving the source pulse and stacking these deconvolutions over the available sites using eq. (9.24). In the next step of the iteration we divide out these source estimates and obtain the estimates of the site factors by stacking over events using eq. (9.26). Subsequently we iterate by removing these site factors to obtain updated values of the source effects. These steps are repeated until the changes become small in the site and source factors; this usually occurs after about four iterations. Without the simultaneous estimation of site effects, the method could be regarded as a multi-channel generalization of the deconvolution technique described by Oldenburg (1981).

In the deconvolution process we utilized only frequency components which had signal-to-noise ratios above 2 and excluded the frequency ranges with poor signal-to-noise ratios from the iterations. This implies that we had to exclude most of the high frequencies from *P* waves observed from low attenuation paths, and this is a fundamental bandwidth limitation imposed on us by the nature of data, which prevents us from resolving very small *pP* or other secondary arrival delays.

The deconvolution procedure has been tested with synthetically constructed models, and it reliably recovers the information fed into it as long as we have reasonable estimates of the instrument, attenuation and, when we decide to remove them, the explosion-source time functions. Therefore, if the process produces complicated time series from actual data, unlike the simple ones assumed in most direct modeling work, the reason must be that actual *P* waves are quite complex.

9.3 Deconvolution of teleseismic P-waves from nuclear explosions

In nuclear monitoring we are primarily interested in estimating the effects of the surface reflection *pP* on the total *P* wave amplitude used for estimating the yield of a nuclear device. The schematic in Figure 9.1 shows the primary *P* wave and its reflection *pP* of the surface. The lower diagram shows the expected idealized time series produced by the primary *P* wave and its reflection *pP*. In this paper, we shall denote by the same '*pP*' the downswing in the deconvolved seismograms immediately following the initial compression associated with the first arrival regardless of the true physical nature of this phase.

If the *pP* delay time, defined as the time lag between the positive and negative arrivals in Figure 9.1, is smaller than the width of the resolution kernel, i.e. the narrowest delta function approximation allowed by the bandwidth (Shumway and Der, 1985), deconvolution will yield progressively smaller amplitudes in the deconvolved $P + pP$ waveforms with decreasing *pP* delay times without seeing any decrease between the positive and negative distances, the only feature that could be used to estimate *pP* delay times from the deconvolved traces. This problem is illustrated in Figure 9.2 by plotting the deconvolutions of the spike sequence in Figure 9.1 for various delay times. Clearly, it is not possible, inspecting normally noisy deconvolved records, to distinguish delay times as small as those shown in Figure 9.2. Of course, for actual nuclear explosions there is a lower limit to the *pP* delay

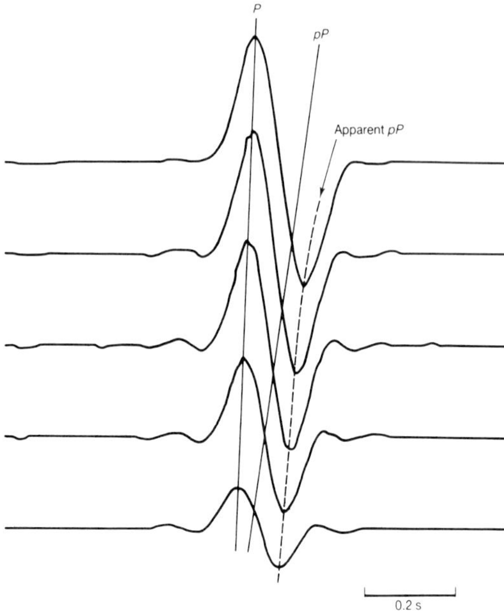

Figure 9.2 Changes in the time lag estimated from apparent *pP* arrivals (dashed line) as opposed to the real ones (solid line) as the *P*–*pP* lag decreases as it applies to data of limited bandwidth.

time, even assuming that everything is linear. This limit is imposed by the minimum depth of containment in the rocks in question.

As a final check on the deconvolution results, we reconstruct each input trace from the convolution of the appropriate source and site terms and the wavelet consisting of attenuation, instrument response, and the source pulse, VSB. Figure 9.3 shows two representative events, one at the Nevada Test Site (NTS) and one at the Russian test site at Shagan. Both events were recorded at the Norwegian Seismic Array (NORSAR). The match between the reconstructions should be good if the factorability condition is valid. In most cases the reconstructions and the original traces have high correlation coefficients in the 0·85–0·95 range (Figure 9.3). Note the site-dependent variability of waveforms for the two events in Figure 9.3, which is a direct obstacle to deriving meaningful source information from a few individual teleseismic short-period *P* wave recordings by waveform modeling.

The same check can also be performed in the frequency domain by computing the ensemble-averaged coherence between the reconstructions and the data given by (9.16). An approximation to (9.16) is the average coherence over the array. Figure 9.4 shows typical coherences obtained by such analyses that indicate the fit is very good, as indicated by coherences over 0·9 over the whole available frequency band for an NTS data set at NORSAR and Shagan data set from the British Atomic Weapons Research

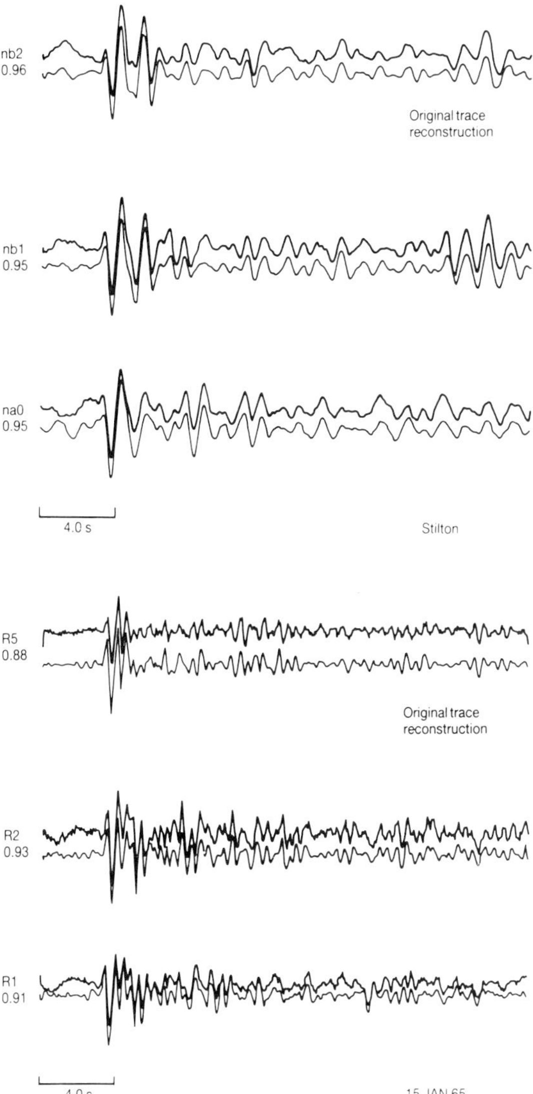

Figure 9.3 Examples of trace reconstructions (bottom) and original traces (top) for some NTS and Shagan events recorded at NORSAR. On the left side of each pair of traces is the name of the sensor and the correlation coefficient between the original and reconstructed traces.

Establishment (AWRE) array (EKA) used for this analysis. These results show that spectral factoring can account for most of the signal power in a broad frequency band. Thus, our multi-channel deconvolution method is truly a broad-band process.

The differences between the original and reconstructed seismograms can

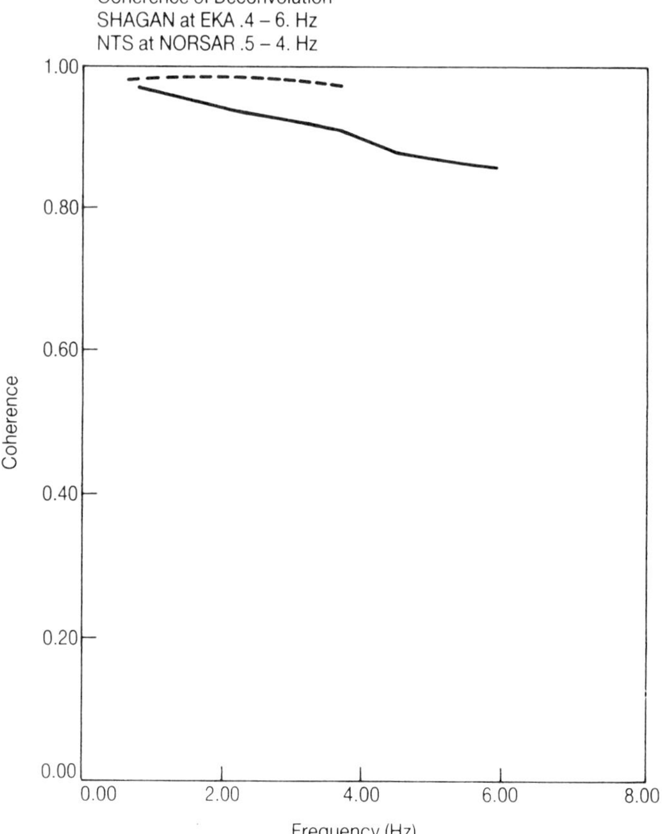

Figure 9.4 Representative ensemble-averaged coherences between data and reconstructed traces for typical Kazakh and NTS data sets. The frequency range is more limited for NTS data since the high frequency energy is much reduced by high mantle attenuation under NTS. The estimated degrees of freedom (DOF) were over 400 for both sets.

be regarded as a measure of the scattering unique to each individual path, i.e. energy unaccounted for by either site or source factors. In the case when eq. (9.1) or (9.2) is not satisfied, the reconstructions may differ considerably from the original traces. As we shall see below, this occurs sometimes when we try to jointly deconvolve recordings of stations or arrays at a wide range of azimuths from the source. This can be explained by differences in the symmetry in the source radiation patterns of at least some of the events jointly analyzed. The success in the reconstructions can thus be an important indicator of the source symmetry or similarity.

The validity of the assumption that the source factor contains only a P and pP sequence (the $P + pP$ model) used in some analyses of P waves from nuclear explosions can also be tested by computing coherences between intercorrelated *P* waves (Lay, 1985; Murphy, 1989). Intercorrelating data from two events at the same receivers should result in identical waveforms if this model is valid (Lay, 1985). Coherences between the intercorrelated traces at different receivers much lower than those expected by considering the prevalent signal-to-ambient-noise ratios would be indicative of the proportion of the signal power not accounted for in the model. Applying this procedure with *pP* parameters derived by Murphy (1989) and Lay (1985) gave values averaging around 0·6 for their Pahute Mesa (Nevada test site) events as opposed to the values in the 0·90–0·95 range obtained from spectral factoring. This indicates that the $P + pP$-model does not fit the data in this case and that the greater complexities in the deconvolved waveforms shown in this paper are required to explain the data.

During the past years, we have deconvolved waveforms for some 245 different source–receiver array pairs, typically using 5–10 array elements for each. In the rest of this paper, we discuss the results of deconvolutions of array data from a number of nuclear test sites – Degelen, Shagan, Novaya Zemlya, Astrakhan, French Sahara, Tuamotu, Sinkiang, Yucca Flats and Pahute Mesa – and of three-component regional and teleseismic data. Due to the large volume of output only a small but typical subset of these deconvolutions will be shown. For additional data regarding the events used, the reader should consult the lists of nuclear explosions published in the books by Bolt (1976) and Dahlman and Israelson (1977) and the reports by Marshall *et al.* (1984, 1985) and standard seismological bulletins. The events are uniquely identified by their dates, names and the test sites.

9.3.1 Deconvolutions at large teleseismic arrays

In the following, we provide an overview of the typical results for the various test sites. The deconvolutions of explosions at the Soviet Shagan River (eastern) test site in Kazakh usually show a well-developed *pP* downswing. The *pP* time delay is typically near 0·3 s. This appears to be valid for most deconvolutions for the British Atomic Weapons Research Establishment AWRE array we have analyzed (Figure 9.5). It is interesting to compare these deconvolutions with those of an earthquake near the Kazakh test site for which the site response function may be assumed to be the same. We had recordings of an earthquake available from three AWRE arrays. The deconvolutions show clear *pP* and *sP* pulses which were not as clear in the original traces (Figure 9.6, see Figure 9.1 for explanation of these arrivals) and some additional *PcP* arrivals (other marks) superposed on some reflections at the crust–mantle boundary (Pooley *et al.*, 1983). We feel that the technique may thus be quite useful for the analyses of earthquake source mechanisms as well if the deconvolved traces, with much reduction of the recording site distortion, are used for reconstructing the temporal and spatial structure of the seismic source.

The second test site we investigated is located in a mountainous area west of the easternmost test area in Kazakh (Degelen Mountain). A distinct characteristic of our deconvolutions from this area is that the downswing of the trace which we term as *pP* is either obscured or absent (Figure 9.7). This

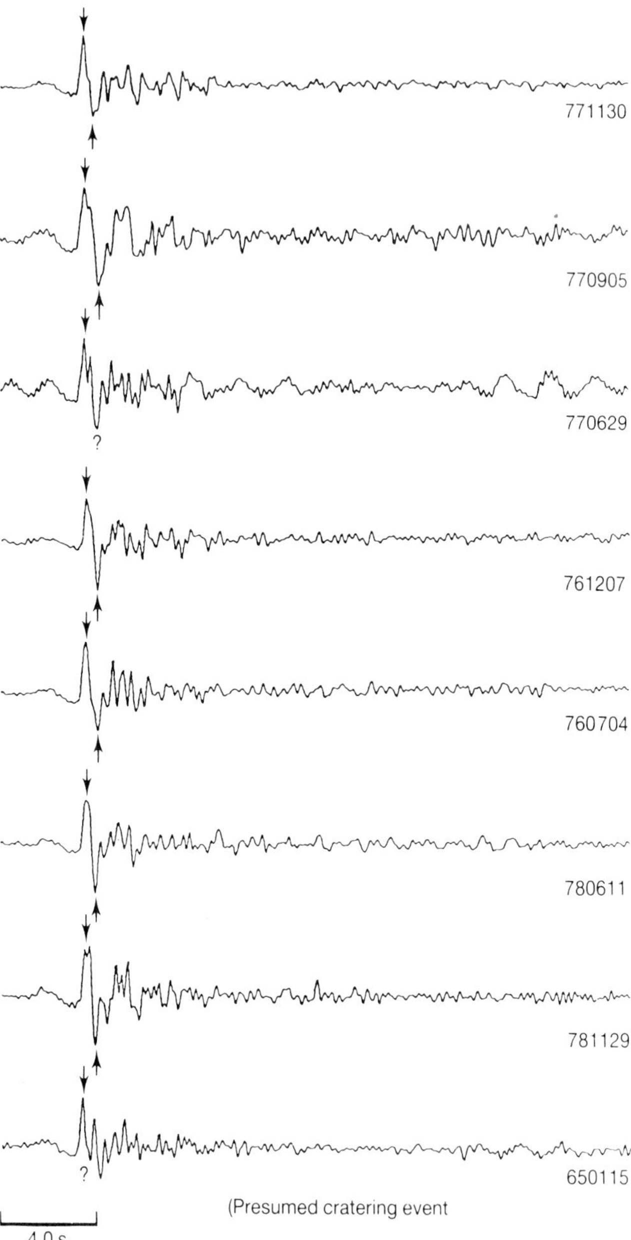

Figure 9.5 Deconvolved source time functions from Shagan data recorded at EKA. Note the difference in the source functions of a cratering explosion as compared to other events.

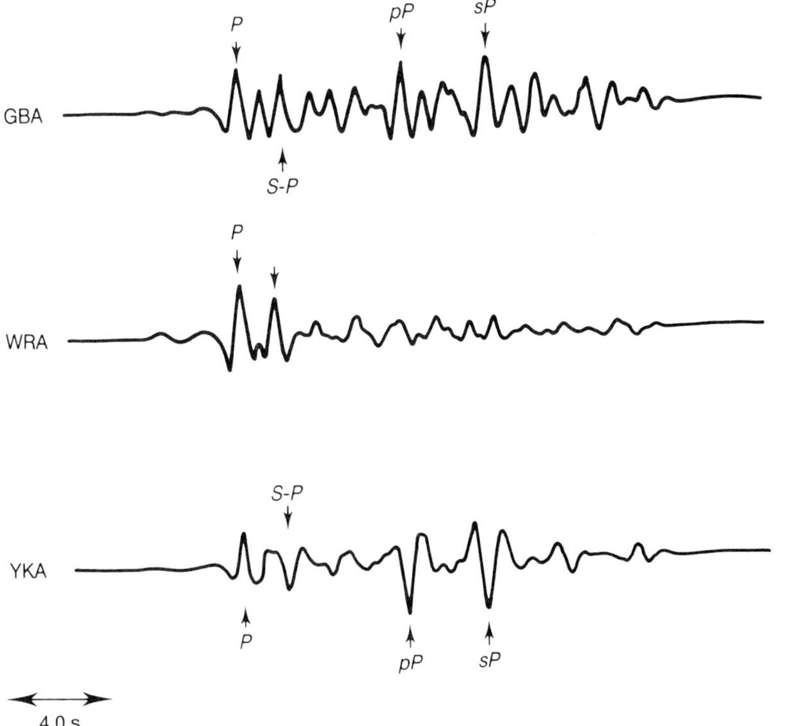

Figure 9.6 Deconvolutions of the 3/20/76 earthquake at various AWRE arrays.

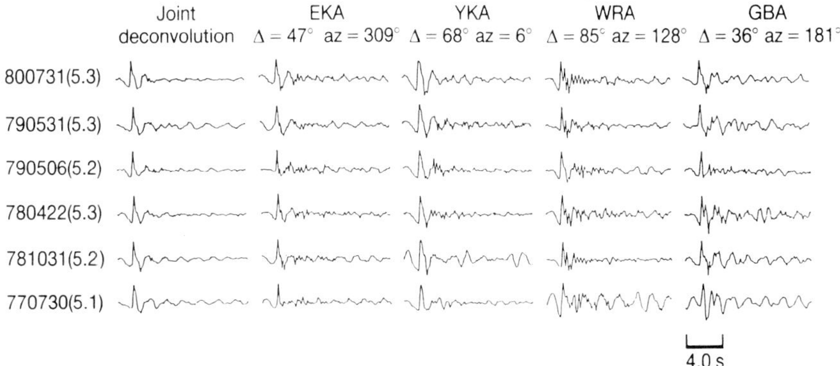

Figure 9.7 Comparison of source deconvolutions for a set of common Degelen events at each of the AWRE arrays and a joint deconvolution using six sensors at each array.

may be the result of the uneven topography above the explosions, and this characteristic is also observed on deconvolutions from similar test sites. The results are also dissimilar at various azimuths. We have selected six sensors at each of the AWRE arrays EKA, GBA, YKA and WRA and treated them as a single array. There are differences among these results, but the most revealing property of the results is that the reconstructions of the joint deconvolutions of the sensors from the four AWRE arrays no longer show a good similarity with the original traces (Figure 9.8). This indicates that the source radiation from the Degelen area explosions has significant azimuthal variations among events and thus the factorability condition is no longer valid for the joint array. This property was also noted by Mellman and Kaufman (1981) for some other Kazakh explosions. We have also performed joint deconvolutions of Shagan and Degelen events at NORSAR which indicate the differences in the pP characteristics, discussed above, between the two test sites. Apparently the site response functions are practically the same for these two closely located test sites at NORSAR.

The deconvolutions of the French tests in Algeria at EKA and YKA shown in the paper by McLaughlin *et al.* (1988) generally show complex waveforms that vary with azimuth considerably. The data for these events are sparse and noisy. The azimuthal variability must be related to the high topographical relief in the test area; the test site itself is a large mountain (Figure 9.9). The French test site in the Tuamotu Archipelago is an atoll, and the devices are exploded in a basalt basement. The mantle attenuation under this test site is high according to evidence of P-wave spectra (Der *et al.*, 1985; Chan and Der, 1988) which show a very low high-frequency content. The attenuation for the paths involved is probably considerably higher than for NTS paths. Deconvolutions of these explosions at the British AWRE stations YKA and RSTN stations show a clear pP on the average about 0·5 s after the upward swing of the direct P wave (Figure 9.10). We

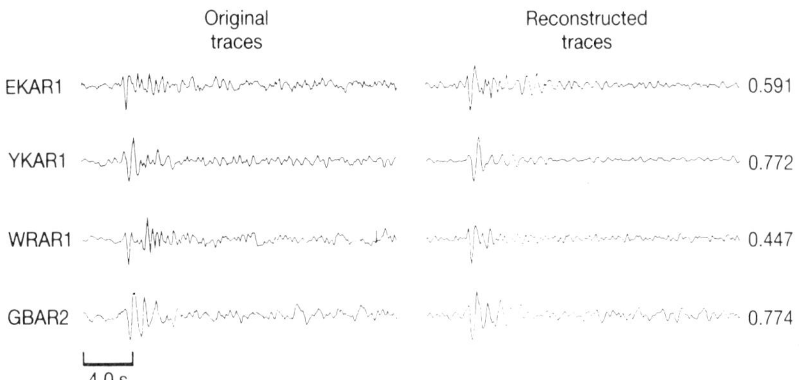

Figure 9.8 Comparison of some original traces and reconstructions obtained from the joint deconvolution of common Degelen events at the four AWRE arrays. The poor quality of reconstructions (unlike those at individual arrays) indicates a considerable asymmetry in the source radiation from Degelen.

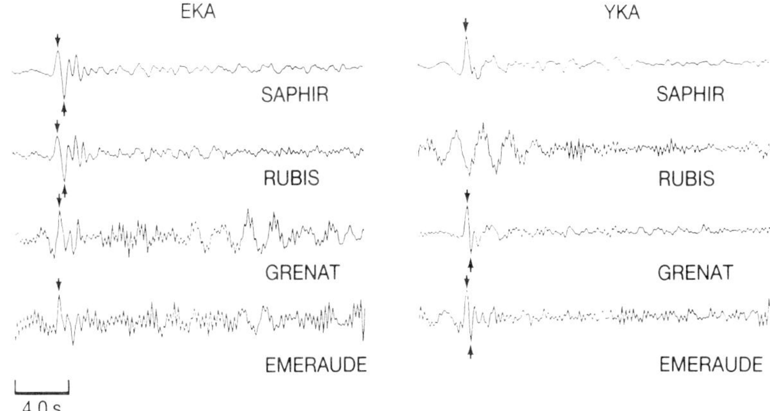

Figure 9.9 Source time function estimates for some French nuclear explosions in the Sahara at the YKA and EKA arrays.

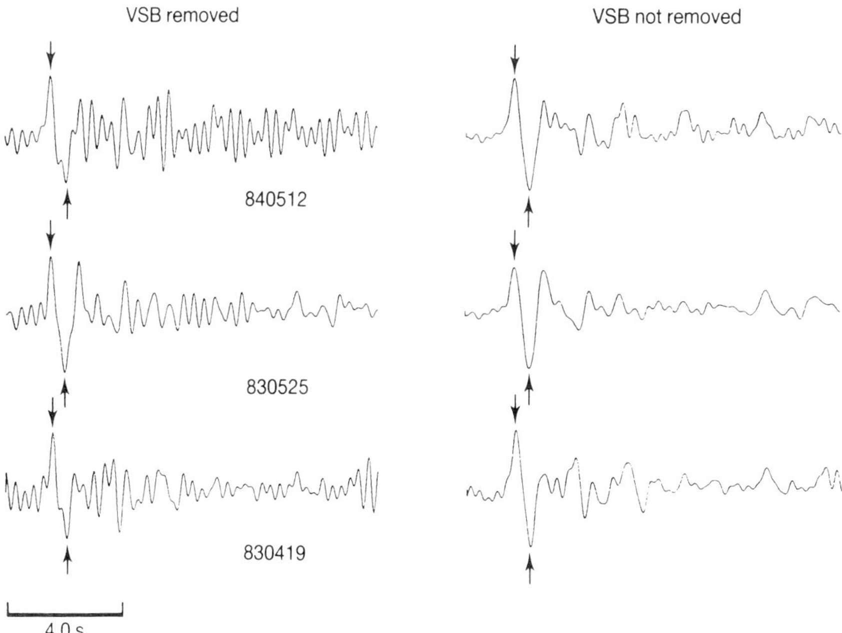

Figure 9.10 Deconvolved source functions for Tuamotu events recorded at the RSTN. To the left, the estimated VSB wavelet has been removed in the deconvolutions, while it has not been removed from the deconvolutions shown on the right.

must note, however, that it would be impossible to resolve much smaller *pP* delays with the high mantle attenuation under this site.

The US nuclear explosions at Pahute Mesa generally show extremely complex waveforms with no clear *pP* arrivals (Figure 9.11) after deconvol-

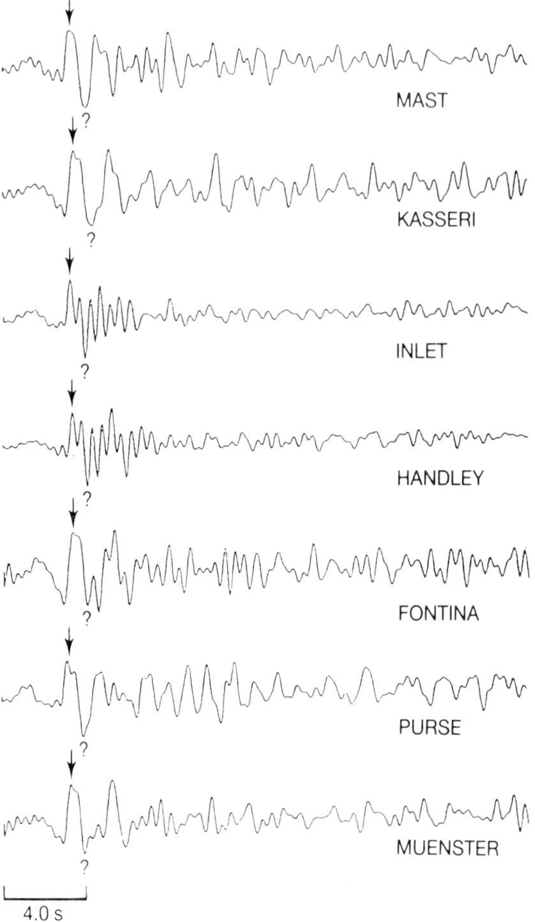

Figure 9.11 Source time function estimates for Pahute Mesa explosions recorded at EKA. VSB wavelets have been removed in these deconvolutions.

ution as noted by Der *et al.* (1987). The absence of clear *pP* in the Pahute Mesa deconvolutions may be due to the combined effects of topography, structural complexity and reverberations in the caldera fill (McLaughlin *et al.*, 1987). We must note, however, that some yet unpublished deconvolutions of Pahute explosions by Douglas (1987), which put more emphasis in the low frequencies below 0·5 Hz, do indicate the presence of *pP* phases. From the viewpoint of estimating yields and the associated magnitude difference in test sites, these low frequencies may not matter for magnitude calculations since magnitude is usually estimated from *P*-waves with dominant periods ranging from 0·5 to1 s.

Despite the fact that there are few direct recordings that indicate what the wave trains from nuclear explosions are like, we are confident that we are looking at the source time functions in some limited bandwidth with much of the recording site effects reduced. In many ways, the results presented above make physical sense. For example, the deconvolution results for test sites with little topographical relief near the shot point tend to show a more pronounced negative pulse (*pP*) following the initial upswing of direct *P* than those from areas of rugged topography. The latter also tend to have more azimuthal variation in the source waveform. The deconvolution of the cratering shot in Figure 9.5 at Shagan lacks this negative pulse, a feature that has been predicted on the basis of numerical elastic–hydrodynamic simulations. Moreover, the deconvolution of the earthquake shown in Figure 9.6 exhibits mostly simple pulses with identical shapes and well-defined polarity as expected for an impulsive source at moderate depth. The late codas of the source time functions seem to be essential for the reconstructions of the events and may represent some (possibly Rayleigh-to-*P*) scattered energy from the source region. We expect many of these results on the basis of our knowledge of the geology and topography of the various test sites, source mechanisms and finite-difference simulations of explosion sources. Since we were removing only the instrument and attenuation operators in our deconvolutions, both of which can be approximated reasonably well (Der *et al.*, 1982; Der and Lees, 1985) we do not depend on questionable, elaborate models of the seismic sources.

We have also explored the consequences of *pP* interference as derived from the deconvolution results and combined with the present best estimates of attenuation variations (Der *et al.*, 1985) by computing *P* waveforms assuming simple linear superposition of *P* and *pP*, a VSB source and a WWSSN (World Wide Standard Seismic Network) instantaneous response. Such calculations showed that the magnitude bias between Soviet Kazakh and the US NTS sites can be enhanced by about 0·3 due to *pP* interference. Moreover, explosions at Degelen may have a smaller magnitude by a few tenths of magnitude units than those at Shagan because of the absence of *pP*. Tuamotu explosions would be similar to NTS explosions in magnitude despite the higher attenuation along the path. Sykes and Ekstrom (1989) estimate the bias to be 0·35 magnitude units based on yield estimates of the Joint Verification Experiment (JVE) explosion published in the New York Times and their own measurements of seismic data. Recently Jih *et al.* (1990) have computed magnitude–yield curves for WWSSN stations using the yield values published by Vergino (1989) and Springer and Kinnaman (1971). They found an average bias between NTS and Shagan in the range of 0·42–0·52 magnitude units and a negative Degelen vs. Shagan offset of 0·03–0·15 magnitude units similar to that predicted from the *pP* differences shown above. All the Shagan–NTS bias values above are higher than the predictions based on attenuation effects alone. Although we are sure that the actual problem is not as simple as the simulations we are referring to, and there are still some discrepancies in the actual numbers, these findings give support to the idea that differences in *pP* characteristics can contribute to the magnitude differences for events of similar yield at various test sites that are above and beyond the effects of mantle attenuation.

9.3.2 Deconvolutions at three-component arrays

An obvious extension of the site-response function idea is the incorporation of the three components of motion in which each of the sensors is treated as a separate site even if they are located in the same place. The particle motions of teleseismic body waves are only in the first approximation linear for body waves *P* and shear waves *S*. Even for simple layered structures of the crust, the motion becomes elliptical in the most complex manner (Su and Dorman, 1965) with prograde and retrograde motions in the various frequency bands. For realistic three-dimensional models of the near-station geology, the motion becomes truly complex and three-dimensional even for supposedly simple, incident *P* waves. While the first few cycles of the *P* wave are reasonably linear-elliptical in the plane of the azimuth of the arrival, the transverse component quickly builds up with time, approaching a respectable fraction of the vertical and radial components in energy. It is a logical extension of the deconvolution work on multiple sites to approach the problem by treating each component of motion as a separate 'site' in the deconvolution process. Each component will have a 'site factor' in the frequency domain, which characterizes the transformation of the motion due to local geology; these factors depend, of course, on the slowness vector (azimuth–distance) of the *P* arrivals and may thus be used to recognize or characterize source regions of *P* waves.

The approach developed here is fundamentally different from practically all of the three-component processing work that has been done thus far (Christofferson *et al.*, 1988). Instead of assuming a simply polarized wave appropriate to a half-space or some other simple, analytically defined mathematical model, we essentially characterize the particle motion as it is, with all its complexities. We believe that source regions may be identified quickly and automatically by recognizing patterns of three-component site factors rather than computing back-azimuths based on simplistic assumptions about polarization patterns. Any systematic deviations between the true azimuths and the back-azimuths can be automatically accounted for in such schemes, as long as the master events used to derive the three-component 'site' factors are accurately located (possibly independently, also including non-seismic information).

To test such concepts, we have assembled a data base of four master events at the Kazakh test site as recorded at the RSTN station network operated by DARPA (Defense Advanced Research Projects Agency) during the mid-1980s. The deconvolved source time functions appear to be quite simple for these events, mainly consisting of *P* and (presumably) *pP* arrivals as marked (Figure 9.12). The site impulse responses tend to be the most complex on the transverse components, building up gradually with time, as expected, but simpler and impulse-like on the vertical and radial components. Nevertheless, all the component site time functions seem to be much more complicated at RSCP and RSSD. The geological structure under these two sites has been shown to be complex by other studies (Owens *et al.*, 1987). Some raw data traces of these events are shown in Figure 9.13, and opposing them, we show the trace reconstructions for easy comparison. The good

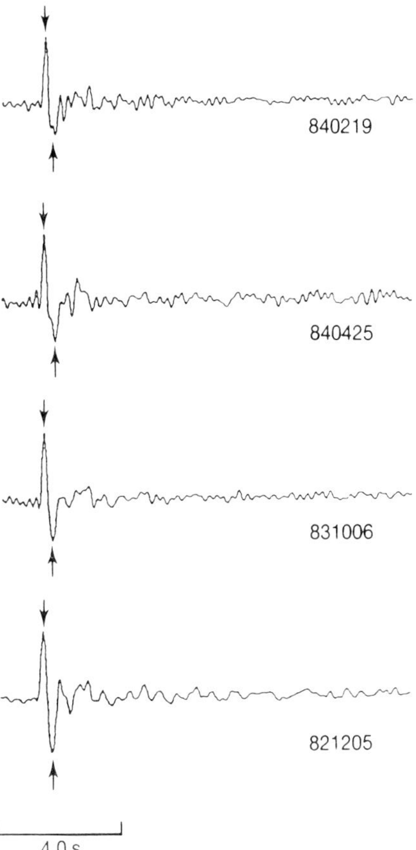

Figure 9.12 Source time function estimates for four Shagan events obtained from the total data set from RSTN three component recordings. VSB wavelets have been removed in these deconvolutions. All of these show the prominent '*pP*' arrival.

similarity of the original and reconstructed traces show that the 'site' approach works well for this multi-site three-component data set and that the complexities in particle motion can be reliably characterized by 'component-site' transfer functions. The fact that the spectral factorability held, i.e. the trace reconstructions were successful, despite the large variation of the back-azimuths from the source region (Shagan River test site) to the individual RSTN stations as compared to that to sensors within a single AWRE seismic array, indicates that the variability in azimuthal asymmetries of the P wave radiation from this test site was not large for the events analyzed.

9.4 Spectral factoring of regional array data

At regional distances the seismogram consists of a number of packets of energy arriving with various velocities. The first arrival is commonly P_n,

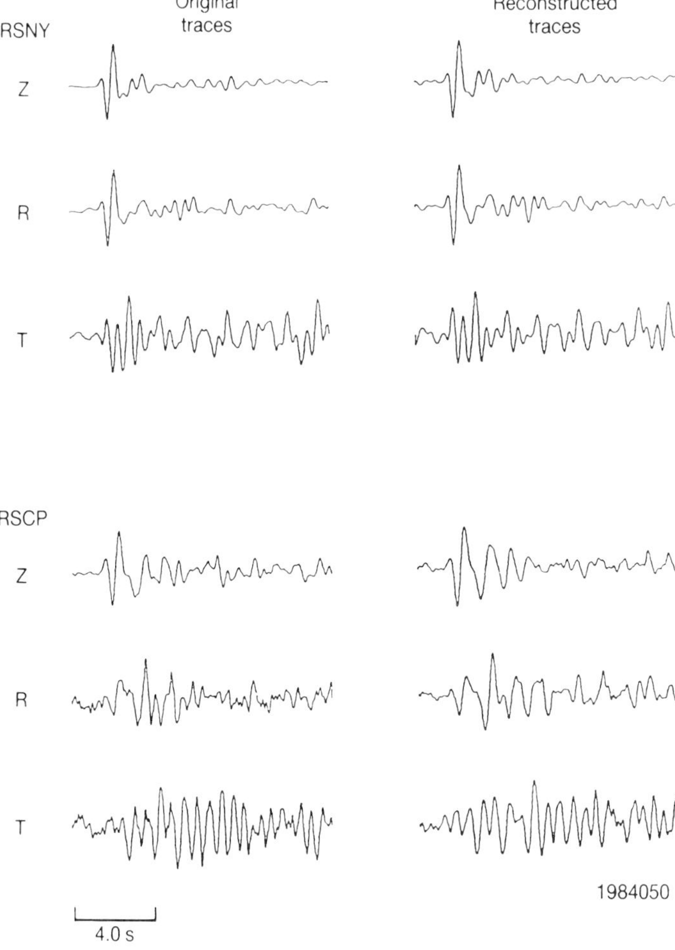

Figure 9.13 Original and reconstructed traces at three component RSTN stations for a Shagan event. These reconstructions are quite good considering the fact that the RSTN stations cover a wide area.

traveling with a subcrustal P velocity of about 8 km/s in the upper mantle, followed by P_g, which travels with crustal P velocities close to 6 km/s, S_n which travels with the subcrustal S velocity, 4·7 km/s, and L_g, which travels with maximum group velocities near 3·7 km/s. The L_g phase consists of numerous higher-mode Rayleigh wave groups and has the highest amplitude of all regional arrivals. All regional arrivals have long codas and often emergent beginnings. Generally it is not possible to model the waveforms of regional arrivals theoretically because of the extreme complexity of the velocity structure of the Earth's crust. Nevertheless, detailed waveform characteristics of regional arrivals can be utilized by spectral factoring methods described here (Der *et al.*, 1990).

The success of spectral factoring depends on similarities among the events in mechanism and source location. Using factoring of a group of events, we can identify events that differ from the rest in location or mechanism (Der *et al.*, 1990) even though they may be difficult to distinguish by conventional methods. Such techniques may be useful for detecting unusual activities in a quarry, such as attempts at hiding nuclear tests in a shot pattern of quarry blasts. We can also test how well the factorability condition accounts for the total energy in various regional arrivals by comparing the similarity between the data and reconstructions.

9.4.1 Factorization of regional data from mine explosions at small arrays

Figures 9.14 and 9.15 show the results of factorization for P_n arrivals from the Titania mine in Norway and L_g arrivals from the Estonian mine E9 using recordings of these phases at NORESS, a very small array of 3 km aperture near Oslo, Norway. In each case four events were used in the factorization involving 12 sensors of the array and the waveforms are nicely reconstructed.

Looking at the same problem in the frequency domain, we can again compute a coherence measure between the data and reconstructions. For the example shown in Figure 9.16 we have factored the L_g wave arrivals from three events at the E9 mine with one from the E4 mine. The factorization will be dominated by the E9 events causing the efficiency of reconstruction for the E4 event to be poor. Since the difference in azimuth of the E9 and E4 mines is only 0·6°, not normally detectable by an array with small aperture similar to NORESS, this example illustrates the capability of such methods for detecting small variations in the azimuth of arrival. We have also found that some of the events at the E9 mine did not factorize well with the group of three at E9 shown in Figure 9.16. We believe that these have different source mechanisms since their L_g envelope shapes are also different. Thus, the modal composition of their L_g phases may also differ.

9.4.2 Spectral factorization of three-component regional data

With the success of deconvolutions of three-component teleseismic data, we then attempted deconvolutions for three-component regional data. The first example is a set of six mine explosions in the Leningrad region recorded at NORESS. These events have complex waveforms, but as Figure 9.17 shows, the original traces and reconstructions for one of the events are very similar for all three components of motion. The excellent quality of the reconstructions shows that the deconvolution procedure is also valid for far-regional signals. We have also obtained similar results for western Norway events at half that distance. The maximum frequencies (up to 6 Hz) utilized in all of our work with regional signals were higher than for the teleseismic deconvolutions.

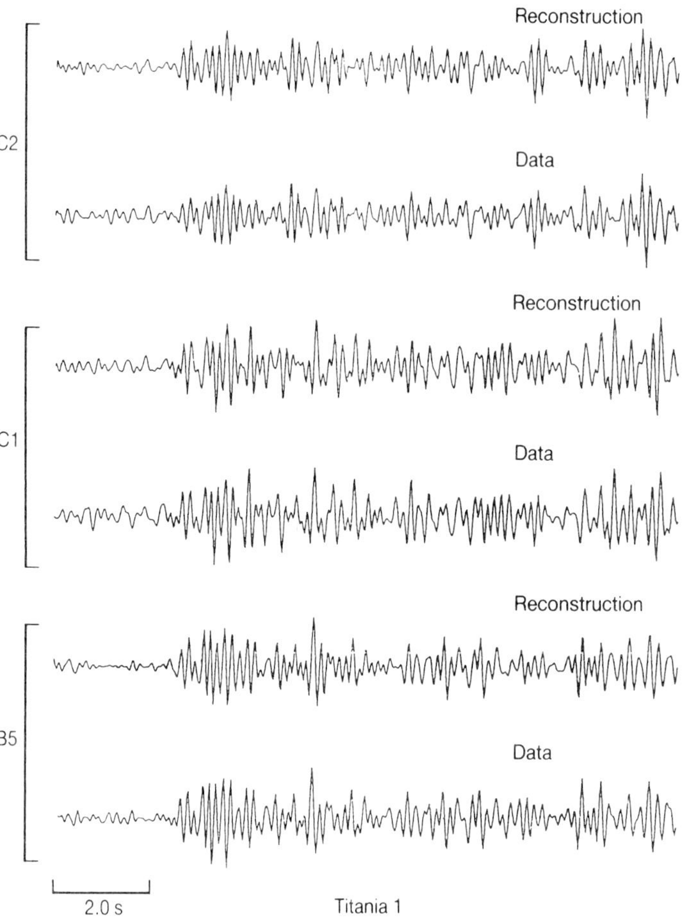

Figure 9.14 Comparisons of data and reconstructed P_n waveforms recorded at NORESS using factorization of four explosions at the Titania mine in Norway.

9.5 Conclusions

A new multi-channel iterative deconvolution method was applied to recordings of P waves from events from a number of test sites. The method used a statistical signal model that resulted from assuming that a matrix of complex spectra of seismograms observed for a suite of events and receivers can be decomposed into fixed receiver responses and random source functions. We derived estimators for both the unknown receiver response and the unknown stochastic source function. The receiver functions were estimated by applying the EM algorithm to the usual Whittle frequency domain likelihood. The sequence of steps involved in producing the receiver estimators also yielded the deconvolved stochastic source functions. The estimated receiver and source functions were exhibited as simple weighted

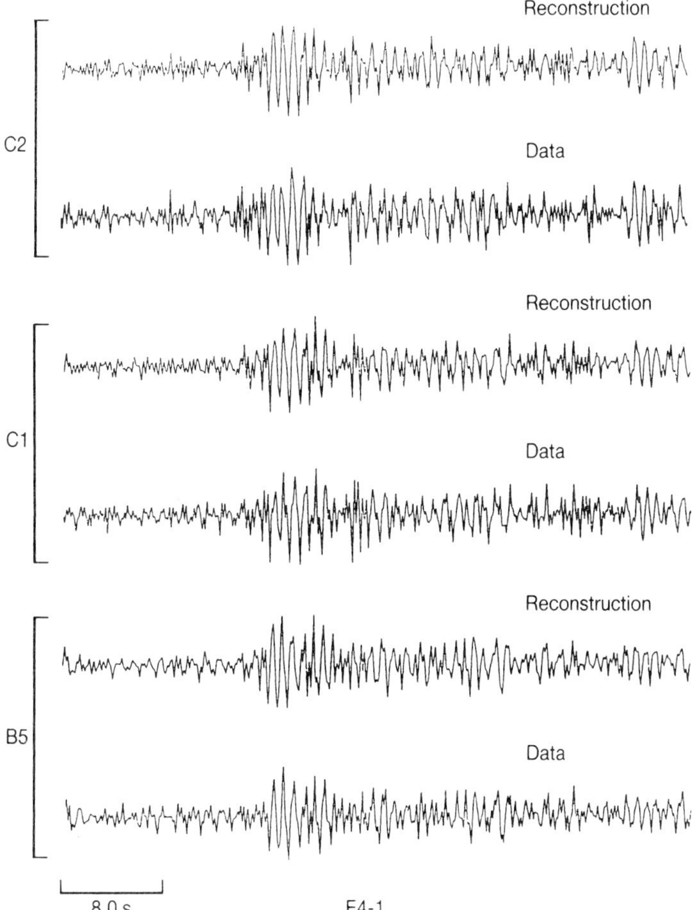

Figure 9.15 Comparisons of data and reconstructed L_g waveforms recorded at NORESS using factorization of four explosions at the E9 mine in Estonia.

stacking operations, carried out over events and sites respectively. We have discussed various tests for model adequacy that arise as extensions of the usual analysis-of-variance approach to the time-series case.

The deconvolved source time functions vary substantially between test sites and sometimes 'within' test sites. Test sites with high topographical relief, shallow source depth and complex geology such as Ahaggar, Degelen and Pahute Mesa are characterized by *P* waves without a clearly identifiable *pP* within the short period band on the deconvolved records. *P* waves from some other test sites, Shagan River, Sinkiang, Tuamotu, Astrakhan, Yucca Flats and Novaya Zemlya, on the other hand, often exhibit indications of a *pP* phase.

Spectral factorability was found to be valid for P_n and L_g arrivals from quarry blasts in Norway and Estonia at the NORESS array. Differences in

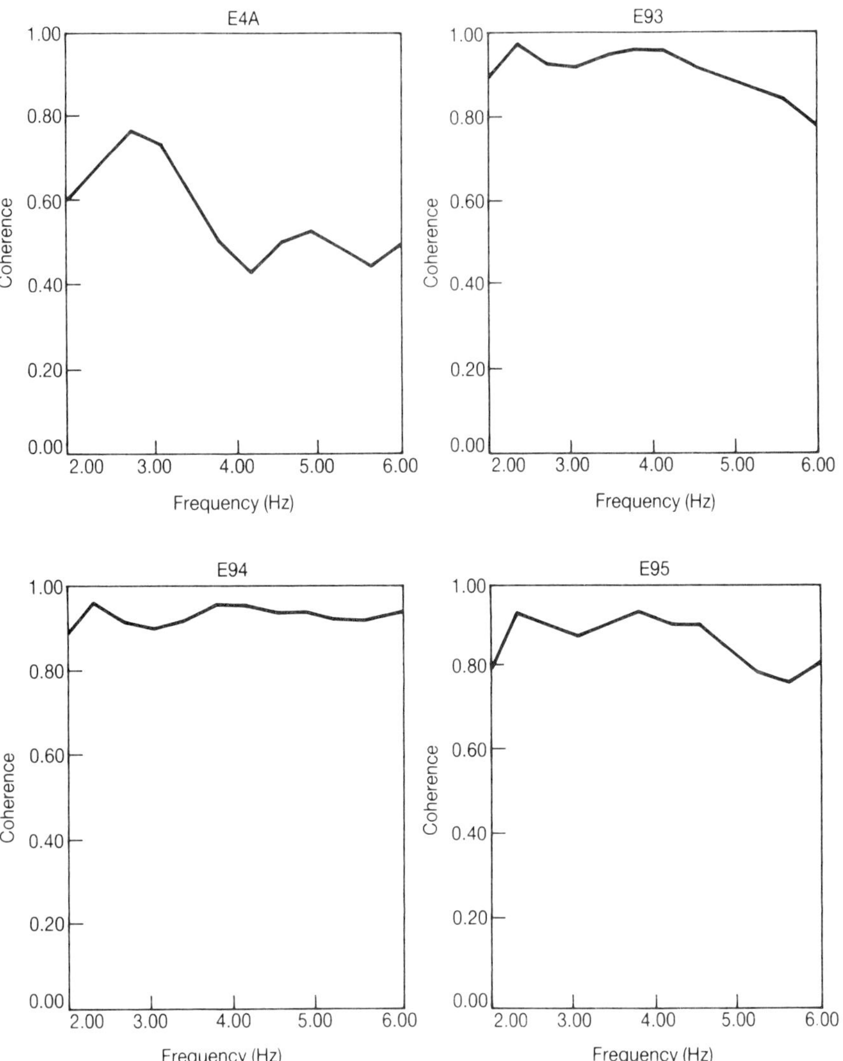

Figure 9.16 Ensemble-averaged coherences between data vs. reconstructed L_g traces for one E4 event jointly factored with three E9 events. The E4 event cannot be reconstructed well using the (dominantly E9) site factors from this factorization. DOF = 250.

factorability of the L_g phases from Estonia can be seen for events differing in azimuth by only 0·6°.

Deconvolutions of teleseismic and regional three-component data showed that the factorability condition also applied to such data. This suggests the possibility of particle motion processing using the intersensor transfer functions. Regional P_n source functions are very complex compared to the

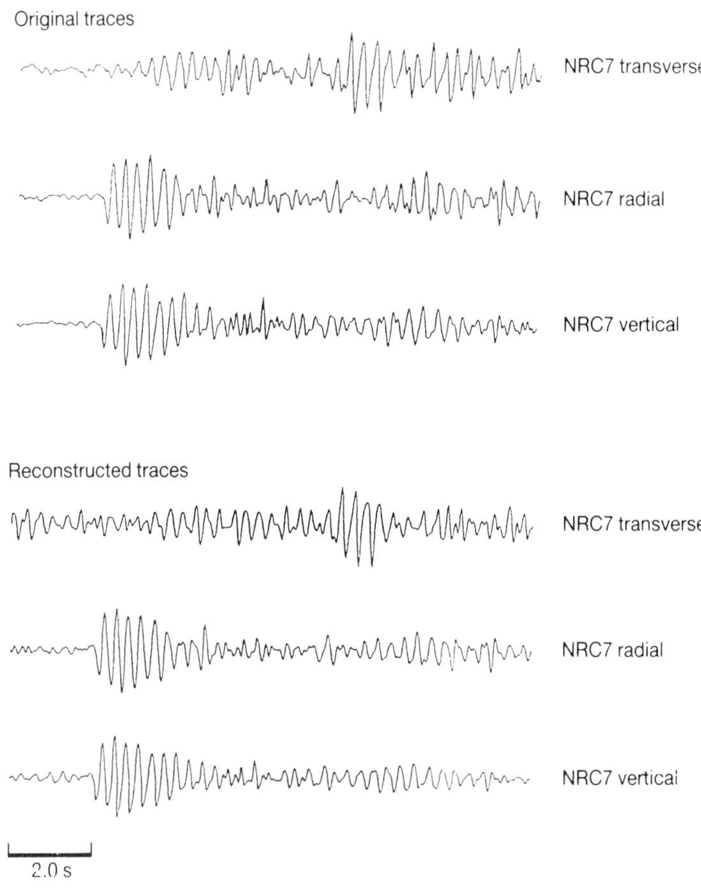

Figure 9.17 Reconstruction of regional P_n arrivals from mine explosion near Leningrad at NORESS.

teleseismic source functions. For *P* waves, the vertical and radial site terms are fairly impulsive, while the transverse site term is more complex, building gradually with time as expected with energy being scattered into the transverse component.

The effects of the lateral variations in anelastic attenuation combined with the effects of the absence or presence of *pP* will cause significant differences in the estimates of event sizes among test sites. Constructive interference of *P* and *pP* increase the m_b values for Shagan, Novaya Zemlya and Tuamotu explosions. Such constructive interferences appears to be absent at Pahute Mesa, Yucca Flats, Ahaggar and Degelen explosions. The associated *pP*-related biases are superposed on the ones caused by attenuation differences (Der *et al.*, 1985), coupling and site geology. The multi-channel deconvolution procedure appears to be a useful tool for a variety of studies related to source function determination at high frequencies.

Acknowledgements

Mike Hirano of Ensco, Inc. assisted with the computations using regional array data. Several colleagues contributed to the development of ideas in this paper through discussions and constructive criticism, most notably R. R. Blandford, D. R. Baumgardt, I. N. Gupta, and A. Douglas. G. Smart has done some early work on stacked single-channel deconvolutions at arrays, R. Wagner and M. Marshall did much of the data handling and preliminary analysis. The work was performed under the DARPA contract F08606-86-C-0006.

References

Bakun, W.H. and Johnson, L.R. (1973). The deconvolution of teleseismic P waves from explosions Milrow and Cannikin. *Geophys. J. Roy. Astron. Soc.*, **34**, 321–42.

Bogert, R.B., Healy, M.J., and Tukey, J.W. (1962). The frequency analysis of time series for echoes. *Proc. Symp. Time Series Analysis*, Rosenblatt, M. (ed.), Wiley, New York.

Bolt, B.A. (1976). *Nuclear Explosions and Earthquakes: The Parted Veil*, Freeman, San Francisco.

Brillinger, D.R. (1969). Asymptotic properties of spectral estimates of second order. *Biometrika*, **56**, 375–90.

Chan, W. and Der, Z.A. (1988). Attenuation of multiple ScS in various parts of the world. *Geophys. J. Roy. Astron. Soc.*, **92**, 303–14.

Christofferson, A., Husebye, E.S., and Ingate, S.F. (1988). Wavefield decomposition using ML probabilities in modeling single-site 3-component records. *Geophys. J. Roy. Astron. Soc.*, **93**, 197–213.

Dahlman, O. and Israelson, H. (1977). *Monitoring Underground Nuclear Explosions*, Elsevier, Amsterdam.

Der, Z.A. and Lees, C.A. (1985). Methodologies for estimating $t^*(f)$ from short-period body waves and regional variations of $t^*(f)$ in the United States. *Geophys. J. Roy. Astron. Soc.*, **82**, 125–40.

Der, Z.A., McElfresh, T.W., and O'Donnell, A. (1982). An investigation of the regional variations and frequency dependence of anelastic attenuation in the mantle under the U.S. in the 0·5–4 Hz band. *Geophys. J. Roy. Astron. Soc.*, **69**, 67–99.

Der, Z.A., McElfresh, T.W., Wagner, R., and Burnetti, J. (1985). Spectral characteristics of P waves from nuclear explosions and yield estimation. *Bull. Seism. Soc. Amer.*, **75**, 379–90, and Errata, **75**, 1222.

Der, Z.A., Shumway, R.H., and Lees, A.C. (1987). Multichannel deconvolution of P waves at seismic arrays. *Bull. Seism. Soc. Amer.*, **77**, 195–211.

Der, Z.A., Hirano, M.R., and Shumway, R.H. (1990). Coherent processing of regional signals at small seismic arrays. *Bull. Seism. Soc. Amer.*, **80 B**, 2161–76.

Deregowski, S.M. (1971). Optimum digital filtering and inverse filtering in the frequency domain. *Geophys. Prospecting*, **19**, 729–68.

Donoho, D.L. (1981). On minimum entropy deconvolution. In *Applied Time Series Analysis*, 2, Findley, D.F. (ed.), Academic Press, New York.

Douglas, A. (1987). Differences in mantle attenuation between Nevada and Shagan River test sites: Can the effects be seen in P wave seismograms? *Bull. Seism. Soc. Amer.*, **77**, 270–86.

Filson, J. and Frasier, C.W. (1972). Multisite estimation of explosive source parameters. *J. Geophys. Res.*, **77**, 2045–61.

Hannan, E.J. and Thomson, P.J. (1974). Estimating echo times. *Technometrics*, **16**, 77–84.

Hunt, B.R. (1972). Deconvolution of linear systems by constrained regression and its relationship to the Wiener theory. *IEEE Trans. Autom. Control*, **1**, 703–5.

Jih, R.-S., Shumway, R.H., and Rivers, D.W. (1990). Maximum likelihood magnitude yield regression with censored information. Submitted to *Bull. Seism. Soc. Amer.*

Kormylo, J. and Mendel, J. (1983). Maximum likelihood seismic deconvolution. *IEEE Trans. Geosci. and Rem. Sensing*, **21**, 72–82.

Lay, T. (1985). Estimating explosion yield by analytical waveform comparison. *Geophys. J. Roy. Astron. Soc.*, **82**, 1–30.

Lii, K.S. and Rosenblatt, M. (1982). Deconvolution and estimation of transfer function, phase and coefficients for non-Gaussian linear processes. *Ann. Statist.*, **10**, 1195–1208.

McLaughlin, K.L., Anderson, L.M., and Lees, A.C. (1987). Effects of local geologic structure on Yucca Flats, NTS, explosion waveforms: 2-dimensional finite difference calculations. *Bull. Seism. Soc. Amer.*, **77**, 1211–22.

McLaughlin, K.L., Lees, A.C., Der, Z.A., and Marshall, M.E. (1988). Teleseismic spectral and temporal M_0 and ψ_0 estimates for four French explosions in southern Sahara. *Bull. Seism. Soc. Amer.*, **78**, 1580–91.

Marshall, P.D., Bache, T.C., and Lilwall, R.C. (1984). *Body Wave Magnitudes and Locations of Soviet Underground Explosions at the Semipalatinsk Test Site.* AWRE Report No. 0 16/84, Atomic Weapons Research Establishment, MOD(PE), Aldermaston, Berkshire, England.

Marshall, P.D., Lilwall, R.C., and Warburton, P.J. (1985). *Body Wave Magnitudes and Locations of French Underground Explosions at the Muroroa Test Site.* AWRE Report No. 0 12/85, Atomic Weapons Research Establishment, MOD(PE), Aldermaston, Berkshire, England.

Mellman, G.R. and Kaufman, S.K. (1981). *Relative Waveform Inversion.* SGI-R-81-048, Sierra Geophysics, Redmond, WA, U.S.A.

Murphy, J.R. (1989). Network averaged teleseismic P wave spectra for underground explosions. Part II. Source characteristics of Pahute-Mesa explosions. *Bull. Seism. Soc. Amer.*, **79**, 156–71.

Oldenburg, D.W. (1981). A comprehensive solution of the linear deconvolution problem. *Geophys. J. Roy. Astron. Soc.*, **65**, 331–58.

Owens, T.J., Taylor, S.R., and Zandt, G. (1987). Crustal structures at regional seismic test network stations determined from inversion of broadband teleseismic P-waveforms. *Bull. Seism. Soc. Amer.*, **77**, 631–62.

Pawitan, Y. and Shumway, R.H. (1989). Spectral estimation and deconvolution for a linear time series model. *J. Time Series Anal.*, **10**, 115–29.

Pooley, C.I., Douglas, A., and Pierce, R.G. (1983). The seismic disturbance of 1976 March 20 in east Kazakhstan: earthquake or explosion? *Geophys. J. Roy. Astron. Soc.*, **74**, 621–31.

Shumway, R.H. (1988). *Applied Statistical Time Series Analysis*, Prentice Hall, Englewood Cliffs.

Shumway, R.H. and Der, Z.A. (1985). Deconvolution of multiple time series. *Technometrics*, **27**, 385–93.

Springer, D.L. and Kinnaman, R.L. (1971). Seismic source summary for U.S. underground explosions, 1961–1973. *Bull. Seism. Soc. Amer.*, **61**, 1073–98.

Stewart, R.C. and Douglas, A. (1983). Seismograms for phaseless seismographs. *Geophys. J. Roy. Astron. Soc.*, **72**, 123–52.

Su, S.S. and Dorman, J. (1965). The use of leaking modes in seismogram interpretation and in studies of crust-mantle structure. *Bull. Seism. Soc. Amer.*, **55**, 989–10121.

Sykes, L.R. and Ekstrom, G. (1989). Comparison of seismic and hydrodynamic yield determinations for the Soviet Joint Verification Experiment of 1988. *Proc. Nat. Acad. Sci.*, **86**, 3456–60.

Vergino, E.S. (1989). Soviet test yields. *EOS Trans. Amer. Geophys. Union*, November 28.

von Seggern, D.H., Blandford, R.R. (1972). Source time functions and spectra for underground nuclear explosions. *Geophys. J. Roy. Astron. Soc.*, **31**, 83–97.

Walden, A.T. (1985). Non-Gaussian reflectivity, entropy and deconvolution. *Geophysics*, **50**, 2862–88.

Walden, A.T. (1988). Robust deconvolution by modified Wiener filtering. *Geophysics*, **53**, 186–92.

Wiggins, R.A. (1978). Minimum entropy deconvolution. *Geoexploration*, **16**, 21–35

Chapter 10

Envelope estimation for quasi-periodic geophysical signals in noise: a multitaper approach

J. Park

10.1 Introduction

Many processes in nature are quasi-periodic, and can be represented as having a dominant oscillation with carrier frequency f_0 that suffers amplitude and phase variations on a time scale T_1 much longer than the oscillation period $T_0 = 1/f_0$. Quasi-periodicity is characteristic of certain deterministic classical mechanical systems, such as fluctuations in the Earth's orbital parameters, and resonant systems subject to stochastic excitation, such as the atmospheric southern oscillation. Systems with many periodic oscillations that are closely spaced in frequency, such as the Earth's free oscillation spectrum, are often observed as quasi-periodic. Data collected from these processes take the form $u_n = \Re\{A_n e^{-2\pi i f_0 n \Delta t}\}$, $n = 1, 2, \ldots, N$ for duration $T = N\Delta t$. The slowly varying complex-valued envelope* $A_n = A(n\,\Delta t)$ can be represented as a narrow-band function $\bar{A}(f)$ in the frequency domain: in particular,

$$A(n\,\Delta t) = A_n = \int_{-f_N}^{f_N} \bar{A}(f) e^{-2\pi i f n \Delta t}\,df, \tag{10.1}$$

where $f_N = 1/(2\,\Delta t)$ is the Nyquist frequency. We assume that $\bar{A}(f) \approx 0$ for $|f| > f_w$ for some half-bandwidth f_w, so that it has minimum time scale of variation $T_1 = 1/f_w$. In some situations, however, the envelope of a quasi-periodic signal may undergo sudden changes, e.g. the impulsive excitation of an oscillatory system, or a bifurcation in a chaotic system. Such behavior 'broadens' $\bar{A}(f)$. In most practical situations, however, the quasi-periodic signal is one component of a 'mixed' spectrum containing additional broad-band stochastic processes, e.g. noise. Often $|\bar{A}(f)|$ is comparable to the

* The 'amplitude' of the envelope – the quantity that is most easy to visualize – is the modulus of the complex-valued envelope.

broadband spectrum outside some narrow bandwidth $[-f_w, f_w]$, and one is faced with reconstructing the envelope A_n, $n = 1, 2, \ldots, N$, from spectral information inside $[-f_w, f_w]$ only.

Narrow-band filtering is a common method for analysing quasi-periodic signals. In this chapter a new technique based on multiple-taper spectrum analysis (Thomson, 1982; Park *et al.*, 1987b; Walden, 1990) is presented. The multitaper approach allows the analyst to model the envelope function $A(t)$ using the tools of inverse theory. This approach allows one to solve for the envelope function that fits the time-series data while optimizing some property of the envelope. The use of the term 'inverse theory' in this context can be misleading. Geophysicists differ over what comprises inverse theory, especially over how to deal with the fact that, in most examples, infinitely many models can fit a given finite data set. The approach of Tarantola (1987) seeks the model that is most consistent with *a priori* statistical information about model parameters. Others take a 'strict bounds' approach (e.g. Stark *et al.*, 1986), in which one seeks characteristics common to all models that satisfy a finite data set and a set of *a priori* constraints. The approach of this report is less sophisticated than either of the above examples, but closer in spirit to the former. Discrimination between different models often involves criteria specific to the application, so that the proper choice of *a priori* constraints is part of the research problem to be solved. We develop a flexible set of tools to test hypotheses about quasi-periodic signals in time series, in particular, to investigate the effect of possible discontinuous changes in amplitude and phase.

We use the techniques of multiple-taper spectrum analysis, as these offer spectrum estimators that are optimally band-limited. In the next section we derive inversion algorithms that find the 'smallest' and 'smoothest' envelopes that fit data from a given time series, and show results from synthetic test examples. We generalize the algorithms to model sudden changes in the envelope, and compensate for background 'noise' and data gaps. We show analyses of time series from both long-period seismograms and analysis of sediment cores. Section 10.3 examines seismic data from the great Macquarie Ridge earthquake of 23 May 1989. Section 10.4 examines quasi-periodic changes in Earth climate from the last 2.7 My as recorded by oxygen isotope fluctuations in an oceanic sediment core.

10.2 The inverse problem for an oscillation envelope

If $\bar{A}(f) \approx 0$ for $|f| > f_w$, it is natural to seek an expansion for $\bar{A}(f)$ in terms of a basis set of functions $W(f)$ over $[-f_w, f_w]$. A basis set of the discrete prolate spheroidal wavefunctions $W_k(f)$ (Thomson, 1982; Slepian, 1983) possesses the useful feature that the $W_k(f)$ are orthogonal both on the frequency interval $[-f_w, f_w]$ and the interval $[-f_N, f_N]$. The $W_k(f)$ are the discrete

Fourier transform pairs of time-domain data 'eigentapers' $w_n^{(k)}$, $n = 1, \ldots, N$:

$$W_k(f) = \sum_{n=1}^{N} w_n^{(k)} e^{2\pi i f n \Delta t}$$

$$w_n^{(k)} = \int_{-f_N}^{f_N} W_k(f) e^{2\pi i f n \Delta t} \, df = \lambda_k^{-1} \int_{-f_w}^{f_w} W_k(f) e^{2\pi i f n \Delta t} \, df \qquad (10.2)$$

and satisfy the condition that the functional

$$\lambda(W) = \frac{\int_{-f_w}^{f_w} |W(f)|^2 \, df}{\int_{-f_N}^{f_N} |W(f)|^2 \, df} \qquad (10.3)$$

is stationary. Larger values of the functional occur for functions that are concentrated in the narrow band $[-f_w, f_w]$. Note that $\lambda < 1$ if $f_w < f_N$, because any taper is time-limited, and therefore cannot have a totally band-limited spectrum (Slepian, 1983). The eigentapers satisfy an $N \times N$ matrix eigenvalue problem

$$\mathscr{A} \cdot \mathbf{w}_k = \lambda_k \mathbf{w}_k, \qquad (10.4)$$

where $\mathscr{A}_{nm} = 2\,\Delta t\, f_w \operatorname{sinc}(2\pi f_w (n-m)\,\Delta t)$ and $\mathbf{w}_k$ is an N-vector containing the kth eigentaper. From eq. (10.4) there are N orthogonal eigentapers for a time series with N data, with eigenvalues λ_k, equivalent to the stationary values of (10.3). We take the normalization convention $\mathbf{w}_k \cdot \mathbf{w}_l = \delta_{kl}$, so that $\mathbf{w}_k \cdot \mathscr{A} \cdot \mathbf{w}_k = \int_{-f_w}^{f_w} |\mathrm{W}_k(f)|^2 \, df = \lambda_k$, and order $1 > \lambda_0 > \lambda_1 > \lambda_2 > \cdots > \lambda_{N-1} \geqslant 0$. The eigenvalues λ_k measure the spectral leakage resistance of the eigentapers $w_n^{(k)}$. Eigentapers with $\lambda \approx 1$ can be used to construct spectrum estimates that are resistant to spectral leakage. If $f_w = pf_R$, where $f_R = 1/T = 1/(N\,\Delta t)$ is the Rayleigh frequency of the time series, the first $2p-1$ eigentapers have sufficient spectral leakage resistance to be useful in spectrum analysis. Tapers constructed for $f_w = pf_R$ (p need not be an integer) are called $p\pi$-prolate Slepian tapers (Figure 10.1), following the terminology of Thomson (1990). Methods for using Slepian tapers to construct general-purpose spectrum estimators are described by Thomson (1982), Park *et al.* (1987b), Walden (1990) and Thomson (1990).

If $A(f) = 0$ outside the narrow band $[-f_w, f_w]$, the behavior of $\bar{A}(f)$ that is resolvable with the N-point data series is an expansion of the prolate spheroidal wavefunctions:

$$\bar{A}(f) = \sum_{k=0}^{N-1} a_k W_k(f) \qquad (10.5)$$

for $f \leqslant f_w$, so that

$$A(n\,\Delta t) = A_n = \sum_{k=0}^{N-1} \int_{-f_w}^{f_w} a_k W_k(f) e^{-2\pi i f n \Delta t} = \sum_{k=0}^{N-1} \lambda_k a_k w_n^{(k)}. \qquad (10.6)$$

The coefficient a_k is related directly to the discrete Fourier transform (DFT)

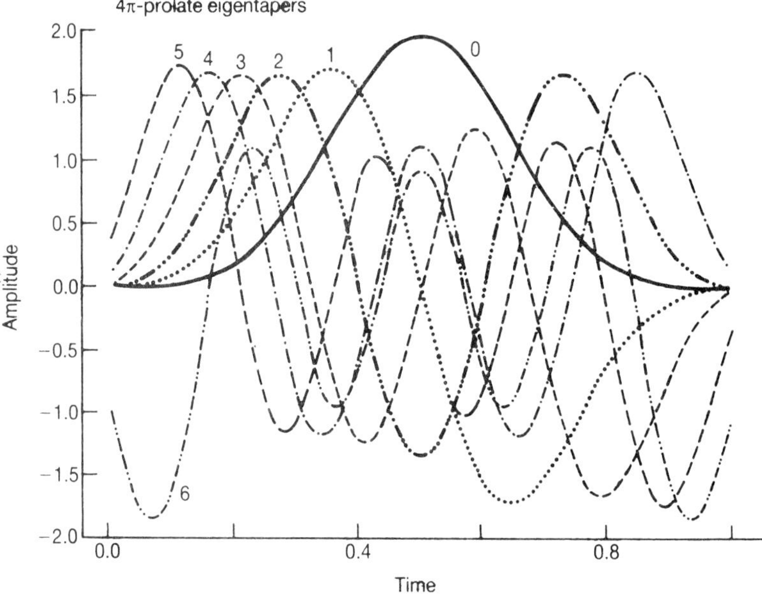

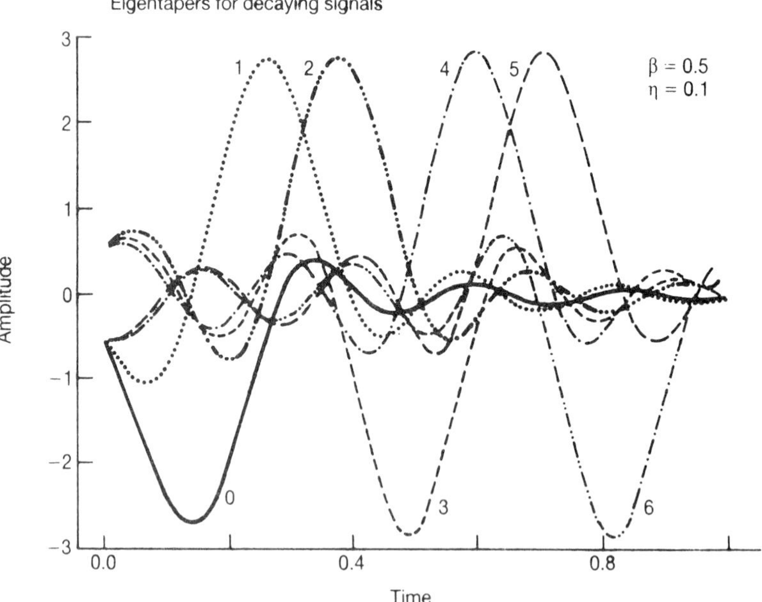

Figure 10.1 (a) The first seven 4π-prolate Slepian tapers. The tapers are constructed with time–bandwidth product $p = 4$, which corresponds to a half-width of $f_w = 4f_R = 4/T$ for the associated convolution kernels in the frequency domain. (b) The first seven Slepian tapers with time–bandwidth product $p = 4$, designed for signals that decay by $\beta = 0{\cdot}5Q$ cycles (a factor of $e^{-\pi/2}$) within a data series, and white-noise weighting parameter $\eta = 0{\cdot}1$.

of the data series $u_1, u_2, \ldots, u_N$ tapered by the kth Slepian taper $w_n^{(k)}$. The complex-valued DFT evaluated at f_0 is the kth eigenspectrum estimate $Y_k(f_0)$, where

$$Y_k(f_0) = \sum_{n=1}^{N} w_n^{(k)} u_n e^{2\pi i f_0 n \Delta t} = \sum_{n=1}^{N} w_n^{(k)} A_n = \lambda_k a_k. \tag{10.7}$$

If $\bar{A}(f) \neq 0$ outside $[-f_w, f_w]$, spectral leakage will bias values of a_k, especially for eigenspectra calculated with higher-order Slepian tapers with λ_k significantly less than unity. In practical situations one must truncate (10.5) at some $K \lessapprox 2p-1$ to minimize spectral leakage bias. For some chosen truncation K, we can estimate the envelope A_n with B_n, $n = 1, \ldots, N$, where

$$B_n = \sum_{k=0}^{K-1} \lambda_k^{-1} Y_k(f_0) w_n^{(k)}. \tag{10.8}$$

This is the multi-taper complex demodulate at the carrier frequency f_0 (Thomson, 1990). In this formula, the tapers $w_n^{(k)}$ are real-valued, while the eigenspectra $Y_k(f_0)$ and therefore the complex demodulate B_n are complex-valued. The proper choice of K will depend on the particular data series, as the analyst seeks to balance resistance to spectral leakage against obtaining a maximum number of data constraints. Thomson (1982) uses 4π-prolate Slepian tapers and chooses $K = 5 < 2p - 1 = 7$ eigentapers for multi-taper estimation of highly colored spectra. In most situations involving envelope estimation, however, a quasi-periodic signal in a data series attracts scrutiny because the signal appears as a prominent peak in the estimated Fourier spectrum. In this case, spectral leakage is less problematic and $K = 2p - 1$ is an appropriate choice. It is possible to downweight the influence of the higher-order eigenspectra without discarding them; this is discussed near the end of this section (eq. (10.21)).

Equation (10.8) is a valid model for the oscillation envelope A_n, but it is not the only valid model. For instance, a constant envelope $A_n \equiv A_0$ cannot be represented by any weighted sum of the first few Slepian tapers, but is often a most reasonable physical model. More flexibility is gained by treating the estimation of A_n as a standard linear inverse problem (Parker, 1977; Menke, 1984). Let $\mathbf{B} = (B_1, B_2, \ldots, B_N)$ be the 'model' for the oscillation envelope $\mathbf{A} = (A_1, A_2, \ldots, A_N)$. We have K 'data constraints'

$$Y_k(f_0) = \sum_{n=1}^{N} w_n^{(k)} A_n = \mathbf{g}_k \cdot \mathbf{A}, \tag{10.9}$$

which identify K data kernels $\mathbf{g}_k = \mathbf{w}_k$. It is a standard consequence of the projection theorem in inner-product spaces that the model $\mathbf{B}$ that fits the K data constraints while minimizing $\|\mathbf{B}\|^2 = \mathbf{B} \cdot \mathbf{B}$ is a linear combination of the data kernels $\mathbf{g}_k$. This minimum-size solution for the envelope is given by (10.8), a linear combination of the Slepian tapers. Define a data vector $\mathbf{y} = (Y_0(f_0), \ldots, Y_{K-1}(f_0))$, a coefficient vector $\mathbf{a} = (a_0, \ldots, a_{K-1})$ and a $K \times N$ kernel matrix $\mathbf{G}$ whose rows are the data kernels $\mathbf{g}_k$. If we form a Gram matrix $\mathbf{\Gamma} = \mathbf{G} \cdot \mathbf{G}^{\mathrm{T}}$, where $\Gamma_{kl} = \mathbf{g}_k \cdot \mathbf{g}_l$, the formal solution for this expansion is $\mathbf{a} = \mathbf{\Gamma}^{-1} \cdot \mathbf{y}$. In the minimum-size solution (10.8), the orthogonality of the $\mathbf{g}_k$ renders the Gram matrix $\mathbf{\Gamma}$ diagonal.

Once the linear inverse problem is formulated, other solutions $\mathbf{B}$ can be found that minimize other quadratic functionals. In addition, one can solve for an envelope that deviates least from some preferred model $\mathbf{w}^{(0)}$, e.g. a constant envelope, or an exponentially decaying envelope. We can construct an envelope solution of the form

$$\mathbf{B} = b\mathbf{w}^{(0)} + \sum_{k=0}^{K-1} a_k \mathbf{g}_k = b\mathbf{w}^{(0)} + \mathbf{G}^{\mathrm{T}} \cdot \mathbf{a}, \tag{10.10}$$

where b and the K-vector $\mathbf{a}$, both complex-valued, are to be determined. A solution of this form obtains oscillation envelopes that are closest to a scalar multiple of $\mathbf{w}^{(0)}$ in the sense of minimizing $\|\mathbf{B} - b\mathbf{w}^{(0)}\|^2$. To obtain a unique solution we specify that the terms on the right sides of (10.10) are mutually orthogonal, leading to the equations

$$\begin{aligned} b\mathbf{z} + \mathbf{\Gamma} \cdot \mathbf{a} &= \mathbf{y} \\ \mathbf{z} \cdot \mathbf{a} &= 0, \end{aligned} \tag{10.11}$$

where $z_k = \mathbf{g}_k \cdot \mathbf{w}^{(0)}$. Multiply the first equation by $\mathbf{z} \cdot \mathbf{\Gamma}^{-1}$ to obtain the formal solution

$$\begin{aligned} b &= (\mathbf{z} \cdot \mathbf{\Gamma}^{-1} \cdot \mathbf{y})/(\mathbf{z} \cdot \mathbf{\Gamma}^{-1} \cdot \mathbf{z}) \\ \mathbf{a} &= \mathbf{\Gamma}^{-1} \cdot (\mathbf{y} - b\mathbf{z}). \end{aligned} \tag{10.12}$$

Numerical efficiency can be gained by avoiding the calculation of $\mathbf{\Gamma}$ and its inverse by using the QR decomposition (Lawson and Hanson, 1974) of the kernel matrix $\mathbf{G}$. Let $\mathbf{G}^{\mathrm{T}} = \mathbf{Q} \cdot \mathbf{R}$, where $\mathbf{Q}$ is an $N \times K$ matrix with orthonormal columns and $\mathbf{R}$ is a $K \times K$ upper triangular matrix. The $K \times K$ identity matrix $\mathbf{I} = \mathbf{Q}^{\mathrm{T}} \cdot \mathbf{Q}$, and the $N \times N$ matrix $\mathbf{Q} \cdot \mathbf{Q}^{\mathrm{T}}$ is an orthogonal projection matrix from R^N onto the subspace spanned by the kernel vectors $\mathbf{g}_k$. After some manipulation, the solution for an oscillation envelope of the form (10.10) is

$$\mathbf{B} = b\mathbf{w}^{(0)} + \mathbf{Q} \cdot ((\mathbf{R}^{\mathrm{T}})^{-1} \cdot \mathbf{y} - b\mathbf{Q}^{\mathrm{T}} \cdot \mathbf{w}^{(0)}), \tag{10.13}$$

where $b = (\mathbf{w}^{(0)} \cdot \mathbf{Q} \cdot (\mathbf{R}^{\mathrm{T}})^{-1} \cdot \mathbf{y})/(\mathbf{w}^{(0)} \cdot \mathbf{w}^{(0)})$. Since $\mathbf{R}$ is upper triangular, $(\mathbf{R}^{\mathrm{T}})^{-1} \cdot \mathbf{y}$ can be calculated efficiently by back-substitution.

To obtain smooth envelopes consistent with measured eigenspectra we use a variation of Occam's inversion (Constable *et al.*, 1987) suggested by Parker (1990). Consider the approximate first-difference operator $\mathbf{D} = D_{ij}$, an $N \times N$ matrix with $D_{11} = \varepsilon \ll 1$, $D_{ii} = 1$ for $i > 1$, $D_{i-1,i} = -1$ and $D_{ij} = 0$ otherwise. The inverse of this matrix is a crude integration operator $\mathbf{D}^{-1}$, where $D_{1i}^{-1} = \varepsilon^{-1}$, $D_{ij}^{-1} = 1$ for $1 < i \leqslant j$, and $D_{ij}^{-1} = 0$ for $j < i$. The data constraints (10.9) can be rewritten as

$$Y_k(f_0) = \mathbf{w}_k \cdot \mathbf{D}^{-2} \cdot \mathbf{D}^2 \cdot \mathbf{A} = \tilde{\mathbf{g}}_k \cdot \tilde{\mathbf{A}}, \tag{10.14}$$

where $\tilde{\mathbf{g}}_k = \mathbf{w}_k \cdot \mathbf{D}^{-2}$ and $\tilde{\mathbf{A}} = \mathbf{D}^2 \cdot \mathbf{A}$. An envelope $\mathbf{B}$ that satisfies the 'data constraints' while minimizing

$$\|\mathbf{D}^2 \cdot \mathbf{B}\|^2 = \sum_{n=2}^{N-1} (B_{n+1} - 2B_n + B_{n-1})^2 + (B_2 - (1+\varepsilon)B_1)^2 + \varepsilon^4 B_1^2 \tag{10.15}$$

can be constructed from $\mathbf{D}^{-2} \cdot \tilde{\mathbf{B}}$, where $\tilde{\mathbf{B}}$ is a linear combination of the $\tilde{\mathbf{g}}_k$. The first term on the right-hand side of (10.15) measures the roughness of the envelope. The extra terms in (10.15) complicate the interpretation of $\mathbf{B} = \mathbf{D}^{-2} \cdot \Sigma_{k=0}^{K-1} \tilde{a}_k \tilde{\mathbf{g}}_k$ as the 'smoothest' model for the oscillation envelope $\mathbf{A}$. The extra terms constrain envelope models for which $\Sigma_{n=2}^{N-1} (B_{n+1} - 2B_n + B_{n-1})^2 = 0$, e.g. constant vectors and linear ramps. In numerical tests (performed with 8-byte floating-point-precision VAX/VMS Fortran) their influence on the solution roughness appears negligible for $N > 100$ data points and $\varepsilon \lessapprox 10^{-4}$. However, the sensitivity to ε may depend somewhat on the computer precision and on the values of N, K and p.

An alternative approach is seminorm minimization (Parker *et al.*, 1987), which can solve for constant and linear-ramp envelopes explicitly, while applying a roughness penalty on the remainder of the solution. We adopt one aspect of the seminorm approach and solve for a 'fixed element' $\mathbf{w}^{(0)}$ as well as a solution that minimizes (10.15). We can construct an estimate for $\tilde{\mathbf{A}}$ of the form

$$\tilde{\mathbf{B}} = \tilde{b}\tilde{\mathbf{w}}^{(0)} + \sum_{k=0}^{K-1} \tilde{a}_k \tilde{\mathbf{g}}_k = \tilde{b}\tilde{\mathbf{w}}^{(0)} + \tilde{\mathbf{G}}^{\mathrm{T}} \cdot \tilde{\mathbf{a}}, \tag{10.16}$$

where $\tilde{\mathbf{w}}^{(0)} = \mathbf{D}^2 \cdot \mathbf{w}^{(0)}$, $\tilde{\mathbf{G}}$ is a $K \times N$ matrix whose kth row is $\tilde{\mathbf{g}}_{k-1}$ and $\tilde{b}$, $\tilde{\mathbf{a}}$ are to be determined. The oscillation envelope $\mathbf{B} = \mathbf{D}^{-2} \cdot \tilde{\mathbf{B}}$ is the solution closest to a scalar multiple of $\mathbf{w}^{(0)}$ in the sense of minimizing $\|\mathbf{D}^2 \cdot (\mathbf{B} - \tilde{b}\mathbf{w}^{(0)})\|^2$. The 'smooth' data kernels $\tilde{\mathbf{g}}_k$ are not orthogonal, and the associated Gram matrix $\tilde{\mathbf{\Gamma}}$ is full, though symmetric. Since the 'data' constraints can be written as $\tilde{\mathbf{G}} \cdot \tilde{\mathbf{B}} = \mathbf{y}$, formulae for $\tilde{b}$, $\tilde{\mathbf{a}}$ and $\tilde{B}$ are straightforward generalizations of (10.11)–(10.13).

Four generalizations of the above development are useful:

(1) In many physical processes an oscillation suffers a sudden shift in amplitude and phase, but is otherwise slowly varying. In long-period seismology, this would occur if a second seismic event (e.g. an aftershock) re-excited the free oscillations. The first-difference matrix $\mathbf{D}$ can be partitioned at a chosen breakpoint $\tilde{N} < N$:

$$\tilde{\mathbf{D}} = \begin{bmatrix} \mathbf{D}^{(\tilde{N})} & \mathbf{0} \\ \mathbf{0} & \mathbf{D}^{(N-\tilde{N})} \end{bmatrix}, \tag{10.17}$$

where $\mathbf{D}^{(M)}$ is an $M \times M$ first-difference matrix. This partitioned matrix and its inverse can be used to set up the linear inverse problems, similar to (10.11), for envelopes that can have discontinuities at the chosen breakpoint, but are constrained to be smooth elsewhere.

(2) Seismic data records often have segments of bad or missing data due to instrument failure or nonlinearity. Therefore edited data records used for analysis often have gaps. The effect of gaps on data spectrum estimates can be modeled by introducing identical gaps in the data tapers $\mathbf{w}_k$. An envelope estimate derived from (10.16) will interpolate the gap with a sum of a third-order polynomial and the fixed element $\mathbf{w}^{(0)}$.

(3) The third generalization applies when the desired carrier frequency f_0

does not equal one of the discrete frequencies of the numerical fast Fourier transform. If we choose eigenspectra $Y(\hat{f}_0)$ at a nearby FFT frequency $\hat{f}_0$ for analysis, the K real-valued data kernels $\mathbf{g}_k$ in (10.9) become complex-valued vectors and eq. (10.7) becomes

$$Y_k(\hat{f}_0) = \sum_{n=1}^{N} w_n^{(k)} u_n e^{2\pi i \hat{f}_0 n \Delta t} = \sum_{n=1}^{N} w_n^{(k)} A_n e^{2\pi i(\hat{f}_0 - f_0) n \Delta t} = \mathbf{g}_k \cdot \mathbf{A}, \tag{10.18}$$

where $(\mathbf{g}_k)_n = w_n^{(k)} e^{2\pi i(\hat{f}_0 - f_0) n \Delta t}$. The Gram matrix $\mathbf{\Gamma}$ becomes complex-Hermitian rather than real-symmetric, with $\Gamma_{kl} = \mathbf{g}_k \cdot \mathbf{g}_l^*$ (the asterisk denotes complex conjugation), and the QR decomposition involves complex-valued matrices. Otherwise the solution algorithm is similar.

(4) Park *et al.* (1987a) and Lindberg and Park (1987) extended the definition of the bandwidth concentration functional (10.4) to design Slepian tapers for decaying signals in the presence of stationary noise. These tapers satisfy the generalized eigenvalue problem

$$\tilde{\mathscr{A}} \cdot \mathbf{w}_k = \tilde{\lambda}_k \tilde{\mathscr{C}} \cdot \mathbf{w}_k, \tag{10.19}$$

where $\tilde{\mathscr{A}}_{nm} = 2\,\Delta t\, f_w \operatorname{sinc}(2\pi f_w (n-m)\,\Delta t) e^{-\alpha(n+m)\Delta t}$ and $\tilde{\mathscr{C}}_{nm} = (e^{-\alpha(n+m)\Delta t} + \eta)\delta_{nm}$. The factor α is the decay rate of the oscillation and η is a weighting parameter that scales the sensitivity of the tapers to stationary noise. The amplitude of an oscillation with quality factor Q will decay by a factor of $e^{-\beta\pi}$ after βQ cycles. Tapers are constructed for chosen parameters η and a total decay of βQ oscillation cycles. If background noise is ignored, $\eta = 0$, and the optimal tapers will be the stationary-signal data tapers (Figure 10.1(a)) multiplied by an increasing exponential. Examples of 4π-prolate tapers designed for $\beta = 0{\cdot}5$ and $\eta = 0{\cdot}1$ are shown in Figure 10.1(b). Additional help in modeling envelopes that suffer exponential decay can be obtained by removing βQ-cycles of exponential growth from the specialized tapers before constructing the data kernels $\mathbf{g}_k$. Formally, this involves a renormalization of the data constraints similar to (10.14). Define an $N \times N$ matrix $\mathbf{E} = \operatorname{diag}\{e^{\beta\pi/N}, e^{2\beta\pi/N}, \ldots, e^{\beta\pi}\}$ and rewrite the data constraints $Y_k(f_0) = \mathbf{w}_k \cdot \mathbf{E}^{-1} \cdot \mathbf{E} \cdot \mathbf{A} = \tilde{\mathbf{g}}_k \cdot \tilde{\mathbf{A}}$ and modify the roughness penalty function (10.15) accordingly. If one solves for $\mathbf{E} \cdot \mathbf{B}$ rather than $\mathbf{B}$ (in practice this is achieved by skipping the last renormalization of the inverse solution $\tilde{\mathbf{B}}$ from (10.16)), we obtain a model for the envelope of the quasi-periodic signal with the decay removed.

We show two synthetic examples to contrast different approaches for estimating an oscillation envelope $A(t)$. We compare solutions of the form (10.10) and (10.16), that is, size versus roughness constraints. Figure 10.2 shows two test oscillations with $f_0 = \frac{1}{6}$ mHz. For test series A both the amplitude and phase of $A(t)$ vary as a second-degree polynomial in t. The second test series (B) suffers an abrupt change in amplitude and phase at its midpoint, but varies smoothly elsewhere. We construct solutions using seven 4π-prolate Slepian tapers $\mathbf{w}_k$ and a constant amplitude fixed element $\mathbf{w}^{(0)}$. Figure 10.3 graphs the modeled amplitude and phase of $A(t)$ against the

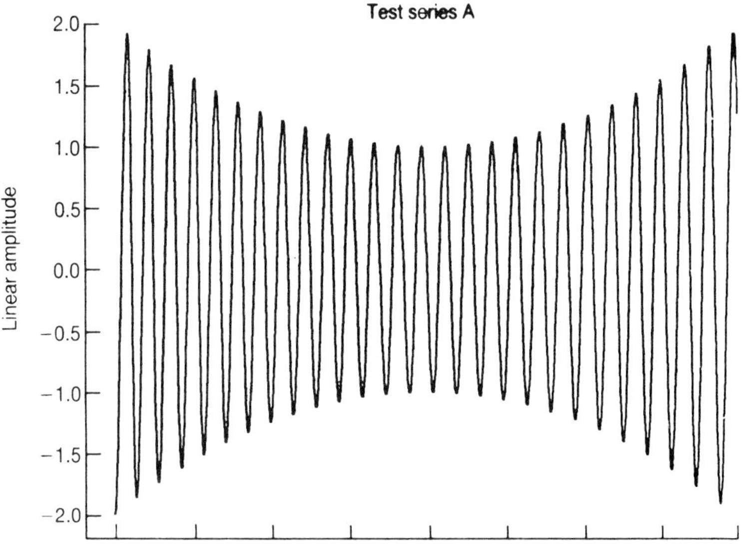

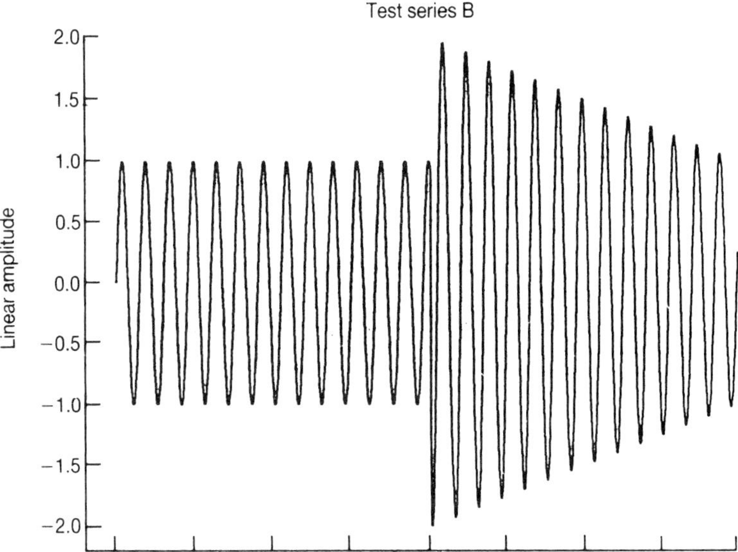

Figure 10.2 Synthetic test series for envelope estimation.

theoretical values for the first synthetic. The smooth model $\mathbf{D}^{-2} \cdot \tilde{\mathbf{B}}$ for the envelope is more successful at following the smooth variation in $A(t)$. Figure 10.4 graphs three models for the envelope of the second synthetic. In addition to the two solutions (10.10) and (10.16), a third model partitions $\mathbf{D}$ at the midpoint of the time series, allowing a discontinuity. All models detect

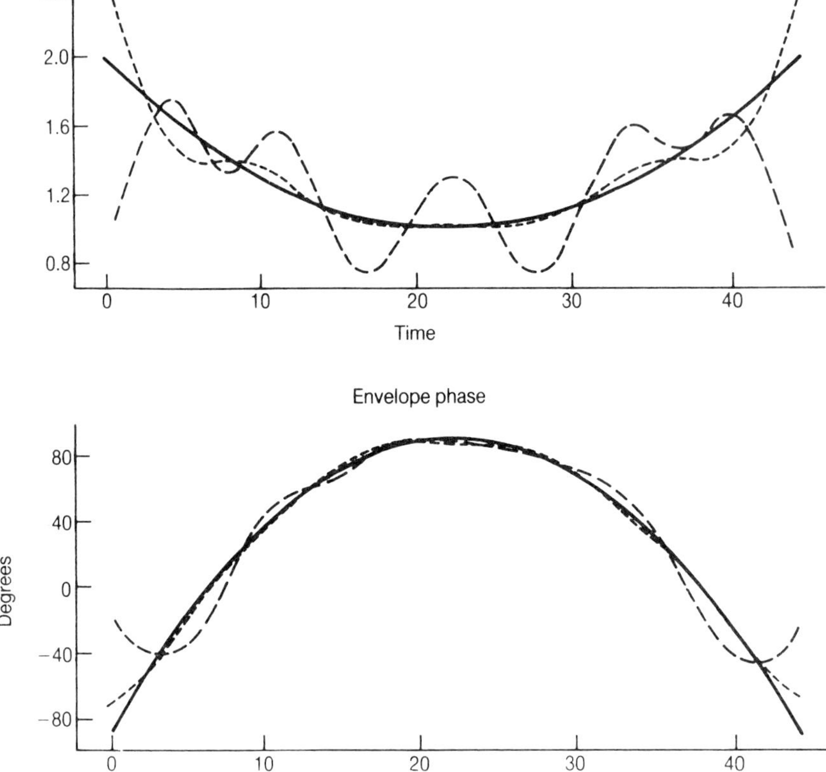

Figure 10.3 Envelope estimators for test series A, using eigenspectra computed with the seven 4π-prolate tapers shown in Figure 10.1. The solid line is the exact amplitude and phase. The coarse-dashed line is a solution of form (10.10), the envelope that is the smallest deviation from constant amplitude and phase. The fine-dashed line is a solution of form (10.16), the envelope that is the smoothest deviation from constant amplitude and phase.

a rapid change in phase, but only the inversion that allows a breakpoint follows well the abrupt change in $|A(t)|$. An incorrect choice of breakpoint leads to much more complicated models (Figure 10.5).

All of the above solutions fit the 'data' (the eigenspectra $Y_k(f_0)$) within numerical precision. Infinitely many envelopes will fit the 'data' equally well, because the inverse problem for $A(t)$ is non-unique. A large proportion of this infinity of models, however, corresponds to $\bar{A}(f)$ with large amplitude outside $[-f_w, f_w]$. In practical situations it is prudent to use for interpretation the smoothest model that fits the data, for it suppresses features that are not required by the data, and leads to models that are reasonably band-limited.

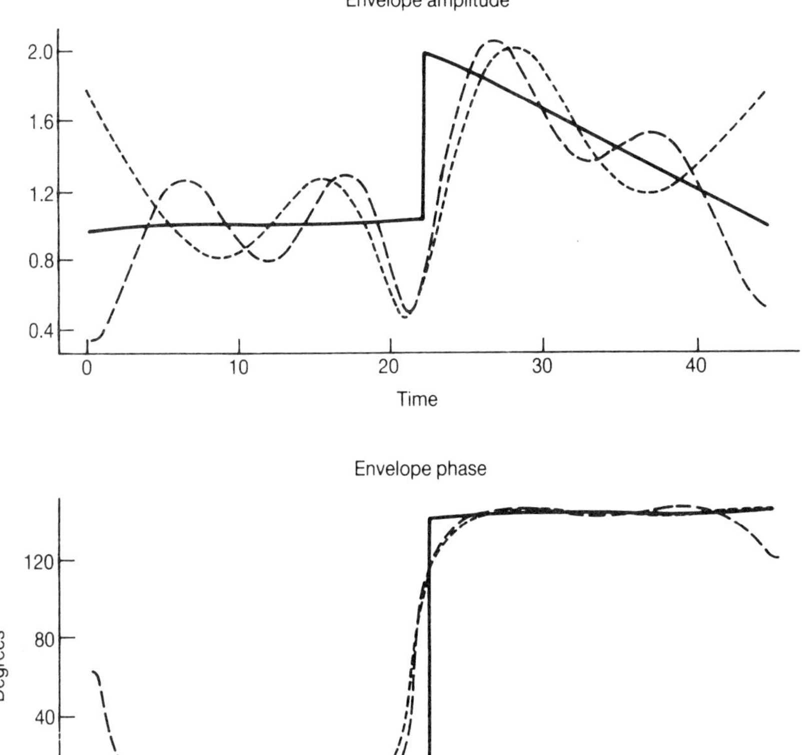

Figure 10.4 Envelope estimators for test series B, using eigenspectra computed with the seven 4π-prolate tapers shown in Figure 10.1. The coarse-dashed line is a solution of form (10.10), the envelope that is the smallest deviation from constant amplitude and phase. The fine-dashed line is a solution of form (10.16), the envelope that is the smoothest deviation from constant amplitude and phase. The solid line is a solution of form (10.16), but with a discontinuity allowed at the series midpoint.

Unusual behavior, such as abrupt changes in amplitude and phase, can suggest rapid change or discontinuities in the envelope.

Two further extensions of the above treatment are useful. The first involves situations where two carrier frequencies f_1, f_2, with independent slowly varying envelopes $\mathbf{A}_1, \mathbf{A}_2$, are an appropriate model for the data. If $|f_1 - f_2| > 2pf_R$, there is minimal correlation between eigenspectra calculated using $p\pi$-prolate Slepian tapers, and the envelopes can be modeled separately. If the two carrier frequencies are more closely spaced, the envelopes should be modeled simultaneously to avoid bias. Formally, for K tapers and an N-point data series, one augments the data vector $\mathbf{y} =$

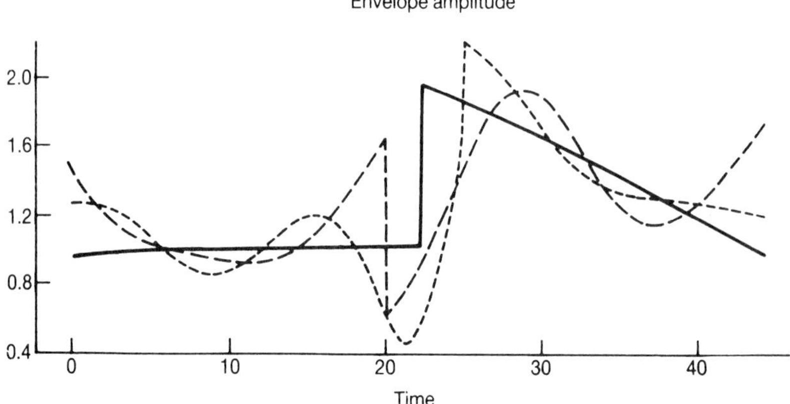

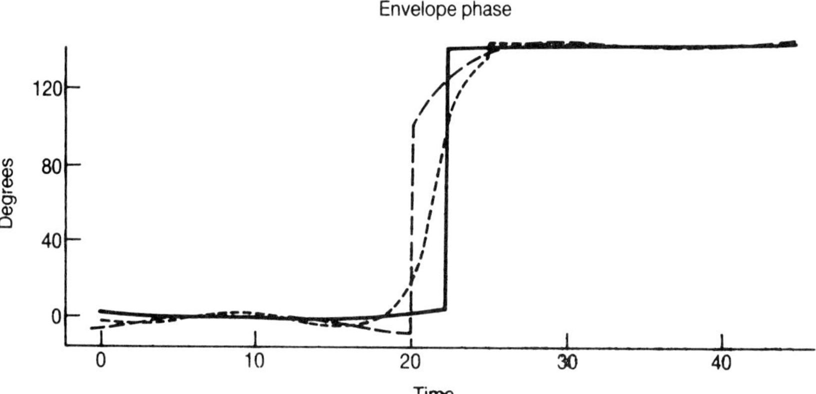

Figure 10.5 The effect of an incorrect choice of envelope discontinuity. The solid line is a solution of form (10.16) for test series B from Figure 10.4. The dashed lines are a solution of form (10.16), but with discontinuities allowed at points in the series away from the series midpoint.

$(Y_0(\hat{f}_1), \ldots, Y_{K-1}(\hat{f}_1), Y_0(\hat{f}_2), \ldots, Y_{K-1}(\hat{f}_2))$ into a complex-valued $2K$-vector, where $\hat{f}_1, \hat{f}_2$ are FFT frequencies near f_1 and f_2, respectively. One solves for a $2N$-vector $\mathbf{B} = (\mathbf{B}_1, \mathbf{B}_2)$, where $\mathbf{B}_1$ and $\mathbf{B}_2$ are estimates of $\mathbf{A}_1$ and $\mathbf{A}_2$, respectively. The first K data kernels are $2N$-vectors

$$\mathbf{g}_k = (\mathbf{w}_1^{(k)} e^{2\pi i(\hat{f}_1 - f_1)\Delta t}, \ldots, w_N^{(k)} e^{2\pi i(\hat{f}_1 - f_1)N\,\Delta t}, w_1^{(k)} e^{2\pi i(\hat{f}_2 - f_1)\Delta t}, \ldots, w_N^{(k)} e^{2\pi i(\hat{f}_2 - f_1)N\,\Delta t}), \tag{10.20a}$$

and the second K data kernels are

$$\mathbf{g}_{K+k} = (w_1^{(k)} e^{2\pi i(\hat{f}_1 - f_2)\Delta t}, \ldots, w_N^{(k)} e^{2\pi i(\hat{f}_1 - f_2)N\,\Delta t}, w_1^{(k)} e^{2\pi i(\hat{f}_2 - f_2)\Delta t}, \ldots, w_N^{(k)} e^{2\pi i(\hat{f}_2 - f_2)-\Delta t}). \tag{10.20b}$$

Solutions can be obtained in a manner similar to a single-envelope inversion, with the proviso that the first-difference matrix $\mathbf{D}$ used to obtain smooth solutions must be partitioned at the midpoint of the solution vector to avoid an unwanted correlation between the endpoints of the envelope estimates $\mathbf{B}_1$ and $\mathbf{B}_2$. Note that fixed elements $\mathbf{w}_0$ used in the inversion are also $2N$-vectors in a form similar to (10.20). Care should be exercised in interpreting double-envelope inversions, because the algorithm may associate an envelope fluctuation with the wrong carrier frequency in order to satisfy the optimization condition. Roughly speaking, the likelihood of misinterpretation varies inversely with the spacing $|f_1 - f_2|$. Narrow frequency spacings also can lead to poorly conditioned kernel matrices $\mathbf{G}$.

The second extension is useful if background noise is known (or is suspected) to contribute significantly to the eigenspectrum estimates $Y_k(f)$ used as data in the above inversion schemes. A standard tactic is to allow a specified misfit to the data $S^2 = \Sigma_k(Y_k(\hat{f}_0) - \tilde{Y}_k)^2$, where $\tilde{Y}_k$ is the kth model prediction, while minimizing the penalty functional $\|\mathbf{B}\|^2$ or (10.15). Details of this method are given by Shure *et al.* (1982), in which the constrained optimization is set up with a Lagrange multiplier μ. The effect of a locally white background is roughly constant for eigenspectra for leakage-resistant Slepian tapers, so that the data misfits can be assumed to have equal variance to first order. The formal solution for real-valued kernels $\mathbf{g}_k$ is a modification of (10.12):

$$b = (\mathbf{z} \cdot \mathbf{\Gamma}^{-1} \cdot \mathbf{y})/(\mathbf{z} \cdot \mathbf{\Gamma}^{-1} \cdot \mathbf{z})$$

$$\mathbf{a} = (\mathbf{\Gamma} + \mu \mathbf{I})^{-1} \cdot (\mathbf{y} - b\mathbf{z}) \tag{10.21a}$$

subject to the constraint

$$\|\mathbf{y} - b\mathbf{z} - \mathbf{\Gamma} \cdot \mathbf{a}\|^2 = F(\mu) = S^2. \tag{10.21b}$$

The misfit variance S^2 can be chosen to represent the background noise level. In this manner the DFT of the data series near f_0 would be modeled as a quasi-periodic 'signal' embedded in a locally white background process. The value of μ consistent with (10.21) is found interatively, using the singular value decomposition of the kernel matrix $\mathbf{G} = \mathbf{V} \cdot \mathbf{\Psi} \cdot \mathbf{U}^{\mathrm{T}}$ (Lawson and Hanson, 1974). $\mathbf{V}$ is a $K \times K$ matrix whose columns are the orthonormal left-eigenvectors $\hat{\mathbf{v}}_k$ of $\mathbf{G}$, $\mathbf{U}$ is an $N \times K$ matrix whose columns are the orthonormal right-eigenvectors $\hat{\mathbf{u}}_k$ of $\mathbf{G}$, and $\mathbf{\Psi} = \mathrm{diag}\{\psi_1, \ldots, \psi_K\}$ contains the singular values of $\mathbf{G}$. The residual data vector can be decomposed $\mathbf{y} - b\mathbf{z} = \Sigma_k \mathcal{Y}_k \hat{\mathbf{v}}_k$ and the solution vector expressed as $\mathbf{a} = \Sigma_k a_k \hat{\mathbf{u}}_k$, where $a_k = \mathcal{Y}_k \psi_k/(\psi_k^2 + \mu)$. Using this decomposition, $F(\mu) = \Sigma_k \mu^2 \mathcal{Y}_k^2/(\psi_k^2 + \mu)^2$, and the iterative solution of (10.21) using Newton's method can be performed using scalar functions. If the roughness constraint (10.15) is used as a penalty functional, the contribution of the more-oscillatory higher-order Slepian tapers to the solution will be diminished relative to that of the low-order tapers. The extension to complex-valued kernels $\mathbf{g}_k$ is straightforward.

The implementation of the above algorithms on a computer relies heavily on routines from the double-precision-complex version of the LINPACK library (Dongarra *et al.*, 1979). Routines ZQRDC and ZSVDC calculate and manipulate the QR and SVD decompositions, respectively, of a general rectangular complex-valued matrix. The Slepian tapers $\mathbf{w}_k$ can be calculated directly from (10.4), but a more efficient method uses the fact that the tapers are also the eigenvectors of a symmetric tridiagonal matrix given by a recursion relation in eq. (10.14) of Slepian (1978). The computational burden can be decreased significantly by decimating the tapers from N to N' points and dividing the eigenspectra $Y_k(f_0)$ by the decimation factor N/N' in order to normalize the data constraints (10.9). The inaccuracy introduced by decimation is small, only affecting the solution if the kernel matrix $\mathbf{G}$ is very poorly conditioned.

10.3 Envelopes of coupled long-period seismic oscillations

The major portion of research in seismology involves the analysis of the first few wave arrivals that travel from the seismic source (earthquakes, explosions, volcanic eruptions) to the seismic receiver through the intervening Earth. These signals appear principally as transient broadband pulses – although many localities suffer extended shaking due to natural resonances in the underlying surface sediments. If the seismic source is large enough,* the ground motion excited by the source persists above the noise level long enough to enable observation of the long-period (3600 s $> T >$ 100 s) natural resonances of the whole Earth. The spectral peaks of the whole Earth's seismic free oscillations have been studied by a variety of time-series techniques (e.g. Gilbert and Dziewonski, 1975; Buland *et al.*, 1977; Buland and Gilbert, 1978; Bolt and Brillinger, 1979; Geller and Stein, 1979; Chao and Gilbert, 1980; Dahlen, 1982; Hansen, 1982a, b; Masters and Gilbert, 1983; Giardini *et al.*, 1987; Ritzwoller *et al.*, 1986; Park *et al.*, 1987a; Lindberg and Park, 1987; Hori *et al.*, 1989; Lindberg and Thomson, 1990; Chao, 1990) – a short history of the subject can be found in Lindberg (1986).

When analysing the Earth's free oscillations, seismologists are aided by the fact that the Earth's interior is nearly 'spherical', that is, its internal properties, expressed as a function of spherical polar coordinates r, θ, ϕ, depend dominantly on the radial coordinate r. Dependence on the angular variables θ and ϕ, associated with the Earth's hydrostatic ellipticity-of-figure and lateral structure in seismic velocities (compressional and shear), density and internal friction, affect the Earth's long-period free oscillations weakly. The oscillations associated with spectral peaks in Fourier-transformed seismic data can be correlated with the oscillations of an idealized, non-rotating, spherically averaged, elastic Earth model. The equations of motion for such a body separate into distinct systems of equations that govern spheroidal and toroidal motions in the Earth. Each vibrational degree of freedom is a free oscillation 'singlet' whose angular dependence can be derived from a single fully normalized spherical harmonic $Y_l^m(\theta, \phi)$. The

*Typically, Richter-scale magnitude $M \gtrsim 6{\cdot}5$.

$2l+1$ singlets with azimuthal orders $m=-l, -l+1, \ldots, l-1, l$ that share angular degree l and the same dependence on radius r form a 'multiplet' with degenerate frequency f. Spheroidal-motion multiplets are denoted by ${}_nS_l$, where n indicates the radial overtone number. Toroidal-motion multiplets are denoted by ${}_nT_l$. Using Rayleigh's variational principle (Aki and Richards, 1980), a model of weak spherically averaged anelasticity can be used to calculate the quality factor, or Q, of each multiplet oscillation. Each vibrational singlet is represented by a vector-valued function $\mathbf{s}_k(\mathbf{r})$, defined at positions $\mathbf{r}$ within the Earth, which describes the particle displacement associated with the oscillation. The response $\mathbf{u}(\mathbf{r}, t)$ of a spherically symmetric, nonrotating Earth to a seismic event at $t=0$ with step-function time dependence can be expressed as an infinite sum of vibrational free oscillations $s_k(\mathbf{r})$ (Gilbert and Dziewonski, 1975),

$$\mathbf{u}(\mathbf{r}, t) = \sum_{k=1}^{\infty} \mathbf{s}_k(\mathbf{r})\aleph_k(\cos \omega_k t e^{-\alpha_k t} - 1)H(t) \qquad (10.22)$$

where $\aleph_k$ is the excitation scalar for the kth singlet, $\omega_k = 2\pi f_k$ is the radial frequency of oscillation, $H(t)$ is the stepfunction and $\alpha_k = \omega_k/2Q_k$, where Q_k is the quality factor of the kth singlet. If we neglect internal-wave motions in the liquid outer core, the frequencies $\omega_k \rightarrow \infty$ as $k \rightarrow \infty$.

The effect of deviations from a spherical reference model on the free oscillations can be treated with perturbation theory – see Park (1988) for a review. On a weakly aspherical, slowly rotating Earth, the singlets of a spherical reference Earth model couple and hybridize, and their frequencies split into a cluster in narrow bands in the neighborhood of the spherical-Earth multiplet frequencies. An analogy can be made between coupled-mode singlet frequencies and splitting of the electron energy states for atoms and molecules, as the matrix equations that govern splitting have similar structure. For some multiplets, the individual singlet oscillations are resolvable and can be 'stacked' using a global array (e.g. Buland *et al.*, 1977; Ritzwoller *et al.*, 1986). Most often the singlets are unresolvable and a free oscillation multiplet is modeled by $u(t) = \Re\{A(t)e^{-2\pi i f_0 t}\}$, where $A(t)$ is a slowly varying complex-valued envelope of a cosinusoid with carrier frequency f_0. The carrier frequency f_0 is typically, but not always, the spherical-Earth frequency of the multiplet.

The slow variation of $A(t)$ is equivalent to narrow-band concentration of energy in its Fourier transform $\tilde{A}(f)$ about $f=0$. Attempts to extract information from $A(t)$ pertaining to Earth asphericity have taken the form of narrow-band integrals of $\tilde{A}(f)$ (the 'multiplet moments' of Jordan (1978)) and Born-series expansions of $A(t)$ (Woodhouse, 1983; Smith and Masters, 1989). However, the potential for complex coupling interaction, such as that between spheroidal and toroidal multiplets due to rotation (Masters *et al.*, 1983; Park, 1986), makes a direct estimate of $A(t)$ useful in some situations. Narrow-band filtering is an option for retrieving $A(t)$ from long-period seismic records, e.g. Geller and Stein (1979), Hansen (1982a, b). However, an extremely long filter would be necessary to isolate the behavior of most free oscillation spectral peaks, making the onset of the oscillation difficult to resolve.

It is conventional to distinguish 'self-coupling' among the $2l+1$ degenerate singlets within an isolated multiplet ${}_nS_l$ or ${}_nT_l$ and the more general coupling between singlets belonging to distinct multiplets. The perturbation of the initial amplitude and phase of the multiplet envelope $A(t)$ caused by self-coupling is negligible. The coupling of a multiplet with nearby multiplets will perturb its initial amplitude (Park, 1987), and coupling with multiplets that possess different Q will perturb the initial phase of the complex-valued oscillation envelope. To see this, consider the coupling of two oscillation singlets $\mathbf{s}_1, \mathbf{s}_2$ with radial frequencies ω_1, ω_2, quality factors Q_1 and Q_2 and cross-coupling interaction term ε. Assume the oscillations are unit-normalized, ε is real-valued, and neglect other coupling interaction to simplify the algebra. The hybrid singlets and their oscillation frequencies are found from the eigenvectors and eigenvalues of the matrix

$$\begin{bmatrix} \omega_1^2(1-\mathrm{i}Q_1^{-1}) & \varepsilon \\ \varepsilon & \omega_2^2(1-\mathrm{i}Q_2^{-1}) \end{bmatrix}. \tag{10.23}$$

If we define

$$\begin{aligned} \Lambda &= (\omega_1^2(1-\mathrm{i}Q_1^{-1})+\omega_2^2(1-\mathrm{i}Q_2^{-1}))/2 \\ \gamma &= (\omega_1^2(1-\mathrm{i}Q_1^{-1})-\omega_2^2(1-\mathrm{i}Q_2^{-1}))/2\varepsilon, \end{aligned} \tag{10.24}$$

the hybrid eigenfrequencies satisfy $\tilde{\omega}_\pm^2 = \Lambda \pm \sqrt{(\gamma^2+1)}$. The hybrid singlets and excitation scalars are

$$\begin{aligned} \tilde{\mathbf{s}}_\pm &= \mathbf{s}_1 + (\gamma \pm \sqrt{(\gamma^2+1)})\mathbf{s}_2 \\ \tilde{\aleph}_\pm &= (\aleph_1 + (\gamma \pm \sqrt{(\gamma^2+1)})\aleph_2)/\eta_\pm, \end{aligned} \tag{10.25}$$

where $\eta_\pm = 2(\gamma^2+1 \pm \gamma\sqrt{(\gamma^2+1)})$. In general, the hybrid singlets $\tilde{\mathbf{s}}_\pm$ will be excited with initial phase $\phi_0 \neq 0^\circ$ or 180°. If the frequency difference $\Re\,\Delta\tilde{\omega} = \Re(\tilde{\omega}_1 - \tilde{\omega}_2)$ is too small to resolve with a given seismic record length, the observed oscillation envelope is the sum of $\tilde{\mathbf{s}}_1$ and $\tilde{\mathbf{s}}_2$. The amplitude $|A(t)|$ of the envelope will beat with period $2\pi/\Re\,\Delta\tilde{\omega}$. The beating will weaken with time as the oscillation with lower Q decays relative to its partner. A beating envelope may be observed even if coupling is absent, if both $\mathbf{s}_1$ and $\mathbf{s}_2$ are excited and are observed on the same component of ground motion.

Coupled seismic free oscillations can be observed on three separate ground-motion components, which aids the detection of coupling between spheroidal and toroidal multiplets. The east and north horizontal ground motion, as measured by the seismometer, can be rotated into radial (R) and transverse (T) components, parallel and perpendicular, respectively, to the great circle connecting source and receiver on the Earth's surface. Toroidal motion is absent from the vertical component and, for $l \gg 1$, only weakly present on the radial component of horizontal ground motion. Likewise, for $l \gg 1$, spheroidal motion is only weakly present on the transverse horizontal component. Therefore beating observed in the oscillation envelope, especially on the vertical component, can be taken as strong evidence of mixed-type modal coupling. In particular, we expect to observe a behavior common to all weakly coupled systems: motion excited on, say, the toroidal component of motion should rotate slowly onto the spheroidal motion component and back again. The strongest coupling observed between spheroidal and toroidal multiplets is caused by the Coriolis force associated with the Earth's rotation. The strength of mixed-type rotational coupling varies with the latitude of the source–receiver great-circle pole (Park, 1986), with strongest interaction on polar (north–south) propagation paths.

The 5/23/89 Macquarie Islands earthquake is nearly ideal for the study of spheroidal–toroidal coupling. The earthquake is one of the largest (surface-wave magnitude $M_s = 8{\cdot}1$, seismic moment $M_0 \approx 2{\cdot}0 \times 10^{21}$ N m) events to occur since the installation of digital seismic networks (Ekstrom and Romanowicz, 1990; Braunmiller and Nabelek, 1990), and has an unusually short inferred rupture duration (Tichelaar and Ruff, 1990; Houston, 1990). The epicentre is at high latitude (52·5°S), so that all propagation paths are more polar than equatorial in orientation. The source geometry is almost pure strike-slip (fault strike $\approx 35°$ west of north, right-lateral motion), which enhances toroidal-motion excitation.

Horizontal components from most stations are fairly noisy at long periods, except from a handful of new and recently upgraded stations with quiet sites and modern broadband-response seismometers. Horizontal-component data from such stations can have signal-to-noise nearly equal to that of the vertical component. Strong mixed-type rotational coupling is predicted from synthetic seismograms for the Macquarie event. The behavior of this interaction is evident in smooth-envelope inversions of hybrid *Ts–St* multiplet pairs. Figures 10.6 and 10.7 show spectrum estimates and smooth envelope inversions for the multiplet pair ${}_0S_{11}$–${}_0T_{12}$ on station PAS (Pasadena, California), part of the Global Seismic Network operated by the US Geological Survey and the Incorporated Research Institutions for Seismology (IRIS). The data analyzed is collected on the VP-channel of this seismometer, which samples ground motion at $0{\cdot}1\ \mathrm{s}^{-1}$, low-passed and decimated to $\Delta t = 40$ s. Figure 10.6 shows 'high-resolution' spectra (Thomson, 1982), calculated from 40-hour records with five 3π-prolate Slepian tapers appropriate for $0{\cdot}9Q$-cycles of decay and $\eta = 0$. The decay rate is chosen to be larger than that experienced by the modes ${}_0S_{10}$, ${}_0S_{11}$ and ${}_0S_{12}$, but less than the decay experienced by the modes ${}_0T_{11}$, ${}_0T_{12}$ and ${}_0T_{13}$. The frequency band shown is a cross-over between the ${}_0S_l$ and ${}_0T_{l+1}$ dispersion branches, for which the angular selection rules for Coriolis-force coupling predict interaction.

Results of smooth-envelope inversions for the V (vertical), R and T ground motion components of the spectral peaks corresponding to ${}_0S_{11}$–${}_0T_{12}$ are shown in Figure 10.7. The complex-valued envelopes are corrected for instrument response, and are given in units of ground displacement. The envelopes are boosted by $0{\cdot}9Q$-cycles of exponential growth to appear relatively decay-free for display, so the displacement units are accurate only at $t = 0$ h. Envelopes inferred for the vertical and radial components share a minimum at $t = 22$ h, suffering a phase shift approaching 180°. The transverse-component envelope varies oppositely, with a maximum near $t = 23$ h and minima where the other components are larger. The phase of the envelopes jumps by $\approx 180°$ where the envelope amplitude is close to zero. The phase drift suggests that the chosen carrier frequency $f_0 = 1{\cdot}860$ mHz differs slightly

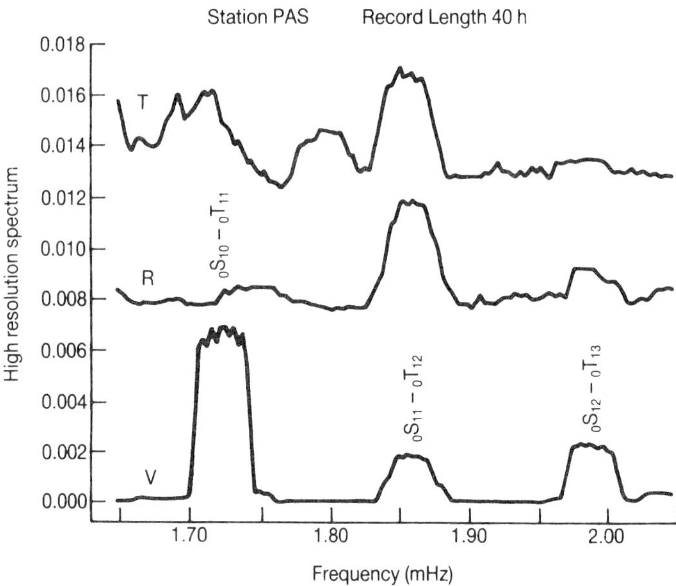

Figure 10.6 High-resolution spectrum estimates using long-period seismic data from station PAS (Pasadena, California) recorded after the $M_S = 8{\cdot}1$ Macquarie Ridge earthquake of 23 May 1989. Vertical (V), radial–horizontal (R) and transverse–horizontal (T) spectrum estimates were computed with five 3π-prolate Slepian tapers with noise-weighting $\eta = 0$ and exponential-decay parameter $\beta = 0{\cdot}9$. Peaks are indicated that correspond to spheroidal–toroidal free oscillation multiplet pairs known to couple through rotational Coriolis force.

from the dominant frequency in the data. The overall behavior is consistent with the simple coupled-oscillation model discussed above, in which vibrational motion alternates slowly between spheroidal and toroidal ground motion. Observed in a narrow frequency passband at the receiver site, ground motion tends to rotate like a Foucault pendulum – although the period of this rotation depends on the Earth's interior properties and not simply on latitude. Figure 10.8 shows the envelope estimates for multiplet pairs $_0S_{31}–_0T_{30}$ and $_0S_{37}–_0T_{35}$ for VP-channel seismograms collected by stations WMQ (Urumchi, China) and KMI (Kunming, China), respectively, in the Chinese Digital Seismic Network. Analysis is similar to that applied to the PAS record, though the more-rapid decay of the higher-frequency oscillations limits analysis to 20-hour records. Angular selection rules suggest that $_0S_{31}–_0T_{30}$ is sensitive to Coriolis-force coupling, and is supported by the observed rotation of vibration motion in the narrow passband around the carrier frequency $f_0 = 3{\cdot}90$ mHz. Coriolis-force coupling is forbidden for $_0S_{37}–_0T_{35}$, so the likely cause of the apparent rotation of vibrational motion between the radial and transverse components is related to lateral structure in the Earth's internal properties.

The detailed structure of long-period seismic motion may help constrain internal variations in Earth properties that govern long-period scattering in

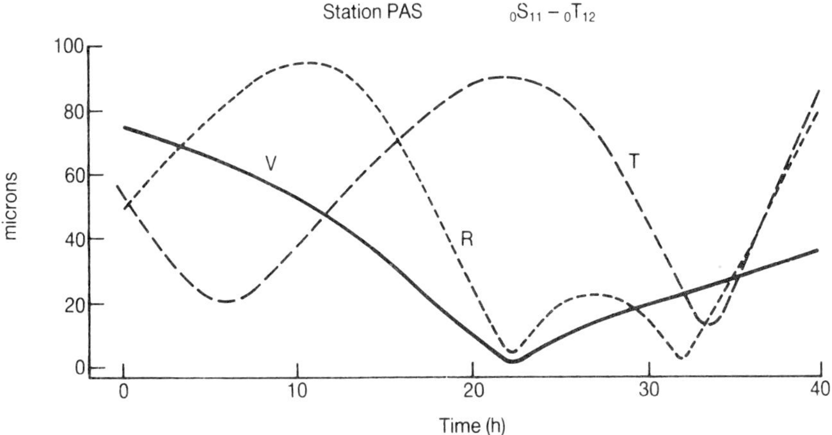

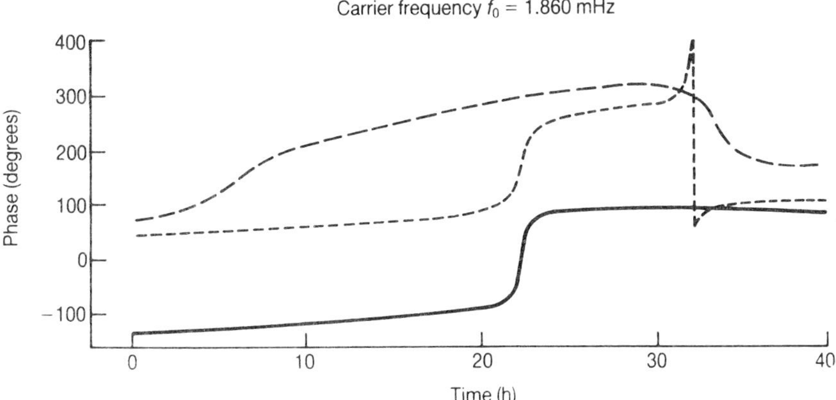

Figure 10.7 Smooth-envelope estimates for PAS data at a carrier frequency appropriate for the free oscillations ${}_0S_{11}$ ($f = 1{\cdot}8611$ mHz, $Q = 340$) and ${}_0T_{12}$ ($f = 1{\cdot}8599$ mHz, $Q = 178$). Frequencies are obtained from spherically averaged Earth model 1066A (Gilbert and Dziewonski, 1975) and attenuation rates from the model of Masters and Gilbert (1983). Plots indicate the smoothest deviation from a simple decaying sinusoid with $0{\cdot}9Q$-cycles of decay ($Q \approx 298$) in the duration of the data record. Decay is removed from the envelopes for display.

the Earth. The particle-motion associated with free-oscillation multiplets varies on the Earth's surface according to associated spherical harmonics $Y_l^m(\theta, \phi)$, $-l \leqslant m \leqslant l$. Therefore it should be possible to extend the single-station analysis presented here to a multiple-station inversion using a global network of seismometers. Observations of envelope fluctuations for spheroidal–toroidal multiplet pairs may be useful in this regard, as modeling studies suggest that source–receiver paths on which mixed-type coupling is strong are associated with considerable lateral structure and/or anisotropy

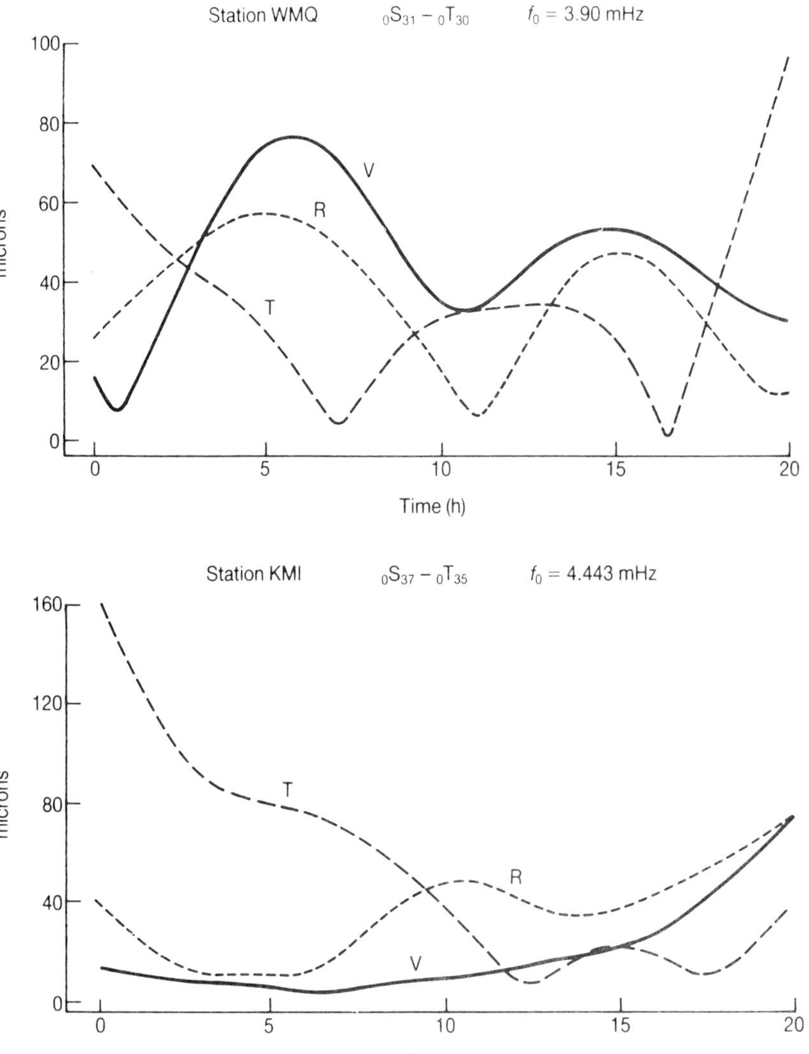

Figure 10.8 Smooth-envelope estimates using long-period seismic data recorded after the $M_S = 8{\cdot}1$ Macquarie Ridge earthquake of 23 May 1989. Vertical (V), radial–horizontal (R) and transverse–horizontal (T) spectrum estimates were computed with five 3π-prolate Slepian tapers with noise-weighting $\eta = 0$ and exponential-decay parameter $\beta = 1{\cdot}5$. Decay is removed from the envelopes for display. (a) Envelopes for data from station WMQ (Urumchi, China) at a carrier frequency appropriate for the free oscillations $_0S_{31}$ ($f = 3{\cdot}9031$ mHz, $Q = 198$) and $_0T_{30}$ ($f = 3{\cdot}8834$ mHz, $Q = 133$). Plots indicate the smoothest deviation from a simple decaying sinusoid with $1{\cdot}5Q$-cycles of decay ($Q \approx 187$) in the duration of the data record. (b) Envelopes for data from station KMI (Kunming, China) at a carrier frequency appropriate for the free oscillations $_0S_{37}$ ($f = 4{\cdot}4382$ mHz, $Q = 172$) and $_0T_{35}$ ($f = 4{\cdot}4324$ mHz, $Q = 129$). Plots indicate the smoothest deviation from a simple decaying sinusoid with $1{\cdot}7Q$-cycles of decay ($Q \approx 188$) in the duration of the data record.

in seismic velocity. Significant observational hurdles exist, not the least of which is that the number of sufficiently-low-noise long-period seismograms is limited by the number of large earthquakes ($M \geqslant 7{\cdot}5$ is best) and the number of high-quality seismic stations. Over time both limitations should recede. Contamination of envelope inversions by neighboring multiplets (e.g. fundamental-mode envelope observations contaminated by overtones) is a potential problem in all records. For instance, the local maximum at $t \approx 10$ h in the radial-component envelope appears related to the transfer of vibrational motion from the T-component, but the likely cause of the growth in the R- and V-envelopes at the record's end is contamination by a relatively high-Q spheroidal overtone multiplet. The latter's weak excitation by the earthquake source is compensated, after 20 h, by greater resistance to exponential decay. With these caveats in hand, smooth-envelope analyses of free oscillation may prove a useful tool in the investigation of the Earth's interior.

10.4 Orbital quasi-periodicities in paleoclimate data

Study of Pleistocene and Pliocene deep-sea cores has demonstrated the sensitivity of the Earth's climate system to small variations in insolation caused by cyclic changes in the geometry of the Earth's orbit and tilt (Hays *et al.*, 1976; Imbrie *et al.*, 1984; Berger, 1988). Cycles have been observed in δ^{18}O measurements, which record variations in global ice volume and ocean temperature, δ^{13}C measurements, a proxy for the supply of nutrients to the deep ocean, and sea–surface temperatures as inferred from foraminiferal assemblages. Rhythmic sedimentary oscillations of parameters related to global climate continue into the more remote geologic past, suggesting a long history of orbital modulation of Earth climate (Herbert and Fischer, 1986; Olsen, 1986). The solar system celestial mechanics problem is quasi-periodic, and numerical solutions are available that are reliable for the last 1–2 My (Berger, 1984; Bretagnon, 1984). The astronomical forcing series that governs long-term insolation variation consists of several groups of closely spaced lines near 413, 126, 96, 41, and 19–23 ky period (Berger, 1978).

Significant anomalies in the solar insolation received by the Earth are associated with the Earth's precession (19–23 ky) and obliquity (41 ky) cycles. Studies with atmospheric global circulation models suggest that these insolation variations can cause profound variations in average Earth climate (Kutzbach and Guetter, 1986; Prell and Kutzbach, 1987). The longer-period eccentricity cycles (96, 126, 413 ky and longer) are associated with much smaller insolation anomalies, but appear to dominate oxygen-isotope records of ice-sheet growth since roughly 1 Ma, as well as climate proxy variables from certain Cretaceous sequences (Herbert and Fischer, 1986; Park and Herbert, 1987). The eccentricity of the Earth's orbit modulates the seasonal insolation anomaly associated with the Earth's precession, so that the eccentricity cycle is the envelope of the precession cycle. Therefore the presence of eccentricity cycles in older sediments may be caused in part by the smearing of thin precession cycles by bioturbation or diagenesis. This explanation is untenable in the late Pleistocene, and the presence of a 100 ky ice-age cycle has been attributed to other factors in the Earth's climate system, such as a nonlinear response to precession and obliquity forcing (Birchfield and Weertman, 1978) or an internal oscillation unrelated to Milankovitch forcing (Saltzman and Sutera, 1984; Maasch and Saltzman, 1990).

Our understanding of the 100 ky ice-age cycle is hampered by the difficulty of modeling its evolution through time, particularly the timing and nature of its onset, thought to have occurred sometime in the interval 0·5–1·0 Ma. Was the onset of the 100 ky cycle gradual or sudden? A sudden onset would be characteristic of a bifurcation in a nonlinear system (Maasch and Saltzman, 1990), and would suggest the existence of a fundamental instability in the terrestrial climate system. A gradual increase in ice-age cycle intensity could, by contrast, suggest a shift in an otherwise stable climate regime or the steady growth of an unstable oscillation. Maasch (1988) analyzed several δ^{18}O and proxy sea surface temperature records for the Pleistocene, and found evidence for a sudden jump in the mean (towards greater average ice mass and cooler average global temperatures) at 0·9 Ma. Ruddiman *et al.* (1989) and Raymo *et al.* (1989) analyzed climate proxy data from DSDP cores 607 and 609, both from the North Atlantic Ocean, to study cyclic climate response. Ruddiman *et al.* (1989) applied bandpass filters to the data series to study the evolution of the 100 ky ice-age near the inferred mid-Pleistocene climate transition. This analysis suggested an increase in amplitude between 0·9 and 0·4 Ma, with the most rapid increase between 0·7 and 0·6 Ma.

Narrow-band filters cannot model discontinuities in the envelope of a quasi-periodic signal, and so are of little help in discriminating between sudden and gradual onset of the 100 ky ice-age cycle. The multi-taper envelope estimation procedure outlined in this chapter can model envelope discontinuities, and offer some aid in this respect. It must be stressed, however, that the technique cannot by itself discriminate between continuous and discontinuous models for the 100 ky cycle, since models of both types can be constructed that fit the spectral 'data' exactly. Rather, the analyst must use other criteria in choosing to accept or reject different models. We demonstrate such use of the algorithm using the 'Site 607' δ^{18}O data series analyzed by Ruddiman *et al.* (1989) and Raymo *et al.* (1989), a composite of data from sediments drilled at DSDP Site 607 and two other piston cores. This data set extends from the present through the Pleistocene into the Pliocene, and is inferred to represent roughly 2·7 My of deposition. The δ^{18}O data is sampled at an average spacing of 15 cm, corresponding to an average time spacing of 3·4 ky. We used the time scale derived by Ruddiman *et al.* (1989) by tuning the series against astronomical obliquity and precession variations. The unevenly spaced data set was interpolated with a cubic spline to obtain an evenly spaced series ($\Delta t = 2$ ky) for analysis. A small amount of misfit to the data was allowed when constructing the cubic spline interpolator (Reinsch, 1967), in order to avoid spurious oscillations in the neighborhood of a handful of closely spaced pairs of data points.

Multi-taper spectrum estimation was performed on the Site 607 record and the corresponding 2·7 My segment of the astronomical eccentricity series (Berger, 1978) using seven 4π-prolate Slepian tapers. Following standard convention, the sign of the $\delta^{18}O$ data was reversed so that lighter isotopic values (i.e. decreased ice volume) correlate with maxima in orbital eccentricity. The spectrum of the astronomical eccentricity series (Figure 10.9) is dominated by quasi-periodic oscillations at 96, 126 and 413 ky (10·4, 7·94 and 2·42 cyc/My, respectively). The complex-valued envelopes of the 96 and 126 ky signals (Figure 10.10), modeled simultaneously with seven 4π-prolate Slepian tapers, a constant-amplitude fixed element and roughness penalty (10.15), vary in tandem, sharing a minimum and a phase jump near 2 Ma. The high-resolution multi-taper spectrum estimate of the 'Site 607' series (Figure 10.11) shows a peak centered near 10·4 cyc/My, with a shoulder at lower frequency. Statistical tests for phase-coherence (Thomson, 1982; Park and Herbert, 1987) show weak peaks (< 99% confidence for nonrandomness) near 96 and 126 ky period, as well as periods not associated with primary Milankovitch cycles. When this peak and shoulder are modeled in terms of envelopes of oscillations with carrier frequencies 7·94 and 10·4 cyc/My (Figure 10.12), poor correlation with the astronomical signal is found before 1 Ma – note the lack of a clear minimum near 2·0 Ma. The phase of both envelopes agrees fairly well with the astronomical series for ages less than 1 Ma. The growth of envelope amplitude for the 126 ky quasi-period appears

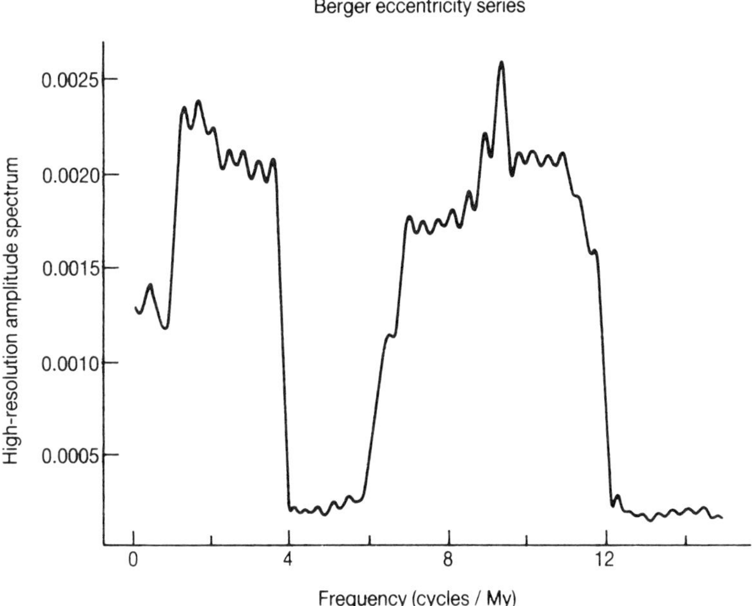

Figure 10.9 High-resolution spectrum estimate of the astronomical eccentricity series from Berger (1978), evaluated from the present to 2·7 Ma. Seven 4π-prolate Slepian tapers were used.

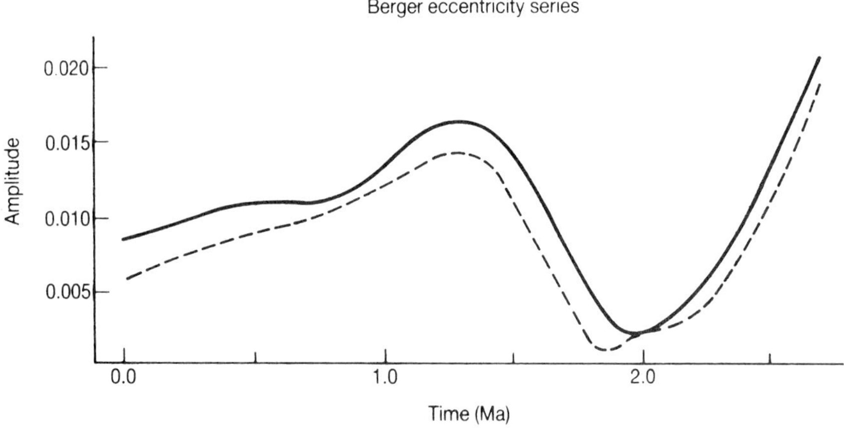

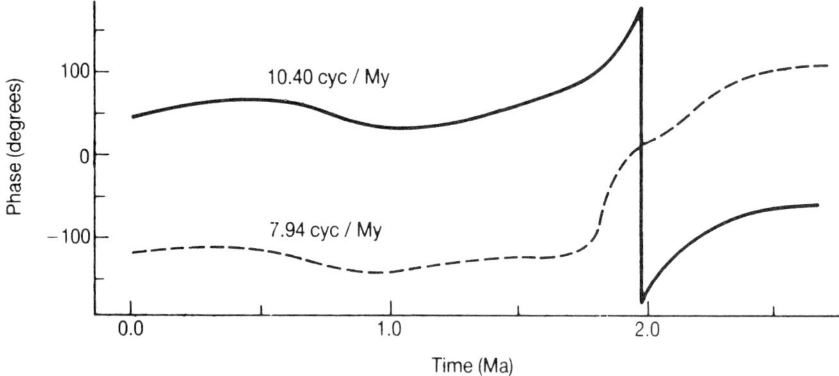

Figure 10.10 Envelope models for the quasi-periodic signal centered at carrier frequencies 7·94 and 10·4 cyc/My, using the astronomical eccentricity series. Seven 4π-prolate Slepian tapers were used, with the roughness penalty function (10.15) and a constant-amplitude fixed element.

to start at 1·25 Ma. This growth is dwarfed by the growth of the 96 ky envelope amplitude, which rises from zero at 1·0 Ma to a maximum at 0·45 Ma, then declines. The behavior of the 96 ky envelope is broadly consistent with that reported by Ruddiman *et al.* (1989) for the 100 ky bandpass. Multi-taper envelope estimation allows the simultaneous modeling of the envelope of signal at the nearby 126 ky eccentricity period.

Although the above smooth-envelope inversion suggests a gradual increase in the 100 ky cycle, abrupt onsets are also consistent with the data. Figure 10.13 shows results of smooth-envelope inversions at $f_0 = 10{\cdot}4$ cyc/My with discontinuities allowed at 0·7, 0·85, and 1·0 Ma. A discontinuity allowed at 1·0 Ma leads to an envelope similar to that shown in Figure 10.12. A discontinuity allowed at 0·7 Ma leads to an envelope with a sharp ampli-

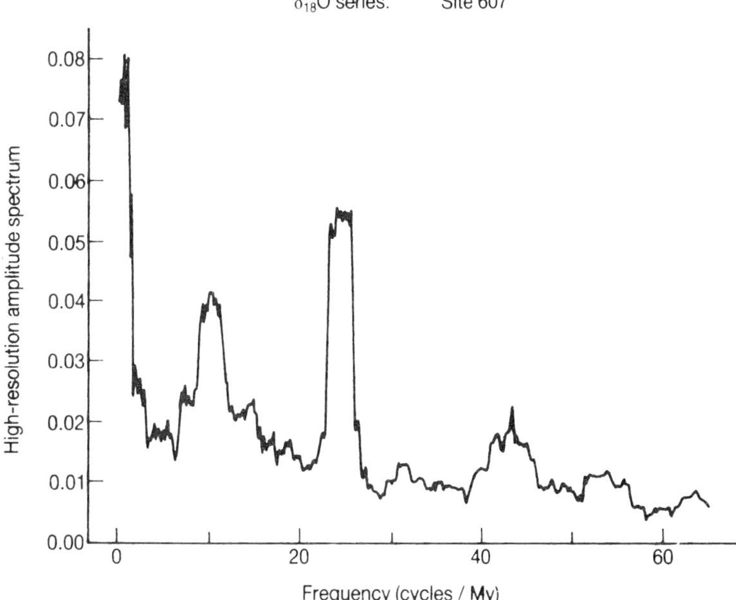

Figure 10.11 High-resolution spectrum estimate of the Site 607 $\delta^{18}O$ data series reported by Ruddiman *et al.* (1989) and Raymo *et al.* (1989), evaluated from the present to 2·7 Ma. Seven 4π-prolate Slepian tapers were used.

tude maximum at the discontinuity. Note that the phase of this envelope (coarse-dashed line in the figure) attempts to match phase across the discontinuity, suggesting that some 100 ky signal with phase close to that of the astronomical cycle exists at ages greater than 0·7 Ma. By contrast, a discontinuity allowed at 0·85 Ma leads to an envelope with roughly constant amplitude at ages younger than 0·85 Ma, and no anomalous deviation in phase at times older than the discontinuity. Based on this comparison, a sudden onset of the ice-age cycle at 0·85 Ma, but not significantly earlier or later, appears a reasonable model for the narrow-band 100 ky signal.

Firmer conclusions about the evolution of the Pleistocene ice-age cycle cannot be reached without a more thorough analysis of this and other climate proxy records. Such an analysis is beyond the scope of the present chapter. We note that time-scale uncertainties in paleoclimate time series create difficulties in interpretation, especially for the quasi-periods used to tune the time scale. For example, smooth envelope inversion for the obliquity signal ($f_0 = 24{\cdot}35$ cyc/My) suggests amplitude variations by a factor of two, with a maximum between 0·9 and 1·5 Ma. Phase variations for this envelope, however, have been suppressed by the use of obliquity as a tuning tool. The time scale given by Ruddiman *et al.* (1989) was engineered to minimize phase fluctuations in the obliquity signal, using an assumed 10 ky phase lag between obliquity and glacial maxima.

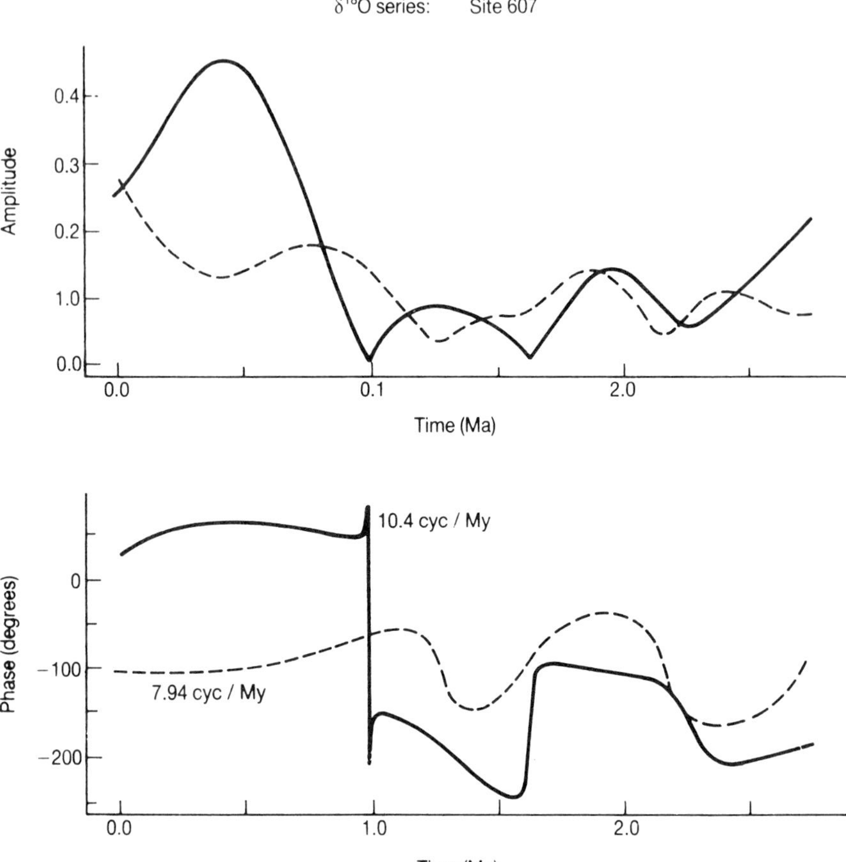

Figure 10.12 Envelope models for the quasi-periodic signal centered at carrier frequencies 7·94 and 10·4 cyc/My, using the Site 607 $\delta^{18}O$ data series. Seven 4π-prolate Slepian tapers were used, with the roughness penalty function (10.15) and a constant-amplitude fixed element.

10.5 Summary

It is often useful to model the time-evolution, or envelope, or quasi-periodic signals in geological and geophysical time series. Multi-taper envelope estimation is a generalization of narrow-bandpass filtering which allows *a priori* constraints to be imposed on the envelope of such a signal, while requiring that the amplitude and phase of the complex-valued envelope be consistent with spectral information in the neighborhood of a carrier frequency f_0. In particular, we use the eigenspectrum estimates $Y_k(f_0)$ (complex-valued), obtained via the discrete Fourier transform of the data multiplied by one of a set of K $p\pi$-prolate Slepian tapers, as 'data' in a linear inverse problem to construct models for the complex-valued envelope. The Slepian tapers serve

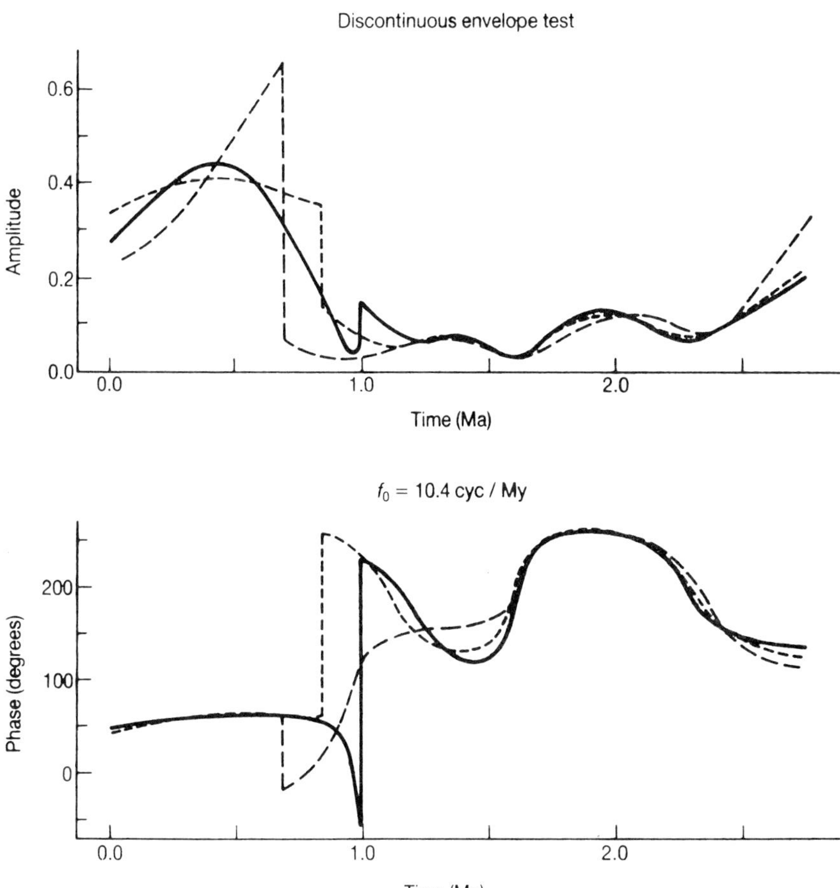

Figure 10.13 Envelope models for carrier frequency $f_0 = 10{\cdot}4$ cyc/My, using the Site 607 δ^{18}O data series. Seven 4π-prolate Slepian tapers were used, with a constant-amplitude fixed element. A roughness penalty is applied everywhere save a single point in time, where a discontinuous jump in the envelope is allowed. Break points shown are 1·0 Ma (solid line), 0·85 Ma (fine-dashed line) and 0·7 Ma (coarse-dashed line).

as kernel functions, or data representers, in the inversion. The Slepian tapers are orthogonal when summed over the series, so that, for locally white data processes, the eigenspectra are statistically uncorrelated. This orthogonality carries over into the manner that the eigenspectra $Y_k(f_0)$ sample the frequency-domain representation $\bar{A}(f)$ of the quasi-periodic signal over the frequency range $[f_0 - pf_R, f_0 + pf_R]$, where $f_R = 1/T$ and T is the duration of the time series. Therefore spectrum estimates using a set of Slepian tapers offer a natural and efficient way to investigate the narrow-band behavior of $\bar{A}(f)$. Envelopes constructed as linear combinations of Slepian tapers can represent the envelope near the ends of the time series, whereas the use of

M-point bandpass filters discards $M - 1$ data points from each end of the series.

All linear inverse problems with finite data have non-unique solutions, but unique models can be constructed that minimize a suitably defined penalty function. A roughness penalty of the form (10.15) is a plausible constraint, as it tends to preserve the bandlimited nature of a quasi-periodic signal. Abrupt changes in an otherwise smoothly varying envelope can be modeled by a suitable modification of the roughness penalty function. Other constraints, data misfit tolerance, and additional model parameters (e.g. a second carrier frequency) can be incorporated into the inverse problem. Discrimination between envelope models often involves criteria specific to the analyzed data set. However, spectral information from outside the narrow band centered on f_0 sampled by the Slepian tapers may be useful in constraining the problem if a suitable case can be made that such spectral information is not contaminated by signals with dominant frequency $f \neq f_0$.

Acknowledgements

I thank Robert Parker for providing a draft copy of his unfinished text on inverse theory. I thank the staff at the IRIS DMC for help in accessing data from the Macquarie Ridge earthquake. I thank Kirk Maasch for helpful comments and for providing a computer file of the Site 607 $\delta^{18}O$ data series, which was provided to him by M.E. Raymo, to whom I also express gratitude. This work has been supported by NSF grant EAR-8657206.

References

Aki, K. and Richards, P.G. (1980). *Quantitative Seismology*, Vol. 1. Freeman, San Francisco.

Berger, A. (1978). A simple algorithm to compute long-term variations of daily or monthly isolation. *Inst. Astron. Geophys. G. Lemaitre*, Contrib. 18, 17 pp.

Berger, A. (1984). Accuracy and frequency stability of the earth's orbital elements during the Quaternary. In *Milankovitch and Climate*, Part 1, Berger, A., Imbrie, J., Hays, J., Kukla, G., and Saltzman, B. (eds). D. Reidel, Hingham, Mass.

Berger, A. (1988). Milankovitch theory and climate. *Rev. Geophys.*, **26**, 624–57.

Birchfield, G.E. and Weertman, J. (1978). A note on the spectral response of a model continental ice sheet. *J. Geophys. Res.*, **83**, 4123–5.

Bolt, B. and Brillinger, D.R. (1979). Estimation of uncertainties in eigenspectral estimates from decaying geophysical time series. *Geophys. J. Roy. Astron. Soc.*, **59**, 593–603.

Braunmiller, J. and Nabelek, J. (1990). Rupture process of the Macquarie Ridge earthquake of May 23, 1989. *Geophys. Res. Lett.*, **17**, 1017–20.

Bretagnon, P. (1984). Accuracy of long-term planetary theory. In *Milankovitch and Climate*, Part 1, Berger, A., Imbrie, J., Hays, J., Kukla, G., and Saltzman, B. (eds), D. Reidel, Hingham, Mass.

Buland, R. and Gilbert, F. (1978). Improved resolution of complex eigenfrequencies in analytically continued seismic spectra. *Geophys. J. Roy. Astron. Soc.*, **52**, 457–70.

Buland, R., Berger, J., and Gilbert, F. (1977). Observations from the IDA network of attenuation and splitting during a recent earthquake. *Nature*, **277**, 358–62.

Chao, B.F. (1990). Commentary on 'A new method of spectral analysis and its application to the Earth's free oscillations: The "Sompi" method'. *J. Geophys. Res.*, **95**, 19 789–90.

Chao, B.F. and Gilbert, F. (1980). Autoregressive estimation of complex eigenfrequencies in low frequency seismic spectra. *Geophys. J. Roy. Astron. Soc.*, **63**, 641–57.

Constable, S.C., Parker, R.L., and Constable, C.G. (1987). Occam's inversion: A practical algorithm for generating smooth models from electromagnetic sounding data. *Geophysics*, **52**, 289–300.

Dahlen, F.A. (1982). The effect of data windows on the estimation of free oscillation parameters. *Geophys. J. Roy. Astron. Soc.*, **69**, 537–49.

Dongarra, J.J., Moler, C.B., Bunch, J.R., and Stewart, G.W. (1979). *LINPACK User's Guide*, Society for Industrial and Applied Mathematics, Philadelphia.

Ekstrom, G. and Romanowicz, B. (1990). The 23 May, 1989 Macquarie Ridge earthquake: A very broad band analysis. *Geophys. Res. Lett.*, **17**, 993–6.

Geller, R.J. and Stein, S. (1979). Time domain attenuation measurements for fundamental spheroidal modes (${}_0S_6$–${}_0S_{28}$) for the 1977 Indonesian Earthquake. *Bull. Seism. Soc. Amer.*, **69**, 1671–92.

Giardini, D., Xi, X.-D., and Woodhouse, J.H. (1987). Three dimensional structure of the Earth from splitting in free oscillation spectra. *Nature*, **325**, 405–11.

Gilbert, F. and Dziewonski, A.M. (1975). An application of normal mode theory to the retrieval of structural parameters and source mechanisms from seismic spectra. *Phil. Trans. Roy. Soc. Lond.*, Ser. A, **278**, 187–269.

Hansen, R.A. (1982a). Simultaneous estimation of terrestrial eigenvibrations. *Geophys. J. Roy. Astron. Soc.*, **70**, 155–72.

Hansen, R.A. (1982b). Observational study of terrestrial eigenvibrations. *Phys. Earth Planet. Int.*, **28**, 29–69.

Hays, J.D., Imbrie, J.I., and Shackleton, N.J. (1976). Variations in the earth's orbit: pacemaker of the Ice Ages, *Science*, **194**, 1121–32.

Herbert, T.D. and Fischer, A.G. (1986). Milankovitch climatic origin of mid-Cretaceous black shale rhythms in central Italy. *Nature*, **321**, 739–43.

Hori, S., Fukao, Y., Kumazawa, M., Furumoto, M., and Yamamoto, A. (1989). A new method of spectral analysis and its application to the Earth's free oscillations: The 'Sompi' method. *J. Geophys. Res.*, **94**, 7535–53.

Houston, H. (1990). Broadband source spectrum, seismic energy, and stress drop of the 1989 Macquarie Ridge earthquake. *Geophys. Res. Lett.*, **17**, 1021–4.

Imbrie, J., Hays, J.D., Martinson, D.G., McIntyre, A., Mix, A.C., Morley, J.J., Pisias, N.G., Prell, W.L., and Shackleton, N.J. (1984). The orbital theory of Pleistocene climate: support from a revised chronology of the marine $\delta(^{18}O)$ record. In *Milankovitch and Climate*, Part 1, Berger, A., Imbrie, J., Hays, J., Kukla, G., and Saltzman, B. (eds). D. Reidel, Hingham, Mass.

Jordan, T.H. (1978). A procedure for estimating lateral variations from low-frequency seismic data. *Geophys. J. Roy. Astron. Soc.*, **52**, 441–55.

Kutzbach, J.E. and Guetter, P.J. (1986). The influence of changing orbital parameters and surface boundary conditions on climate simulations for the past 18,000 years. *J. Atmos. Sci.*, **33**, 1726–59.

Lawson, C.L. and Hanson, R.J. (1974). *Solving Least Squares Problems*, Prentice-Hall, Englewood Cliffs, N.J.

Lindberg, C.R. (1986). *Multiple Taper Spectral Analysis of Terrestrial Free Oscillations*, Ph.D. Thesis, University of California, San Diego.

Lindberg, C.R. and Park, J. (1987). Multiple-taper spectral analysis of terrestrial free oscillations, Part II. *Geophys. J. Roy. Astron. Soc.*, **91**, 795–836.

Lindberg, C.R. and Thomson, D.J. (1990). Commentary on 'A new method of spectral analysis and its application to the Earth's free oscillations: The "Sompi" method'. *J. Geophys. Res.*, **95**, 19 785–8.

Maasch, K.A. (1988). Statistical detection of the mid-Pleistocene transition. *Climate Dynamics*, **2**, 133–43.

Maasch, K.A. and Saltzman, B. (1990). A low-order dynamical model of global climate variability over the full Pleistocene. *J. Geophys. Res.*, **95**, 1955–63.

Masters, G. and Gilbert, F. (1983). Attenuation in the earth at low frequencies. *Phil. Trans. Roy. Soc. Lond.*, Ser. A, **308**, 479–522.

Masters, G., Park, J., and Gilbert, F. (1983). Observations of coupled spheroidal and toroidal modes, *J. Geophys. Res.*, **88**, 10 285–98.

Menke, W. (1984). *Geophysical Data Analysis: Discrete Inverse Theory*, Academic Press, Orlando.

Olsen, P.E. (1986). A 40-million-year lake record of early Mesozoic orbital climate forcing. *Science*, **234**, 842–484.

Park, J. (1986). Synthetic seismograms from coupled free oscillations: the effects of lateral structure and rotation. *J. Geophys. Res.*, **91**, 6441–64.

Park, J. (1987). Asymptotic coupled-mode expressions for multiplet amplitude anomalies and frequency shifts on an aspherical earth. *Geophys. J. Roy. Astron. Soc.*, **90**, 129–69.

Park, J. (1988). Free oscillation coupling theory. In *Mathematical Geophysics*, Vlaar, N.J., Nolet, G., Wortel, M.J.R., and Cloetingh, S.A.P.L. (eds). D. Reidel, Dordrecht.

Park, J. (1990). Observed envelopes of coupled seismic free oscillations. *Geophys. Res. Lett.*, **17**, 1489–92.

Park, J. and Gilbert, F. (1986). Coupled free oscillations of an aspherical, dissipative, rotating earth: Galerkin theory. *J. Geophys. Res.*, **91**, 7241–60.

Park, J. and Herbert, T.D. (1987). Hunting for paleoclimatic periodicities in a geologic time series with an uncertain time scale. *J. Geophys. Res.*, **92**, 14 027–40.

Park, J., Lindberg, C.R., and Thomson, D.J. (1987a). Multiple-taper spectral analysis of terrestrial free oscillations, Part 1. *Geophys. J. Roy. Astron. Soc.*, **91**, 755–894.

Park, J., Lindberg, C.R., and Vernon, F.L., III (1987b). Multiple-taper spectral analysis of high frequency seismograms. *J. Geophys. Res.*, **92**, 12 675–84.

Parker, R.L. (1977). Understanding inverse theory. *Ann. Rev. Earth Plan. Sci.*, **5**, 35–64.

Parker, R.L. (1990). *Inverse Theory*. Unpublished book manuscript.

Parker, R.L., Shure, L., and Hildebrand, J.A. (1987). The application of inverse theory to seamount magnetism. *Rev. Geophys.*, **25**, 17–40.

Prell, W.L. and Kutzbach, J.E. (1987). Variability of the monsoon over the past 150,000 years: Comparison of observed and simulated paleoclimatic time series. *J. Geophys. Res.*, **92**, 8411–25.

Raymo, M.E., Ruddiman, W.F., Backman, J., Clement, B.M., and Martinson, D.G. (1989). Late Pliocene variation in northern hemisphere ice sheets and North Atlantic Deep Water circulation. *Paleoceanography*, **4**, 413–46.

Reinsch, C.H. (1967). Smoothing by spline functions. *Numerische Mathematik*, **10**, 177–83.

Ritzwoller, M., Masters, G., and Gilbert, F. (1986). Observations of anomalous splitting and their interpretation in terms of aspherical structure. *J. Geophys. Res.*, **91**, 10 203–28.

Ruddiman, W.F., Raymo, M.E., Martinson, D.G., Clement, B.M., and Backman, J. (1989). Pleistocene evolution: Northern hemisphere ice sheets and the North Atlantic Ocean. *Paleoceanography*, **4**, 353–412.

Saltzman, B. and Sutera, A. (1984). A model of the internal feedback system involved in late Quaternary climatic variations. *J. Atmos. Sci.*, **41**, 736–45.

Shure, L., Parker, R.L., and Backus, G.E. (1982). Harmonic splines for geomagnetic modelling. *Phys. Earth Plan. Inter.*, **28**, 215–29.

Slepian, D. (1978). Prolate spheroidal wave functions, Fourier analysis and uncertainty – V: The discrete case. *Bell Syst. Tech. J.*, **57**, 1371–1429.

Slepian, D. (1983). Some comments on Fourier analysis, uncertainty, and modeling. *SIAM Rev.*, **25**, 379–93.

Smith, M.F. and Masters, G. (1989). Aspherical structure constraints from free oscillation frequency and attenuation measurements. *J. Geophys. Res.*, **94**, 1953–76.

Stark, P.B., Parker, R.L., Masters, G., and Orcutt, J.A. (1986). Strict bounds on seismic velocity in the spherical earth. *J. Geophys. Res.*, **91**, 13 892–902.

Tarantola, A. (1987). *Inverse Problem Theory*, Elsevier, Amsterdam.

Thomson, D.J. (1982). Spectrum estimation and harmonic analysis. *Proc. IEEE*, **70**, 1055–96.

Thomson, D.J. (1990). Time series analysis of Holocene climatic data. *Phil. Trans. Roy. Soc. Lond.*, **330**, 601–16.

Tichelaar, B.W. and Ruff, L.J. (1990). Rupture process and stress drop of the great 1989 Macquarie Ridge earthquake. *Geophys. Res. Lett.*, **17**, 1001–4.

Walden, A.T. (1990). Improved low-frequency decay estimation using the multitaper spectral analysis method. *Geophys. Prospecting*, **38**, 61–86.

Woodhouse, J.H. (1983). The joint inversion of seismic waveforms for lateral variations in Earth structure and earthquake source parameters. In *Proceedings of the Enrico Fermi International School of Physics, 85*, Kanamori, H. and Boschi, E. (eds), North Holland, Amsterdam.

Chapter 11

Statistical methods for the description and display of earthquake catalogs

D. Vere-Jones

This paper sets out the main statistical techniques which can be used to produce maps of seismicity (intensities) or b-values from catalogs of earthquake data. The emphasis is on techniques which can be used with two space dimensions and with space–time models. There are four main parts, covering respectively basic parameters of catalog data; kernel-smoothing methods; semi-parametric methods (orthogonal functions and splines); stochastic process models. The aim is to provide an overview of the methods, with an indication of advantages, limitations, assumptions, and practical points of implementation. *Ad hoc* methods have long been used in the literature on earthquake risk. In point of fact the topics are closely related to current research in fields such as spatial statistics, space–time models for point processes, density estimation. Many aspects are still only partially resolved, and would repay further investigation at both theoretical and empirical levels.

11.1 Introduction

This paper reviews statistical methods for developing space or space–time models for the data in an earthquake catalog with particular reference to historical catalogs. The emphasis is on models which lead to estimates of seismicity (occurrence intensities) and b-values or mean magnitudes, for graphical display as functions of space or space and time.

Techniques for displaying catalog data in graphical form have a long history in engineering and seismological studies of earthquake risk, where they generally form the first stage of more specialized studies. The aim of the present paper is to set these problems in the context of recent research in topics such as density estimation, point process models, maximum likelihood estimates for spatial processes, etc.

The paper is divided into four main parts. Part I, on the basic parameters of catalog data, is a brief outline of some of the points that need to be considered in regard to the quality and interpretation of catalog data. Part II treats kernel methods for smoothing the data, including the crucial question of choice of bandwidth. Part III looks at methods for developing orthogonal and spline function representations for the data. These are relatively recent methods of considerable flexibility and power, but many questions remain concerning optimal procedures. Finally Part IV deals with estimates based on stochastic process models which begin to incorporate some of the complexities (clustering and stress-release) of the underlying physical processes. These methods are still at an early stage of development, but are of importance in developing forecasts of earthquake risk.

11.2 Part I: The basic parameters of earthquake catalogs

Before embarking on any form of statistical analysis it is important to gain a good understanding of the limitations and special features of the data being analyzed. Nowhere is this more important than in the statistical analysis of earthquake catalogs, above all catalogs extending over historical data. The reliability and quality of the data, the units in which it is measured, and the techniques by which it has been obtained, are all likely to vary significantly and abruptly from one time period or seismic region to another. A careful scrutiny of the data, and wherever possible the first-hand advice of a local expert, are mandatory preliminaries to any meaningful analysis.

The basic parameters in view, which we shall take as the irreducible minimum for any catalog in our sense, are the epicentral latitude and longitude, origin time, depth and magnitude of recorded events. Much additional information is often available, but we shall limit the discussion to the above parameters. Vere-Jones and Smith (1981) give an introduction to some of the statistical problems which arise in estimating those parameters, and to statistical problems in seismology more generally.

Relative to the dimensions of the study, the origin time is generally the most accurately estimated of the parameters mentioned; errors of more than a few seconds in instrumental data, a day in recent historical data, or a year in old historical data, are rare. Epicentral latitudes and longitudes are also estimated rather precisely from instrumental data, though errors in assumed velocities can create systematic biases in some regions, and the very concept of epicentre for the largest events is somewhat questionable. Depth is more difficult to determine. Although depths can reach hundreds of kilometers in certain regions, we shall somewhat beg the question of depths by generally assuming that the catalogs are for shallow events (typically 0–30 km) only. Virtually all recorded historical events are shallow, since with rare exceptions only shallow events produce substantial surface effects. In recent catalogs where depths are more accurately determined, depth can be treated as an additional spatial variable without altering the basic procedures described in the sequel.

One of the least satisfactory but at the same time crucial parameters is the earthquake magnitude. For all data prior to 1900, magnitude estimates are dependent on comparisons of historical records with records of comparable recent events with known magnitudes. Considerable expertise, both historical and technical, is required to make such comparisons reliably; the detailed accounts by Ambraseys and Melville (1982) and in several of the papers in Margottini and Serva (1988) illustrate just how exacting and time-consuming a task this can be. The historical records vary enormously in quality from time period to time period, and in different regions, so that considerable caution is needed in assessing the range of validity of a given catalog.

Several magnitude scales are currently in use. The original magnitudes (M_L) were determined from the logarithm of the maximum amplitude recorded by a standard instrument after adjustment for distance from the source. Later, surface wave and body wave magnitudes (M_s and m_b) were developed from a closer analysis of seismogram traces. In recent years recognition of the important role played by the seismic moment has led to moves, only partially successful, to replace magnitudes by moments. For small events, duration of shaking can often be used as a surrogate measurement in place of magnitude when direct determinations are not feasible. Recent catalogs may cite several or all of these alternatives, and care is needed to ensure that a consistent choice is made throughout the time period. The different kinds of magnitude are closely related, but not identical, since in effect they measure different aspects of a complex physical event.

The crucial importance of the magnitude, and the reason why, despite its manifold defects, it has not been displaced by more recent contenders, is the simple fact that it provides a single number which broadly characterizes the size of the underlying physical event. Here two important characteristics need to be borne in mind.

The first is that, because of its original definition in terms of logarithms, the magnitude tends to be logarithmically related to most variables with more direct physical meaning: the energy release, area of aftershock zone, length of fault trace, etc. The magnitude scale is approximately a decibel scale, with unit increase in magnitude corresponding roughly to a 30-fold increase in energy release.

The second important feature is that, almost universally, magnitudes follow approximately an exponential frequency distribution. There may be arguments about its validity in the extreme upper tail, or possible changes of slope in the log survivor function, but one of the most ubiquitous features of earthquake occurrence, whether of large tectonic earthquakes or micro-fractures in the laboratory, is the approximately linear relation between magnitude and log survivor function. The traditional formulation in the seismological literature (see Gutenberg and Richter, 1954) is

$$\log_{10}\{\text{no. of events with magnitude} \geqslant m\} = a - bm, \tag{11.1}$$

where a is a measure of the total number of events above the threshold magnitude M_0 and the b-value is frequently close to unity.

The maximum likelihood estimate of b is then

$$\hat{b} = \frac{1}{2{\cdot}3026}\hat{\beta}, \qquad \hat{\beta} = (\bar{M} - M_0)^{-1}, \tag{11.2}$$

where $\bar{M}$ is the average of the magnitudes of the events under study. For older data, where the magnitudes may be rather crudely rounded, or clustered around integer values, some corrections to this estimate are desirable (see for example Vere-Jones, 1988) but we shall assume for the sequel that (11.2) is appropriate. One of its consequences is that plotting the b-value as a function of time and space is effectively equivalent to plotting the mean magnitude as a function of time and space: estimates and confidence intervals for the one can be directly converted into estimates and confidence intervals for the other.

Crucial and difficult decisions occur over the choice of boundaries for the region to be studied: geographical boundaries, time interval, and threshold magnitude. The question of completeness is paramount; nothing of value can be expected from a statistical analysis with significant amounts of the data missing. Much has been written about statistical methods of assessing completeness, but in the writer's view simple graphical checks, especially the use of cusum charts, are as effective a means as any of judging the constancy of intensities and b-values over a period of time. Such checks on internal consistency are all that any statistical approach can hope to provide, and their results are bound to confound real physical changes with changes in the efficiency of data capture, so that an element of historical judgment is inescapable. In such circumstances, more sophisticated techniques, such as change-point methods, are of doubtful value, and also hard to apply with confidence because of the effects of clustering on significance levels.

A somewhat analogous problem arises over the treatment of aftershocks and other major clusters such as earthquake swarms. The recording of aftershocks in the older historical data is so haphazard that they are better excluded altogether, but one may well have qualms about cutting out a significant portion of more recent instrumental data. Most aftershocks are indicated as such, or can be largely removed by passing an appropriate series of time–space–magnitude windows across the data. The danger then is that the statistical properties of the data which remain reflect the effects of the windowing procedures as much as they do the properties of the underlying physical phenomena. The most careful approach is to carry out several analyses, before and after applying various 'thinning' techniques, to give some feeling for how far and in what ways the conclusions of interest are affected by the treatment of aftershocks.

11.3 Part II: Kernel methods

The kernel method is the most straightforward and widely used of the standard methods, and for exploratory purposes, in particular, it is hard to see substantial reasons for going beyond it. Perhaps its only major drawback is the difficulty of adapting it for forecasting purposes when substantial fluctuations with time are present. It can be applied to both intensities and mean magnitudes, with many technical points in common. We treat only fixed bandwidth methods.

11.3.1 Kernel estimates for intensities

Smoothing procedures for the related problem of estimating a probability density have been the subject of remarkable development in the last few decades, following a period of equally remarkable neglect. The book by Silverman (1986) can be recommended as an introduction to both theory and practice, and contains an extensive bibliography of the literature prior to 1983. Marron (1988) provides a non-technical review of recent developments, with the emphasis on the problem of bandwidth selection.

In the context of estimating spatial rates of intensities, the basic kernel estimate takes the form

$$\hat{\lambda}(\mathbf{x}) = \sum_{i=1}^{n} k_h(\mathbf{x} - \mathbf{x}_i) = \int k_h(\mathbf{x} - \mathbf{u})\, N(\mathrm{d}\mathbf{u}), \tag{11.3}$$

where N is the counting process, $\mathbf{x}$ is an arbitrary point of the region for which the intensity estimate is required, the $\mathbf{x}_i$ are the spatial coordinates of the events, and n is the number of events in the catalog. The kernel k_h is a probability density function ($k_h \geqslant 0$, $\int k_h = 1$) incorporating a variable scale factor, or bandwidth, h, almost invariably in the form

$$k_h(\mathbf{x}) = h^{-d} k_h(\mathbf{x}/h), \tag{11.4}$$

where d is the dimensionality of the region studied. In space–time problems a product form

$$k(\mathbf{x}, t) = k_{h_1}(\mathbf{x}) k_{h_2}(t) \tag{11.5}$$

seems more appropriate than the isotropic form (11.4) suggested for the purely spatial context. Further variants, for example with elliptical cross-sections, can be derived from the isotropic forms by appropriate changes of scale. Some further comments on the choice of kernel are given below.

If the expected rate of occurrence λ is constant in a region around the point ($\mathbf{x}$), the expected value of the estimate (11.3) can be written as

$$E[\hat{\lambda}(\mathbf{x})] = \int k_h(\mathbf{x} - \mathbf{u})\lambda\, \mathrm{d}\mathbf{u} = \lambda \int k_h(\mathbf{u})\, \mathrm{d}\mathbf{u} = \lambda.$$

If the intensity varies sufficiently slowly in a neighborhood of $\mathbf{x}$, therefore, $\hat{\lambda}(\mathbf{x})$ provides an approximately unbiased estimate of the true intensity. In practice the intensity is not constant, and successful implementation of the method depends on choosing the bandwidth h to obtain the best compromise between bias (h too large) and variability (h too small). Further complications arise from boundary effects, clustering, and the problems of higher dimensionality. These are almost but not identically the same as those arising in the traditional probability density estimation context (where the estimate (11.3) is divided by a factor n), and we briefly summarize some of the main technical points which may need consideration.

11.3.2 Choice of kernel

Conventional wisdom, and our own experiences, suggest that within reasonable bounds the exact choice of kernel is of relatively minor importance compared to the choice of bandwidth. Kernels with finite support have minor computational advantages but may be better avoided in the present context since they may lead to areas of zero intensity and consequent instabilities in the ratio estimates of mean magnitude considered later. Paradoxically, kernels with extremely long tails may produce inadequate smoothing at short or medium distances, tending to produce a uniform background with sharp spikes around clusters. Two generally convenient forms are

$$ {}_1k_h(\mathbf{x}) = c_1 h^{-d}(1 + r^2/h^2)^{-\lambda} \qquad \lambda > \frac{d}{2}, $$

$$ {}_2k_h(\mathbf{x}) = c_2 h^{-d} \exp\left[-(r^2/h^2)^{\lambda}\right], \qquad \lambda > 0, $$

where d is the dimension, $r^2 = \mathbf{x}^{\mathrm{T}}\mathbf{x}$, and c_1, c_2 are the normalization factors.

	c_1^{-1}	c_2^{-1}
1-D	$B(\lambda - \frac{1}{2}, \frac{1}{2})$	$\lambda^{-1}\Gamma(\frac{1}{2}\lambda)$
2-D	$\dfrac{\pi}{(\lambda + 1)}$.	$\pi\Gamma\left(\dfrac{1}{\lambda}\right)$

Values of λ around 1·5–2 and 1–1·5 for the first and second forms respectively can be suggested. The use of the isotropic form (dependence on r^2) assumes the spatial dimensions should be treated equally. In a space–time context, however, it seem more natural to use a product kernel (11.5) where each factor incorporates its own separate choice of bandwidth. A product form can also be used in a spatial context, but leads to a kernel which typically has diamond-shaped contours with the points oriented along the coordinate axes.

In connection with dimensionality, it is worth noting that if the data is assumed roughly uniformly distributed over a square (cubic) region of scale L, then to secure about 10% of the data within the effective support of the kernel, h will need to be taken of the order of $L/10$ in one dimension, $L/3$ in two dimensions, and $L/2$ in three dimensions. In other words, the higher the dimensionality, the greater the smoothing required to produce a density estimate with a given reliability at a single point.

11.3.3 Boundary effects

If no precautions are taken, density estimates near the boundary, and especially the corners, of the observation region will be severely biased downwards by the fact that the integral in (11.3) will be truncated by the boundaries of the observation region. The problem increases in seriousness with dimension.

Wherever possible, this difficulty should be avoided by extending the catalog to include data from a data set extending a distance at least h beyond the edge of the region over which intensity estimates are required. The estimates then incorporate data from this border region but are not themselves extended into that region.

Where this is not possible the data set can be artificially extended by reflecting the original data in each boundary of the region, with higher reflections as needed in the corners. The same procedure can then be applied to this artificially extended data set.

An alternative frequency used in the literature is to weight the integral $k_h(\mathbf{x} - \mathbf{x}_i)$ in (11.3) inversely by the integral of $k_h(\mathbf{x} - \mathbf{u})$ over the observation region. This is computationally more cumbersome and, according to Diggle and Marron (1988), is fractionally less effective in reducing the bias.

11.3.4 Selection of bandwidth

If the time is not explicitly included as a variable in the analysis, a simple procedure is to divide the data into two parts, an initial or training period which is used to produce an intensity estimate, and a final evaluation period in which the goodness of fit is scored in some suitable manner. This approach has the considerable advantages that it provides direct information on the sorts of error that are likely to arise if the intensity estimates are to be used as forecasts for engineering and design purposes, and automatically incorporates the effects of features such as clustering which may invalidate estimates based on the behavior of i.i.d. samples.

Two scoring methods which may be used here are the K–L (Kullback–Leibler) score

$$S_1 = \Sigma \log \hat{\lambda}(\mathbf{x}_i) - \int \hat{\lambda}(\mathbf{x})\, d\mathbf{x} \tag{11.6}$$

and the MISE (mean integrated square error) score

$$S_2 = 2\Sigma\hat{\lambda}(\mathbf{x}_i)w(\mathbf{x}_i) - \int \hat{\lambda}(\mathbf{x})^2 w(\mathbf{x})\, d\mathbf{x}, \tag{11.7}$$

where $\hat{\lambda}(\mathbf{x})$ is the proposed intensity estimate adjusted if necessary for the length of the evaluation period, and the sums are taken over the events $\mathbf{x}_i$ in that period. The weight functions $w(\mathbf{x}_i)$ may be taken as unity in many cases: see also Marron (1987). The scores are derived from estimates of the K–L and MISE distances between $\hat{\lambda}(\mathbf{x})$ and the observed data.

The analysis is repeated for different values of the bandwidth h to determine the value giving the maximum score. Robustness of the choice can be tested by varying the relative lengths of the training and evaluation periods.

An analog to this method in the classical density estimation problem would be to select the training set randomly from the full data set and use the remainder for evaluation. This approach is less used in the literature than 'leave-one-out' cross-validation, where a particular data point $\mathbf{x}_i$ is left out of the analysis, and a value of $\hat{\lambda}(\mathbf{x}_i)$ computed from all remaining data points. The resulting values are then scored as in (11.6) or (11.7), the final integrals in both cases being based on the estimate using all values (and in fact $\int \hat{\lambda}(\mathbf{x})\, d\mathbf{x} = n$ if the estimate is properly normalized).

The cross-validation procedure is computationally more cumbersome, though quick methods of handling the 'leave-one-out' statistics have been developed – see e.g. Silverman (1986) – and is sensitive to the effects of clustering.

Other methods for probability densities are reviewed by Silverman (1986) and Marron (1988), but also seem likely to suffer some lack of robustness in the present context.

In a space–time problem the training-and-evaluation procedure requires some modification because the current estimate of the intensity at time t cannot be expected to be valid at all times $t' > t$. There is also a bias problem if a two-sided kernel is used in time. The following procedure may be suggested to overcome these difficulties. Firstly, use a one-sided kernel for the time factor $k_2(\mathbf{t})$ in the product form of kernel $k_1(\mathbf{x})k_2(\mathbf{t})$. Next select some Δ as a reasonable forecast forward period, say 5–10 per cent of the total time span of the observation period. Then for a sequence of increasing values of t_k use the data up to t_k to produce an estimate $\hat{\lambda}(t_k, \mathbf{x})$, 'freeze' it at time t_k, and evaluate the 'frozen' forecast over the period $(t_k, t_k + \Delta)$. Finally increase t_k to $t_{k+1} = t_k + \Delta$, repeat, and sum the evaluation scores for different k. The resulting scores can then be compared for different bandwidths h and the optimal bandwidth chosen for forecasts of depth Δ.

11.3.5 Error estimates

Order-of-magnitude estimates for confidence intervals or standard errors for $\hat{\lambda}(\mathbf{x})$ can be obtained from a consideration of second-order properties of the underlying point process.

Suppose first that clustering and other dependence effects can be ignored, so that the process can be treated as a nonhomogeneous Poisson process with time intensity $\lambda(\mathbf{x})$, say, and that the intensity can be treated as approximately constant over distances comparable with the bandwidths h. Then from standard formulae for the moment densities (see e.g. Daley and Vere-Jones, 1989, Chapters 5–7),

$$\operatorname{var} \hat{\lambda}(\mathbf{x}) = \int \lambda(\mathbf{x} - \mathbf{u}) k_h^2(\mathbf{u})\, d\mathbf{u} \cong \lambda(\mathbf{x}) \int k_h^2(\mathbf{u})\, d\mathbf{u}. \tag{11.8}$$

Thus the estimate $\hat{\lambda}(\mathbf{x})$ is approximately unbiased with a standard error proportional to $\sqrt{\hat{\lambda}(\mathbf{x})}$. The dispersion index (variance/mean ratio) is approximately independent of $\mathbf{x}$ and given by

$$\rho = \int k_h^2(\mathbf{u})\, d\mathbf{u} \Big/ \int k_h(\mathbf{u})\, d\mathbf{u} = \int k_h^2(\mathbf{u})\, d\mathbf{u}.$$

Approximate confidence bands can be obtained by approximating to the distribution of $\hat{\lambda}$ by a gamma distribution, i.e. by supposing $\hat{\lambda}$ can be written in the form $\hat{\lambda} = A\chi_k^2$, where A is a constant of proportionality and χ_k^2 is a χ^2 random variable with k degrees of freedom, where A and k are to be determined. Equating means and variances from the above we find

$$\hat{A} = \frac{1}{2} \int k_h^2(\mathbf{u})\, d\mathbf{u}$$

$$\hat{k} = \hat{\lambda}(\mathbf{x})/\hat{A},$$

from which the appropriate confidence bounds can be read off from standard tables of the χ^2 distribution.

Although such estimates can be used to provide confidence intervals of $\hat{\lambda}(\mathbf{x})$ at a point, there has yet to be devised a convenient way of graphically displaying the variation of such intervals over a two- or three-dimensional region. A simpler but less informative alternative is to plot contours of the standard error from (11.8).

11.3.6 Effects of clustering

In considering smoothing problems it is important to distinguish between two types of spatial clusters: those such as aftershock sequences which last only for a short time and rarely repeat themselves at the same locations; and spatial concentrations which persist in time at the same locations. Obviously the forecasting implications of these two types of clusters are very different. In the former case, the essential problem is to forecast the location of future clusters, with a further forecast of the expected cluster size. In the latter case, the problem is to delineate the shape of the high-intensity region, and to forecast the fluctuations of intensity in time within that region.

Relatively little seems to have been written about the problem of clustering in the literature on density estimation. Diggle and Marron (1988) discuss smoothing and bandwidth selection for a doubly stochastic Poisson process (Cox process), but their intention is not to smooth over but rather to provide an optimal representation of the clusters within the data set. By contrast Musmeci and Vere-Jones (1986) describe a histogram-type approach which seeks to smooth over the major clusters, and to estimate the intensity of the underlying process of cluster centers.

In so far as bandwidth selection is concerned, the training-evaluation method will broadly adapt to these two types of features, since clusters occurring in unpredicted locations will reward more smoothing while persistent regions of high activity will reward less smoothing. (Of course the use of a fixed bandwidth is a limitation, but not one we shall attempt to remove in the present discussion). Most of the other methods, including cross-validation, are based on the assumption of i.i.d. data, and are likely to undersmooth.

The other point at which clustering is relevant is the estimation of errors, since the clustering will introduce a greater amount of variability per observation than in the Poisson or i.i.d. case. In a cluster model the formula for the variance of the estimate $\hat{\lambda}(\mathbf{x})$ generalizes to

$$\operatorname{var} \hat{\lambda}(\mathbf{x}) \simeq \hat{\theta}\hat{\lambda}(\mathbf{x}) \int k_h^2(\mathbf{u})\, d\mathbf{u}$$

and the estimates $\hat{A}$, $\hat{k}$ for the approximating χ^2 distribution became respectively

$$\hat{A} = (\hat{\theta}/2) \int k_h^2(\mathbf{u})\,\mathrm{d}\mathbf{u}$$

$$\hat{k} = \hat{\lambda}(\mathbf{x})/\hat{A},$$

where $\hat{\theta}$ is the variance/mean ratio for the cluster size. To find an estimate of $\hat{\theta}$, one approximate procedure is to divide the observation region into cells of equal edge length and to consider the variance/mean ratio for the numbers of events in contiguous blocks of four cells (two dimensions) or eight cells (three dimensions). The average of this variance/mean ratio across all disjoint blocks of cells can then be used as an estimate of the variance/mean ratio of the number of points in a typical cluster (see Musmeci and Vere-Jones, 1986).

11.3.7 Kernel estimates for mean magnitudes and *b*-values

If the space and time coordinates are thought of as regressor variables, the problem of estimating mean magnitude as a function of space and time becomes a problem in nonparametric regression. The points (epicentres and origin times) at which magnitude values are taken are randomly rather than systematically located, but this does not introduce a major change in methodology. Since the *b*-value is effectively the reciprocal of the mean magnitude, contour maps and confidence bounds for mean magnitude can be readily converted into contour maps and confidence bounds for *b*-values.

The kernel estimate of the mean magnitude at a point $\mathbf{x}$ is a weighted mean of the magnitudes of events in the neighborhood of $\mathbf{x}$, with weights depending (through the kernel) on the distance of the event from $\mathbf{x}$. This results in an estimate of the form

$$\hat{M}(\mathbf{x}) = \left[\sum_{i=1}^{n} M_i k_h(\mathbf{x} - \mathbf{x}_i)\right] \Big/ \left[\sum_{i=1}^{n} k_h(\mathbf{x} - \mathbf{x}_i)\right]$$

$$= \left[\int k_h(\mathbf{x} - \mathbf{u}) M(\mathbf{u}) N(\mathrm{d}\mathbf{u})\right] \Big/ \left[\int k_h(\mathbf{x} - \mathbf{u})\, N(\mathrm{d}\mathbf{u})\right], \qquad (11.9)$$

where $(\mathbf{x}_i, M_i)$ are the coordinates of the events, and $M(\mathbf{u})$ is a notional 'magnitude field' which is sampled at the points $\mathbf{x}_i$ and whose expected value at $\mathbf{x}$ is estimated by $M(\mathbf{x})$.

Similar problems concerning the choice of kernel and bandwidth, edge effects, etc. arise here as in the estimation of intensities, and can be tackled in similar ways. Härdle (1990) and Härdle *et al.* (1988), provide further background and references.

11.3.8 Scoring methods for magnitudes

The analog of the IMSE score is here the function

$$\sum_{i=1} [M_i - \hat{M}(\mathbf{x}_i)]^2 w(\mathbf{x}_i). \qquad (11.10)$$

The rationale behind the 'leave-one-out' cross-validation procedure is more transparent here, but the scoring can be used equally in a training-and-evaluation scheme.

The analog to the K–L score corresponds to the loglikelihood of a set of i.i.d. exponential random variables with parameters $\lambda_i = \hat{M}(\mathbf{x}_i)$:

$$\Sigma[\log \hat{M}(\mathbf{x}_i) - M_i/\hat{M}(\mathbf{x}_i)]. \tag{11.11}$$

This latter seems more appropriate since (11.10) corresponds rather to the case of normally distributed variables. In general the mean magnitudes appear to vary rather less than the intensities, so that wider bandwidths may be anticipated.

11.3.9 Error estimates for mean magnitudes

Conditionally on the locations $\mathbf{x}_i$, the estimate (11.9) has the form of a weighted average $\Sigma\alpha_i(\mathbf{x})M_i$, where the weights $\alpha_i(\mathbf{x})$ decrease as a function of the distance between $\mathbf{x}$ and $\mathbf{x}_i$. Assuming independent magnitudes, the variance of $\hat{M}(\mathbf{x}_i)$ is then

$$\text{var } \hat{M}(\mathbf{x}) = \Sigma\alpha_i(\mathbf{x})^2 \text{ var } M_i \cong [\Sigma\alpha_i(\mathbf{x})^2]\hat{M}(\mathbf{x})^2,$$

where $\alpha_i(\mathbf{x}) = k_h(\mathbf{x} - \mathbf{x}_i)/[\Sigma k_h(\mathbf{x} - \mathbf{x}_i)]$.

But

$$\Sigma\alpha_i(\mathbf{x})^2 = [\Sigma k_h^2(\mathbf{x} - \mathbf{x}_i)]/[\Sigma k_h(\mathbf{x} - \mathbf{x}_i)]^2 \cong \hat{\lambda}(\mathbf{x})^{-1} \int k_h^2(\mathbf{u})\,d\mathbf{u},$$

so that

$$\text{var } \hat{M}(\mathbf{x}) \cong [\hat{M}(\mathbf{x})^2/\hat{\lambda}(\mathbf{x})] \int k_h^2(\mathbf{u})\,d\mathbf{u}. \tag{11.12}$$

Again rough confidence intervals can be obtained by equating the mean and variance to the corresponding quantities for a χ^2-multiple, $A\chi_k^2$. Estimates of A and k are given by

$$\hat{A} = [\hat{M}(\mathbf{x})/2\hat{\lambda}(\mathbf{x})] \int k_h^2(\mathbf{u})\,d\mathbf{u}$$

$$\hat{k} = \hat{M}(\mathbf{x})/\hat{A}.$$

We mention finally that the intensive space–time clustering of earthquakes does not extend to their magnitudes, which seem close enough to being independent for most practical purposes. In contrast to the intensity estimates, therefore, there is no need to inflate the variance (11.12) by a factor to accommodate clustering.

11.4 Part III: Semiparametric methods: orthogonal function and spline estimates

11.4.1 General issues

The methods to be reviewed in this section depend on representing the intensity or mean magnitude as a linear combination of standard functions with coefficients estimated from the data. Thus the estimates take the general form

$$\hat{Q}(\mathbf{x}) = \sum_{k=1}^{K} \hat{a}_k \phi_k(\mathbf{x}) \tag{11.13}$$

where $\hat{Q}(\mathbf{x})$ is either the intensity or its logarithm in the first case, and the mean magnitude or its reciprocal in the second case. The $\phi_k(\mathbf{x})$ are taken either from one of the standard orthogonal function bases for a finite interval, such as the trigonometric functions, or a spline basis. The coefficients $\hat{a}_k$ are estimated from the data, and the crux of the smoothing problem here, analogous to the choice of bandwidth in the kernel-smoothing context, is the choice of K, or more generally of a weighting scheme for the higher-order terms. In essence, the models fall into the category of generalized linear (regression) models, with the underlying variates having Poisson distributions in the intensity case and exponential distributions in the magnitude case. The special feature is the very large number of terms which may need to be included (50–100 would not be unusual in some situations at least) and the associated problem of weighting the higher-order terms.

The earlier discussions of orthogonal function representations, such as Kronmal and Tarter (1968) typically made use of method-of-moments estimates for the $\hat{a}_k$, often with a weighting term $w_k = (c + k)^{-\alpha}$ multiplying the basic estimates. In the case of trigonometric functions on the interval [0, 1], for example, the moments estimates for intensity are just the Fourier coefficients

$$\hat{\alpha}_0 = N,$$

$$\hat{\alpha}_k = 2 \sum_{i=2}^{N} \cos 2\pi k x_i = 2 \int_0^1 \cos(2\pi k x)\, N(\mathrm{d}x) \qquad (k = 1, 2, \ldots)$$

and in (11.13),

$$\hat{a}_k = w_k \hat{\alpha}_k.$$

Discussions by Wahba (1975), and others in the density estimation context showed that with optimal choices of the weighting factors, similar rates of convergence could be obtained with these as with the kernel methods. The sense in which these results are asymptotically optimal needs to be reformulated in the intensity context (see Diggle and Marron, 1988), but it seems reasonable to assume that the efficiencies would be comparable here also.

Despite these results, there are some obvious disadvantages to the method of moments approach. It cannot be guaranteed without further precautions that the estimates will be non-negative. There is, moreover, an implicit link with techniques which are optimal for normal distributions, while here we have Poisson or negative exponential distributions. More recent investigations have centered around likelihood-based methods, including the use of Bayesian methods and penalized likelihoods, and automatic model selection criteria such as AIC for determination of the order K. It is these likelihood-based approaches that we consider in detail below. Another advantage is their ability to link into more complex models, as will be indicated in Part IV.

Boundary problems arise with these approaches as with the kernel method, and typically manifest themselves in the form of uncontrolled behavior ('flapping wings') near the edges of the region, particularly when polynomial functions are used. Again such effects can be reduced by making use of a boundary zone, augmenting the original data set with additional data where that is available, or by reflecting the original data in the boundaries where it is not.

Clustering also creates difficulties, particularly for the intensity estimates. Should clusters be treated as nuisance features to be smoothed over, or part of the geophysical pattern of interest? For the time being we shall take the latter point of view, assuming that aftershocks and similar features have been either removed or deliberately retained as features of the occurrence pattern which have to be fitted by the Poisson-based modeling procedures. Models which attempt to fit clustering as a separate component of the activity will be described in Part IV.

Finally it should be emphasized that there are many open questions concerning the performance of the techniques described below, and much scope for further investigation.

11.4.2 Orthogonal function estimates of intensity

The underlying assumption throughout this subsection is that the data forms a nonhomogeneous Poisson process in space (space–time, etc.), with rate function $\lambda(\mathbf{x})$. The log likelihood then takes the form

$$\log L = \sum_{i=1}^{N} \log \lambda(\mathbf{x}_i) - \int_A \lambda(\mathbf{x})\,d\mathbf{x}, \tag{11.14}$$

where the $\mathbf{x}_i$ are the coordinates of the events and A is the observation region.

An initial decision is whether to expand $\lambda(\mathbf{x})$ itself or its logarithm. Expanding the intensity directly leads to a simple evaluation of the integral (trivial if the functions are orthogonal with respect to Lebesgue measure over A), but has the major disadvantage that special techniques must be used to ensure $\lambda(\mathbf{x})$ remains non-negative over the entire region, without which constraint totally spurious likelihoods and estimates can be obtained (see Vere-Jones and Musmeci, 1986).

Since, however, the natural parameter for a Poisson distribution is the logarithm of the intensity, it is the logarithm which is the more natural choice. The successive sums $\Sigma\phi_k(\mathbf{x}_i)$ are then sufficient statistics for the respective coefficients $\hat{a}_k$. A unique solution is guaranteed by the convexity of (11.14) as a function of the parameters. The main computational problem is evaluating the integral on the right-hand side of (11.14) at each step in the optimization subroutine. If numerical integration is used, there is a danger that minor errors in the integration can lead to instabilities in the optimizing search routine and cause the latter to fail; higher-order precision for the integration step may therefore be needed. Since the derivatives of (11.14) can be written down fairly simply, for example

$$\frac{\partial \log L}{\partial a_k} = \sum_{i=1}^{N} \phi_k(\mathbf{x}_i) - \int_A \phi_k(\mathbf{x})\lambda(\mathbf{x})\,\mathrm{d}\mathbf{x} \tag{11.15a}$$

$$\frac{\partial^2 \log L}{\partial a_k \partial a_n} = -\int_A \phi_k(\mathbf{x})\phi_n(\mathbf{x})\lambda(\mathbf{x})\,\mathrm{d}\mathbf{x}, \tag{11.15b}$$

a form of optimization which uses these derivaties may yield more stable convergence.

Expression (11.15b) shows up the role of orthogonality in the expressions; it is not crucial, but helps to keep the Hessian close to diagonal form, improving the convergence properties of the optimization routine.

The choice of orthogonal family seems to have no major effect on the performance; the two most obvious choices in one dimension are the trigonometric functions $\cos 2\pi kx$ and the Legendre polynomials, both orthogonal with respect to constant weight functions. Use of these in standard form requires rescaling the observation interval to $[0, 1]$ and $[-1, 1]$ respectively. We have found no substantial reason for going beyond the trigonometric polynomials.

In higher dimensions, orthogonal families on the corresponding square or cube can be built up from products of the one-dimensional functions, say with two space and one time dimension,

$$\phi_k(\mathbf{x}) = \phi_{k_1}(\mathbf{x})\phi_{k_2}(y)\phi_{k_3}(t),$$

where a suitable lexical ordering is imposed on the triple (k_1, k_2, k_3) to obtain from it the single integer k. An immediate difficulty, which affects determination of the maximum order K, is that there is no natural ordering of the terms.

Perhaps the most straightforward approach to determining the value of K, at least in one dimension, is through the use of the AIC criterion (Akaike, 1977)

$$-\tfrac{1}{2}\mathrm{AIC}(K) = \log L(K) - K, \tag{11.16}$$

where $L(K)$ refers to the maximum likelihood obtained with just K parameters fitted. A maximum possible value $K_{\max}$ is fixed *a priori*, and that value of K chosen which maximizes (11.16) in the range $1 < K \leqslant K_{\max}$.

In higher dimensions the problem caused by the lack of a natural ordering can be partially overcome by treating the parameters in blocks; for example with two spatial variables, all terms $\phi_{k_1}(x)\phi_{k_2}(y)$ with $k_1 + k_2 = s$, say, can be treated as one block and included or omitted together. With two space and one time coordinates, a square array can be built up by treating the spatial terms in blocks to give one parameter s, and the time terms individually to give a second parameter k. Then the pair (s, k) maximizing (11.16) within some constraints $1 \leqslant s \leqslant s_{max}$ and $1 \leqslant k \leqslant k_{max}$ can be chosen.

As an alternative or back-up to the use of AIC, a training and evaluation method can be used to determine K, again using the Kullback–Leibler score (11.14).

The probability density estimation routine incorporated in the program 'GALTHY' (Akaike and Arahata, 1978) incorporates not a single estimate for a particular choice of K but a weighted average

$$\hat{\lambda}(\mathbf{x}) = \sum_k W(k)\hat{\lambda}_k(\mathbf{x}),$$

where the weights $W(k)$ are taken proportional to $\exp\{\text{AIC}(k)\}$.

More recently, Akaike and colleagues have developed an 'objective Bayesian approach' based on selecting a very large value of K and maximizing not (11.14) itself but a posterior log likelihood

$$\log L^*(\mathbf{a} \mid \boldsymbol{\theta}) = \log \pi(\mathbf{a} \mid \boldsymbol{\theta}) + \log L(\mathbf{a}; \mathbf{x}), \tag{11.17}$$

where $\log L(\mathbf{a}; \mathbf{x})$ is the log likelihood (11.14) and $\pi(\mathbf{a} \mid \boldsymbol{\theta})$ is a prior density for the parameter set $\mathbf{a}$, the prior itself depending on some parameters $\boldsymbol{\theta}$. The term $\log \pi(\mathbf{a} \mid \boldsymbol{\theta})$ can be thought of as a penalty term which penalizes departures from smoothness, for example by down-weighting the higher-order terms in a trigonometric series expression. In the one-dimensional case a natural choice in this context might be a prior in which the a_k had independent normal distributions with variance proportional to k^θ for some $\theta > 0$, so that (11.17) takes the form

$$\log L^*(\mathbf{a} \mid \boldsymbol{\theta}) = C(\theta, B) + \log L(\mathbf{a}; \mathbf{x}) - \tfrac{1}{2} B \sum_{k=0}^{K} k^\theta a_k,$$

where $C(\theta, B)$ is a normalizing constant and the parameters (θ, B) then remain as 'hyperparameters' to be determined by a further round of maximization (possibly with respect to a prior of their own) or by a training and evaluation routine. Akaike suggests integrating the likelihood in (11.17) to obtain (suppressing the dependence on $\mathbf{x}$ for brevity)

$$\text{ABIC} = L^t(\boldsymbol{\theta}) = \int L^*(\mathbf{a} \mid \boldsymbol{\theta})\, d\mathbf{a} \tag{11.18}$$

and then maximizing (11.18) with respect to $\boldsymbol{\theta}$. The difficulty of the integration in (11.18) is overcome by an approximation suggested by Ishiguro and Sakamoto (1983). Since the main contribution to the integral comes from values in the neighborhood of the value $\hat{\mathbf{a}}$ which maximizes (11.17), $\log L^*(\mathbf{a} \mid \boldsymbol{\theta})$ can be replaced in (11.18) by the quadratic approximation

$$\log L^*(\mathbf{a} \mid \boldsymbol{\theta}) = \log L^*(\hat{\mathbf{a}} \mid \boldsymbol{\theta}) - \tfrac{1}{2}(\mathbf{a} - \hat{\mathbf{a}})^{\mathrm{T}} H(\hat{\mathbf{a}} \mid \boldsymbol{\theta})(\mathbf{a} - \hat{\mathbf{a}}), \tag{11.19}$$

where $H(\hat{\mathbf{a}} \mid \boldsymbol{\theta})$ is the Hessian for (11.17) which can be obtained directly or as part of the numerical optimization routine. Thus (11.18) simplifies to

$$\log L^t(\boldsymbol{\theta}) \cong \log L^*(\hat{\mathbf{a}} \mid \boldsymbol{\theta}) - \log \det H(\hat{\mathbf{a}} \mid \boldsymbol{\theta}) - k \log 2\pi,$$

which can be maximized with respect to $\boldsymbol{\theta}$ in a relatively straightforward manner.

Only a limited amount of investigation has been carried out on these various alternatives. Most of the recent Japanese work has concentrated on the use of spline functions and a different form of smoothness prior (see the discussion below). Vere-Jones and Musmeci (1986) report on two- and three-dimensional applications of the simple AIC approach. The Bayesian variant suggested above does not seem to have been tried in this context.

We conclude by observing that once the parameter vector $\hat{\mathbf{a}}$ has been determined, and the Hessian

$$H(\hat{a}) = \left.\frac{\partial^2 \log L}{\partial \mathbf{a} \partial \mathbf{a}^{\mathrm{T}}}\right|_{\mathbf{a}=\hat{\mathbf{a}}}$$

evaluated, an approximate expression for the standard deviation of $\hat{\lambda}(\mathbf{x})$ can be obtained from

$$\operatorname{var}[\log \hat{\lambda}(\mathbf{x})] \cong \boldsymbol{\phi}^{\mathrm{T}} H^{-1}(\hat{\mathbf{a}}) \boldsymbol{\phi},$$

where $\boldsymbol{\phi}$ is the vector $\{\phi_k(\mathbf{x});\ k = 1, 2, \ldots, K\}$. In the Bayesian case, L^* should replace L in the Hessian.

11.4.3 Spline estimates of the intensity

To obtain a spline representation of the one-dimensional function $Q(x)$ the interval (a, b) over which the estimate is required is first divided into K subintervals (δ_k, δ_{k+1}) of length $\Delta = (b - a)/K$, where $\delta_0 = a$, $\delta_K = b$, and the intervening $\{\delta_k\}$ are referred to as 'knots'. Then, on each sub-interval the function $Q(x)$ is represented as a low-order polynomial, the polynomials being matched so as to satisfy continuity requirements at the knots. We shall consider only cubic splines, in which case the polynomials are cubic, and the function and its first two derivatives are required to be continuous across the knots. The continuity conditions are elegantly incorporated into the form of the representation by writing the polynomial on (δ_k, δ_{k+1}) in the form

$$Q(x) = \sum_{j=0}^{3} a_{k+j} B_{4-j}\left(\frac{x - \delta_k}{\Delta}\right),$$

where the $B_i(x)$ vanish outside the interval (0, 1) and in that interval have the form

$$B_1(u) = u^3/6; \qquad B_2(u) = (-3u^3 + 3u^2 + 3u + 1)/6$$

$$B_3(u) = (3u^3 - 6u^2 + 4)/6 \qquad B_4(u) = (-u^3 + 3u^2 - 3u + 1)/6$$

(see Inoue, 1986; Ogata and Katsura, 1988).

It is easily checked that the continuity conditions are met automatically in passing over a knot, and that overall the function $Q(x)$ has a linear representation

$$Q(x) = \sum_{k=0}^{K+3} a_k \beta_k(x), \tag{11.20}$$

where each $\beta_k(x)$ is a linear combination of terms $B_i((x - \delta_i)/\Delta)$.

The representation can readily be extended to higher dimensions by using products of the $B_i(u)$. In two dimensions, for example, $Q(x, y)$ is written in the form

$$Q(x, y) = \sum_{i,j=0}^{3} a_{k+j,n+i} B_{4-j}\left(\frac{x - \delta_k}{\Delta_x}\right) B_{4-i}\left(\frac{y - \gamma_n}{\Delta_y}\right)$$

on rectangles $(\delta_k, \delta_{k+1}) \times (\gamma_n, \gamma_{n+1})$, with sides of length Δ_x, Δ_y, respectively, the observation region comprising $K \times L$ such rectangles. Again it can be checked that $Q(x, y)$ is continuous in both coordinates across the boundaries of such rectangles and that overall the function can be represented as a linear combination of product terms analogous to (11.20).

From this point the discussion proceeds much as in the previous subsection. For $Q(x)$ we may take $\lambda(x)$ or its logarithm, the latter being preferred. The number of knots K replaces the maximum order as the primary variable controlling the degree of averaging in the estimation. The value of K can be fixed by AIC or by a training and evaluation scheme. Alternatively, as discussed in detail by Ogata and Katsura (1988), the ABIC procedure can be followed. Then the value of K is taken very large (of the order of the number of data points) and the straight likelihood replaced by a penalized form in which the penalty can be interpreted as deriving from a prior distribution on the parameters. The prior they suggest follows earlier work on penalized likelihood estimation of probability densities by Good and Gaskin (1971) (see also Whittle, 1958) and takes the form

$$\log \pi(\mathbf{a} \mid \boldsymbol{\theta}) = C(\theta_1 \theta_2) - \theta_1 \Phi_1(\mathbf{a}) - \theta_2 \Phi_2(\mathbf{a})$$

with $C(\theta_1 \theta_2)$ a normalization constant and

$$\Phi_1(\mathbf{a}) = \int_A \sum \left(\frac{\partial \log \lambda}{\partial x_i}\right)^2 \mathrm{d}x$$

$$\Phi_2(\mathbf{a}) = \int_A \left\{\sum_i \left(\frac{\partial^2 \log \lambda}{\partial x_i^2}\right)^2 + 2 \sum_{i<j} \left(\frac{\partial^2 \log \lambda}{\partial x_i \partial x_j}\right)^2\right\} \mathrm{d}\mathbf{x}$$

It therefore penalizes high values of the slope and curvature of $\log \lambda$. A technical difficulty arises because the resulting Gaussian prior is singular: see Ogata and Katsura (1988) for details. The estimation procedure is completed by minimizing the ABIC criterion (11.18) with respect to the parameters θ_1 and θ_2, using for this purpose the quadratic approximation (11.19).

The spline representation seems well suited for interpolation purposes, having effectively one free parameter per cell in the observation region, and hence disposing the available degrees of freedom uniformly over the observation region. As with kernel methods, some problems arise if the estimates are to be used for extrapolation or forecasting.

11.4.4 *b*-value estimates

The magnitudes can be treated as an extra dimension or 'mark' associated with the spatial or space–time process. The overall process then takes the form of a point process in one higher dimension, with intensity function of the form

$$\lambda^*(\mathbf{x}, M) = \lambda(\mathbf{x}) f(M \mid \mathbf{x}) \tag{11.21}$$

where $\lambda(\mathbf{x})$ is the space or space–time intensity, and $f(M \mid x)$ is the conditional density for magnitudes given the particular location $\mathbf{x}$. The overall likelihood then splits into two terms:

$$\begin{aligned} \log L &= \sum \log \lambda^*(\mathbf{x}_i M_i) - \int_A \int_M \lambda^*(\mathbf{x}, M)\, \mathrm{d}\mathbf{x}\, \mathrm{d}M \\ &= \left\{ \sum \log \lambda(\mathbf{x}_i) - \int_A \lambda(\mathbf{x})\, \mathrm{d}\mathbf{x} \right\} + \sum f(M_i \mid \mathbf{x}_i) \\ &= \log L_1 + \log L_2. \end{aligned} \tag{11.22}$$

The term in braces is the spatial (or space–time) likelihood already treated, while the second sum represents the partial log likelihood for the magnitudes, taking the locations as given.

The factorization (11.21) embodies the strong assumption that the magnitude distribution depends only on the location $\mathbf{x}$ and does not depend on other features of the process, such as, for example, the magnitudes of neighboring or immediately prior events. In this sense it is the appropriate extension to magnitudes of the Poisson assumption already made for the space and time coordinates.

Assuming also that the magnitudes are exponentially distributed, with

$$f(M \mid \mathbf{x}) = \beta(\mathbf{x}) \exp\{-M\beta(\mathbf{x})\}$$

we can state the main problem of the present subsection as that of trying to estimate the '*b*-value field' $\beta(\mathbf{x})$. Note that in fact

$$b = (\log_e 10)^{-1} \beta = \beta / 2{\cdot}3026.$$

To this end we may take $\beta(\mathbf{x})$ or its reciprocal as $Q(\mathbf{x})$, and seek its expansion in terms of orthogonal or spline functions. Note that $\beta(\mathbf{x})$ corresponds to the reciprocal of the mean magnitude treated in (11.9). It is, nevertheless, the more appropriate function to use here, since it corresponds to the natural parameter of the exponential distribution. If, for example, we take

$$\beta(\mathbf{x}) = \sum_{k=0}^{K} \beta_k \phi_k(\mathbf{x}),$$

the partial likelihood for the magnitudes take the form

$$\log L_2 = \sum^{K} \log \beta(\mathbf{x}_i) - \sum M_i \beta(\mathbf{x}_i)$$
$$= \sum \log \beta(\mathbf{x}_i) - \sum_k \beta_k \left(\sum_i M_i \phi_k(\mathbf{x}_i) \right)$$

so that the sums $\Sigma M_i \phi_k(\mathbf{x}_i)$ are sufficient for the $\{\beta_k\}$.

From this point on similar considerations apply to the estimation of the magnitude as to the estimation of intensities. The functions used in the expression can be orthogonal functions or splines, the number of terms can be decided by AIC or by training and evaluation, or by a Bayesian approach based on the ABIC criterion can be used. Variants of the last approach are given by Imoto and Ishiguro (1986) and Ogata and Katsura (1988).

11.5 Part IV: Stochastic process models

11.5.1 Introduction

There is, of course, a huge range of stochastic models that could be applied to the many different facets of earthquake occurrence. We shall restrict attention to just two models, which follow closely the aims and procedures outlined in the preceding sections, but extend these by incorporating the first elements of a dependence structure between events.

A paradoxical feature of earthquakes is that they exhibit two mutually exclusive types of dependence; one leading to clustering, the other leading to the spacing out of events. Even the most cursory examination of a recent instrumental catalog will show the extreme prevalence of clustering, in space as well as time, but most of all in space and time together. Spacing out effects are not so obvious from the data, and their existence is not established beyond all possible doubt. However, historical records from several different parts of the world suggest that the occurrence of very large events seems to inhibit the occurrence of similar large events in the same region for periods of decades or sometimes even centuries. Reconciling these two opposing features within a single comprehensive model is at present an unresolved problem, not so much because of the lack of a suitable mathematical apparatus, but because of the lack of physical insights to guide it.

Both the models we shall describe incorporate an extremely powerful principle in point-process model-building, namely that the framework developed in the previous section for unconditional Poisson processes can be extended to space–time processes with causal dependence relations. The same likelihood structure can be evoked even when the intensity function depends on the past of the process (or indeed on external variables) provided it does not depend on its future. The idea behind this principle is that the iterated product formulae for independent events, which dictates the form of the Poisson likelihood, can be replaced by an iterated product formula for dependent events provided the events are linked in a causal chain through time. Expositions of this theory are given, for example, in Bremaud (1981),

Daley and Vere-Jones (1989) and Karr (1986). The one place where the theory does not apply is to events in a purely spatial context, where the causal character is lost.

With this understanding, the basic task is simply that of selecting an explicit form for the intensity which will produce a model that is at once tractable and capable of reproducing the required qualitative features.

11.5.2 The Hawkes process and its extensions

The clustering model which lends itself most easily to development of this kind is the so-called 'self-exciting' or Hawkes process, first introduced by Hawkes (1971a), initially applied to earthquake modeling by Hawkes and Adamopoulos (1973), and subsequently developed in a series of recent papers by Ozaki, Ogata, Akaike and others; Ogata and Katsura (1986, 1988) give reviews with many further references.

In the simplest context of a sequence of events in time, the intensity of the Hawkes process is of the form

$$\lambda(t) = \lambda_0(t) + \sum_{i:t_i<t} g(t - t_i) = \lambda_0(t) + \int_0^t g(t-u)\, N(\mathrm{d}u). \qquad (11.23)$$

Here $\lambda_0(t)$ (a given function) represents the intensity of a background or 'immigration' point process which is assumed to follow a Poisson process and contributes events independently of all other aspects of the process; it is assumed to include also any residual effects from time $t < 0$. The second term describes the clustering behavior in terms of an 'infectivity' function $g(u)$, where $g(u)\,\mathrm{d}u$ represents the probability that an event at 0 produces a further event within $(u, u + \mathrm{d}u)$. Insofar as these further events can be interpreted as 'offspring' from an initial event ('ancestor') at the origin, the process admits a branching process interpretation of considerable power (Hawkes and Oakes, 1974). In particular, if the process is to settle down to an equilibrium form as $t \to \infty$, the branching process must be 'subcritical', so that the increase due to immigration can be balanced by the losses due to 'deaths'. For this the condition is

$$m = \int_0^\infty g(u)\, \mathrm{d}u < 1;$$

m then represents the mean number of 'offspring' per 'ancestor'.

In order to fit a model of this type it is necessary to give a parametric form to the infectivity function $g(u)$ and the immigration rate $\lambda_0(t)$. Simple choices are

$$g(t) = \sum a_k L_k(u) \mathrm{e}^{-\gamma u}$$

and

$$\lambda_0(t) = A + B\mathrm{e}^{-\delta t},$$

where the $L_k(u)$ are Laguerre polynomials orthogonal with respect to the exponential function e^{-u}. Examples of fitting such models to earthquake data using the likelihood (11.14) are given by Ogata and Akaike (1982) and Vere-Jones and Ozaki (1982), which may be referred to for computational details.

The nature of the clusters produced by this model can be understood from the representation (11.23). As soon as two or three events occur close together, the risk rapidly increases from the sum in the second term of (11.23), exhibiting the positive feedback implied in the 'self-exciting' title. Between events the risk falls back to the background level described by $\lambda_0(t)$. If the quantity m is close to 1, very substantial and extended clusters can develop, while for $m > 1$ the process is unstable and may increase without bound; for m near zero, on the other hand, the process is close to the Poisson model described by $\lambda_0(t)$.

The background term $\lambda_0(t)$ can easily be extended to include periodic or other time-dependent effects, so that the overall process embodies clusters imposed on a time-varying background. Ogata and Katsura (1986) have tackled in this way the problem of decomposing the catalog records into a long-term trend component and a short-term clustering component.

It is usually important to incorporate some dependence of cluster size on magnitude into the model. As in Section 11.4.4, this can be done by including the magnitude as an extra dimension. The intensity function of the extended process can often be written in the factored form

$$\lambda(t, M) = f(M)\lambda(t). \tag{11.24}$$

The magnitude distribution $f(M)$ could be allowed to depend explicitly on time or implicitly on time through a dependence on the history of the process. The factorization (11.24) corresponds to the simplest possible case of independently distributed magnitudes. As in Section 11.4.4 the log likelihood then decomposes into the sum of two terms, each of which can be maximized independently of the other. In place of (11.23), $\lambda(t)$ itself may now include an explicit dependence on magnitudes of past events, say

$$\lambda(t) = \lambda_0(t) + \sum_{i:t_i<t} \psi(M_i)g(t - t_i),$$

where for example $\psi(M) = \exp(\alpha M)$, while the stability condition becomes

$$m^* = E[\psi(M)] \int_0^\infty g(t)\,\mathrm{d}t < 1.$$

Another straightforward extension is to a family of 'mutually exciting' point processes, to use the terminology of Hawkes (1971b). In a bivariate model, for example, (11.23) is replaced by the pair of intensities

$$\lambda_1(t) = \lambda_{01}(t) + \int_0^t g_{11}(t-u)N_1(\mathrm{d}u) + \int_0^t g_{21}(t-u)\,N_2(\mathrm{d}u)$$

$$\lambda_2(t) = \lambda_{02}(t) + \int_0^t g_{22}(t-u)N_2(\mathrm{d}u) + \int_0^t g_{12}(t-u)\,N_1(\mathrm{d}u), \tag{11.25}$$

where explicit parametric representations can be introduced as previously for the functions $g_{ij}(t)$, $\lambda_{0i}(t)$. The joint likelihood can be written in the form

$$\log L(t_1, \ldots, t_N; s_1, \ldots, s_M) = \sum_{i=1}^{N} \log \lambda_1(t_i) + \sum_{j=1}^{M} \log \lambda_2(s_j) - \int_0^t [\lambda_1(t) + \lambda_2(t)]\, dt, \tag{11.26}$$

where $(t_1, \ldots, t_N)$ and $(s_1, \ldots, s_M)$ are the occurrence times of events of the first and second types respectively.

Again an explicit dependence on magnitudes can be incorporated into the functions $g_{ij}(t)$.

This bivariate model has been successfully used by Ogata and Akaike (1982) to demonstrate excitation of inland events by seaward events in Japan.

A final extension of these models is to include space as well as time and magnitude dimensions. In this case $\lambda(t)$ is replaced by a space–time intensity $\lambda(\mathbf{x}, t)$ which (still supposing independence of the magnitudes) again factorizes into a product $\lambda(\mathbf{x}, t)f(M)$ where $\lambda(\mathbf{x}, t)$ can be represented in the form

$$\lambda(\mathbf{x}, t) = \lambda_0(\mathbf{x}, t) + \sum_{i:t_i<t} \psi(M_i)g(\mathbf{x} - \mathbf{x}_i, t - t_i).$$

The infectivity function $g(\mathbf{x}, t)\, d\mathbf{x}\, dt$ now represents the probability that an 'ancestor' at $t = 0$, $\mathbf{x} = 0$ gives rise to an 'offspring' in the time interval $(t, t + dt)$ and the space interval $(\mathbf{x}, \mathbf{x} + d\mathbf{x})$. One possible structure for this function is

$$g(\mathbf{z}, u) = Ae^{-\beta u}k_u(\mathbf{z}), \tag{11.27}$$

where $e^{-\beta u}$ is an overall damping factor and $k_u(\mathbf{z})$ is a two- or three-dimensional kernel such as those described in Section 11.3, for example a diffusion kernel

$$k_u(\mathbf{z}) = (2\pi\sigma^2 u)^{-d/2} e^{-\|\mathbf{z}\|^2/2u\sigma^2}.$$

Then (11.27) represents a kind of diffusion of risk from the epicenter, together with an exponential damping factor. Of course many other structures in place of (11.27) could be suggested. The function $\lambda_0(\mathbf{x}, t)$ represents a possibly time-varying background risk, which could itself be modeled by the procedures of Sections 11.3 or 11.4. Some difficulties with convergence must be anticipated in an overall model because of the inherent ambiguity in attributing fluctuations in activity to clustering or long-term trends. A promising numerical procedure is to alternate procedures for estimating $\lambda_0(\mathbf{x}, t)$ and $g(\mathbf{z}, u)$, holding the parameters for the one fixed while those of the other are estimated.

Some preliminary work on fitting this model to Italian data is reported in Musmeci and Vere-Jones (1991). Here $\lambda_0(\mathbf{x})$ was assumed constant in time and proportional to a preliminary kernel estimate of seismicity. Both Gaussian and Cauchy kernels were used in (11.27) with little difference in performance. In both cases, however, it was found necessary to prevent the scaling factor σ^2 in the kernel from approaching too close to zero, to avoid spurious maxima in the likelihood function, possibly due to near coincidence of events in the historical catalog.

11.5.3 Stress release models

The classical elastic rebound model for earthquake release envisages a slow accumulation of stress in a region followed by its sudden release in the form of an earthquake. A simple stochastic setting for this idea is the theory of jump-type Markov processes with drift, analogous to the Lundberg collective risk model (see Feller, 1966). The stress level $X(t)$ is treated as a scalar Markov process which increases deterministically between events, the events themselves being controlled by a risk function $\psi(x)$, where $\psi(x)\,dt$ gives the chance of an event occurring in $(t, t+dt)$ given that $X(t)=x$, and a family of jump density functions $f(y \mid x)$ for the conditional distribution of the size of event, given that it occurs when $X(t)=x$.

Then if the times and stress drops (inferred approximately from the magnitudes) of the events in the catalog are denoted by (t_i, s_i), the evolution of the stress level $X(t)$ can be represented by the equation

$$X(t) = X(0) + \rho t - \sum_{i:t_i<t} s_i,$$

the last sum representing the accumulated stress release over the period $(0, t)$.

A key to the statistical analysis of the process is the observation that, although the underlying process $X(t)$ is Markovian, the observed data (t_i, s_i) can be treated as a point process in time–stress space with conditional intensity function

$$\lambda(t, s) = \psi[X(t)]f(s \mid X(t)). \tag{11.28}$$

Further analysis can then proceed along the lines of the preceding sections, maximizing the log likelihood (11.14) which, as in (11.22), splits into the sum of two terms,

$$\log L = \left\{\sum_i \log \psi[X(t_i)] - \int_0^t \psi[X(u)]\,du\right\} + \sum_i \log f(s_i \mid X(t_i)). \tag{11.29}$$

In the simplest possible case the size of the jump is independent of the stress level $X(t)$ and the two terms are completely decoupled. Otherwise the first term can be maximized in its own right as a 'partial likelihood' (see Cox, 1975), ignoring any correction due to the interaction with the second term.

To complete the estimation an explicit form for $\psi(x)$ is needed, of which the simplest is perhaps

$$\psi(x) = A \exp \beta x.$$

The stress releases themselves can be rather arbitrarily related to the square roots of the energy releases, themselves loosely related to magnitudes by an equation of the form (Gutenberg and Richter, 1956)

$$\log_{10} E = 1{\cdot}5M + \text{constant},$$

so that very approximately

$$s_i \propto 10^{0{\cdot}75M_i}.$$

Substitutions into (11.28) and (11.29), followed by a standard numerical optimization procedure, leads to estimates of the parameters A, β, ρ and $X(0)$.

Special cases of the model include the simple Poisson process ($\beta = 0$), and the Poisson process with exponential linear trend ($\beta \to 0$, $\beta\rho \to \theta$, say). Simple multivariate extensions allow the data to be separated into classes by region or magnitude or both, with dependence either on a common stress level or a separate stress level for each class. Zheng and Vere-Jones (1991) provide a detailed treatment of the Chinese historical catalog from this point of view, and use the AIC criterion to show that the data is best described by a division into two geographical regions, with perhaps a somewhat different relation between small and large magnitude events in the two regions. Similar analyses are currently under way for historical data for Iran and Japan, and again show significant improvements over the simpler Poisson models. Despite the manifest crudities of both model and data, the fact that positive results are obtained from a number of major historical catalogs suggests that the results are more than can be explained by chance coincidences or biased data selection, though the latter is always a danger since the fit is sometimes extremely sensitive to the inclusion or exclusion of individual events.

Earlier variants on the model are described by Knopoff (1971), Molchan (1984) and Vere-Jones (1978). A comprehensive treatment of ergodic and existence theorems is given by Zheng (1991). Statistical and other aspects are discussed by Ogata and Vere-Jones (1984), and Hayashi (1986).

11.5.4 Concluding remarks

The point process framework developed in Parts III and IV of this paper represents an extremely flexible and powerful approach to the more complex problems of estimating earthquake models. We have illustrated its use first with space and space–time Poisson models (Part III), then with the causal clustering and stress-release models of Part IV. Many other variations are possible. The so-called 'time predictable process' (Shimazaki and Nakata, 1980) is a special case of the stress-release model corresponding to fixed strength threshold above which failure is certain while below it failure is impossible. Combinations of factors are readily allowed for, as is the dependence of risk on external variables such as precursory phenomena. The latter can be incorporated along the same lines as explanatory variables within the 'Cox regression model' for times to failure (Cox, 1972); i.e. simply as additional regressor variables in a representation of the intensity such as (11.14). We have not developed in detail the applications to earthquake risk, but it is just the extrapolation of the intensity function, conditioned by any relevant explanatory variables, which should form the basis of risk forecasts; Kagan and Knopoff (1976, 1977), Vere-Jones (1978) give preliminary expositions of this idea. Another important advantage of these models for forecasting is the simplicity with which they can be simulated; see, for example, Ogata (1981), Zheng and Vere-Jones (1991).

If the development of this framework seems to the author the most significant progress in modeling earthquake occurrence since his review paper (Vere-Jones, 1970), the most significant open problem seems to be that of providing a proper physical context for the discussion of clustering, stress concentration, and stress transfer. Without any proper physical basis, space–time models are forced to stay at a descriptive level, and even these are hampered by the complications of modeling in high-dimensional spaces without any physical guide.

Closely related to this question is that of providing a statistically useful model for the process of rock fracture itself. The literature on fracture processes, or even on rock fracture, is enormous, but it has had very little, if any, impact into the model-building context described in the present paper. Several recent papers by Kagan (e.g. Kagan, 1982), describing what are essentially simulation models for fracture, are of interest here, both for this reason and because of their links with the concept of 'self-similarity'. Despite the importance in the long term of such attempts to bridge the gap between statistical and physical models, for the time being they remain at a relatively speculative and preliminary stage.

Acknowledgements

The first draft of this paper was developed while the author was a visitor at the Stochastic Process Centre at the University of North Carolina in Chapel Hill. Much of the work leans on the collaboration over several years with members of the Department of Environmental and Human Health Protection of the ENEA, Rome, Italy, above all to the skills, assistance and enthusiasm of Fabio Musmeci. The support of both these groups is gratefully acknowledged. The helpful advice of Steven Marron, in guiding the author's footsteps in the realm of density estimation, is also much appreciated.

References

Akaike, H. (1977). On entropy maximisation principle. In *Application of Statistics*, Krishnaiah, P.R. (ed.) North Holland, Amsterdam, pp. 27–41.

Akaike, H. and Arahata, E. (1978). *GALTHY: A Probability Density Estimation*, Computer science monograph 9, Inst. Statist. Maths, Tokyo.

Ambraseys, N.N. and Melville, C.P. (1982). *A History of Persian Earthquakes*. Cambridge University Press, Cambridge.

Bremaud, P. (1981). *Point Processes and Queues: Martingale Dynamics*, Springer, New York.

Cox, D.R. (1972). Regression models and life tables (with discussion) *J. Roy. Statist. Soc.*, **B**, **34**, 187–220.

Cox, D.R. (1975). Partial likelihoods. *Biometrika*, **62**, 269–76.

Daley, D.J. and Vere-Jones, D. (1989). *An Introduction to the Theory of Point Processes*, Springer, New York.

Diggle, P. (1985). A kernel method for smoothing point process data. *Applied Statistics*, **34**, 138–47.

Diggle, P. and Marron, J.S. (1988). Equivalence of smoothing parameter selectors in density and intensity estimation. *J. Amer. Statist. Assoc.*, **83**, 793–800.

Feller, W. (1966). *Introduction to Probability Theory and Applications*, Vol. II, Wiley, New York.

Good, I.J. and Gaskin, R.A. (1971). Nonparametric roughness penalties for probability densities. *Biometrika*, **58**, 255–77.

Gutenberg, B. and Richter, C.F. (1954). *The Seismicity of the Earth*, Princeton University Press.

Härdle, W. (1990). *Applied Non-Parametric Regression*. Econometric Society Monographs 19, Cambridge University Press, Cambridge.

Härdle, W., Hall, P., and Marron, J.S. (1988). How far are automatically chosen regression smoothers from their optimum? (with discussion) *J. Amer. Statist. Assoc.*, **83**, 86–101.

Hawkes, A.G. (1971a). Spectra of some self-exciting and mutually exciting point processes. *Biometrika*, **58**, 83–90.

Hawkes, A.G. (1971b). Point spectra of some mutually exciting point processes. *J. Roy. Statist. Soc.*, **B**, **33**, 438–43.

Hawkes, A.G. and Adamopoulos, L. (1973). Cluster models for earthquakes – regional comparisons. *Bull. Int. Statist. Inst.*, **45**, 454–61.

Hawkes, A.G. and Oakes, D. (1974). A cluster representation of a self-exciting process. *J. App. Prob.*, **11**, 493–503.

Hayashi, (1986). Laws of large numbers in self-correcting point processes. *Stoch. Proc. App.*, **23**, 319–26.

Imoto, M. and Ishiguro, M. (1986). A Bayesian approach to the detection of changes in the magnitude frequency relation of earthquakes. *J. Phys. Earth*, **34**, 441–55.

Inoue, H. (1986). A least squares smooth fitting for irregularly spaced data: finite element approach using cubic B-spline basis. *Geophysics*, **51**, 2051–66.

Ishiguro, M. and Sakamoto, Y. (1983). A Bayesian approach to binary response curve estimation. *Ann. Inst. Stat. Math.* (Tokyo), **35**, 115–37.

Kagan, Y. (1982). Stochastic model of earthquake fault geometry. *Geophys. J. Roy. Astron. Soc.*, **71**, 659–91.

Kagan, Y. and Knopoff, L. (1976). Statistical search for non-random features of the seismicity of strong earthquakes. *Phys. Earth Planet. Inter.*, **12**, 291–318.

Kagan, Y. and Knopoff, L. (1977). Earthquake risk predictions as a stochastic process. *Phys. Earth Planet. Inter.*, **14**, 97–107.

Karr, A.F. (1986). *Point Processes and their Statistical Inference*, Marcel Dekker, New York.

Knopoff, L. (1971). A stochastic model for the occurrence of main sequence earthquakes. *Rev. Geophys. and Space Phys.*, **9**, 175–88.

Kronmal, R. and Tarter, M. (1968). The estimation of probability densities and cumulative distributions by Fourier methods. *J. Amer. Statist. Ass.*, **63**, 925–32.

Margottini, C. and Serva, L. (1988) (eds). *Historical Seismicity of Central-Eastern Mediterranean Region*, ENEA-IAEA, Rome.

Marron, J.S. (1987). A comparison of cross-validation techniques in density estimation. *Ann. Statist.*, **15**, 152–62.

Marron, J.S. (1988). Automatic smoothing parameter selection: a review. *Empir. Econ.*, B, 65–86.

Molchan, G.M. (1984). Some remarks on L. Knopoff's model for an earthquake sequence. *Vychislitel'naya Seismologiya*, **16**, 36–57.

Musmeci. F. and Vere-Jones, D. (1986). A variable-grid algorithm for smoothing clustered data. *Biometrics*, **42**, 483–94.

Musmeci, F. and Vere-Jones, D. (1991). A space–time clustering model for historical earthquakes. *Ann. Inst. Stat. Math.* (to appear).

Ogata, Y. (1981). On Lewis' simulation method for point processes. *IEEE Trans. Inf. Th.*, **27**, 23–31.

Ogata, Y. (1988). Statistical models for earthquake occurrences and residual analysis for point processes. *J. Amer. Statist. Assoc.*, **83**, 9–27.

Ogata, Y. and Akaike, H. (1982). On linear intensity models for mixed doubly stochastic Poisson and self-exciting point processes. *J. Roy. Statist. Soc.*, B, **44**, 102–7.

Ogata, Y. and Katsura, K. (1986). Point process models with linearly parameterized intensity for the application to earthquake catalogue. *J. Appl. Prob.*, **23A**, 231–40.

Ogata, Y. and Katsura, K. (1988). Likelihood analysis of spatial inhomogeneity of mankind point patterns. *Ann. Inst. Stat. Math.*, 40, 29–39.

Ogata, Y. and Vere-Jones, D. (1984). Inference for earthquake models: self-correcting model. *Stoch. Proc. Appns.*, **17**, 337–47.

Ogata, Y., Akaike, H., and Katsura, K. (1982). The application of linear intensity models to the investigation of causal relations between a point process and another stochastic process. *Ann. Inst. Stat. Math.*, **34**, 373–87.

Prakasa Rao, B.L.S. (1983). *Non-Parametric Functional Estimation*, Academic Press, New York.

Shimazaki, K. and Nakata, T. (1980). Time predictable recurrence model for large earthquakes. *Geophys. Res. Lett.*, **7**, 279–82.

Silverman, B. (1986). *Density Estimation for Statistics and Data Analysis*. Chapman and Hall, London.

Vere-Jones, D. (1970). Stochastic models for earthquake occurrence (with discussion). *J. Roy. Statist. Soc.*, B, **32**, 1–62.

Vere-Jones, D. (1978). Earthquake prediction – a statistician's view. *J. Phys. Earth*, **26**, 129–46.

Vere-Jones, D. (1988). Statistical aspects of the analysis of historical earthquake catalogues. In *Historical Seismicity of Central-Eastern Mediterranean Region*, Margottini, C., and Serva, L. (eds), ENEA-IAEA, Rome, pp. 271–95.

Vere-Jones, D. and Musmeci, F. (1986). *Orthogonal Polynomial Estimates for Point Process Densities*. ENEA Technical Report No. RT/PAS/86/2.

Vere-Jones, D. and Ozaki, T. (1982) Some examples of statistical estimation applied to earthquake data in cyclic Poisson and self-exciting models. *Ann. Inst. Statist. Math.*, **34**, 189–207.

Vere-Jones, D. and Smith, E.G.C. (1981). Statistics in Seismology. *Commun. Statist. Theor. Math.*, A**10**, 1559–85.

Wahba, G. (1975). Optimal convergence properties of variable knot, kernel and orthogonal series methods for density estimation. *Ann. Statist.*, **3**, 15–29.

Zheng, X. (1991). Ergodic Theorems for stress-release processes, *Stoch. Proc. Appns.*, **37**, 239–25.

Zheng, X. and Vere-Jones, D. (1991). Applications of stress release models to historical earthquakes from North China. *Pure and Applied Geophys.*, **135**, 559–76.

Chapter 12

Stochastic models for the distribution of rock types in petroleum reservoirs

B. D. Ripley

12.1 Overview

The potential of enhanced oil recovery schemes is assessed by numerical simulation of multi-phase flow through a discrete block representation of the reservoir. Traditionally, the inputs to the simulation package have been expected values of the porosity and permeability for each block. It is clear that using mean values will result in a 'smoother' picture of the reservoir than is likely to be realistic. This has serious repercussions for the estimated production curve for the reservoir. Heterogeneities in the flow properties of the reservoir encourage instabilities in the flow, hence 'fingering' patterns of the injection fluid will occur and lead to break-through of the injection fluid. This suggests that any departures from the expected 'smooth' model will lead to early break-through and less oil being recovered. In other words, average values for the flow parameters lead to extreme (optimistic) values for oil recovery.

The only way known to assess the size of this effect is to simulate the lithology of the reservoir, and to run numerical flow simulations on the randomly generated synthetic reservoirs. The range of production curves produced by such simulations will indicate the effects of likely heterogeneity. Indeed, it allows recovery schemes to be designed to minimize the risk of a low recovery rather than to maximize oil recovery in the most optimistic view of the reservoir.

Readers should note that *simulation* is used in two senses in the field (and in this paper). There is both *numerical* simulation of flows, and *stochastic* simulation of the rock fabric. Since we are primarily concerned with synthesizing reservoirs, simulation will be used here to mean synthesis of rock types, and where flow simulation is meant, this will be stated specifically.

Simulation of a reservoir is not done in the complete absence of information, so the stochastic simulation algorithm has to take account of what is known. In decreasing order of reliability we have

- rock types from core samples in wells;
- classification of lithofacies from well logs;

- indications of density, from seismic studies;
- geologists' world knowledge of the rock structures and morphogenesis.

Our simulations must take account of these data and knowledge, and so are known as *conditional simulations*.

At least four different approaches have been tried in conventional models for reservoirs. Three of these are illustrated in Figure 12.1:

(1) levels of a continuously varying random field;
(2) fitting the correlation structure of blocks, directly;
(3) Markov random fields for blocks;
(4) randomly placed geometric objects.

To be successful, a class of models must allow:

- sufficient flexibility;
- efficient simulation methods;
- conditional simulation techniques;
- estimation of parameters;
- encapsulating 'real-world' knowledge.

Our survey and experiments suggest that the last two classes are the most promising.

The major technical problems with stochastic simulations are to find reasonably efficient ways to produce realizations (the synthetic reservoirs) within an acceptable amount of computer time, and to find ways to choose the parameters of the models to match the data and background knowledge. Significant progress has been made in these areas in the last five years; the bulk of this paper is devoted to explaining these results in some detail.

Haldorsen *et al.* (1988) review the need for stochastic models of reservoir structure.

12.2 Experimentation with stochastic models

Classical statistical techniques work effectively only for linear or nearly linear systems. For highly non-linear systems the mean of the output need bear no resemblance to the output for the mean input (Figure 12.2). For such systems, the only general method of resort is (stochastic) simulation. By using, say, 100 random inputs in Figure 12.2(b) we can recover a very good idea of the output distribution. Simulation allows us to estimate not only the mean and variance of the output distribution, but also the frequency of extreme events.

The flow properties of the gas/oil/water mixtures in heterogeneous rocks are not at all well understood, but both theoretical considerations and

Figure 12.1 (opposite) Approaches to modeling rock types in a reservoir: (a) discretizing the levels of a continuous random function, here illustrated for a one-dimensional section; (b) dividing the reservoir into small blocks; (c) including rock types as 'object' against a homogeneous 'background'.

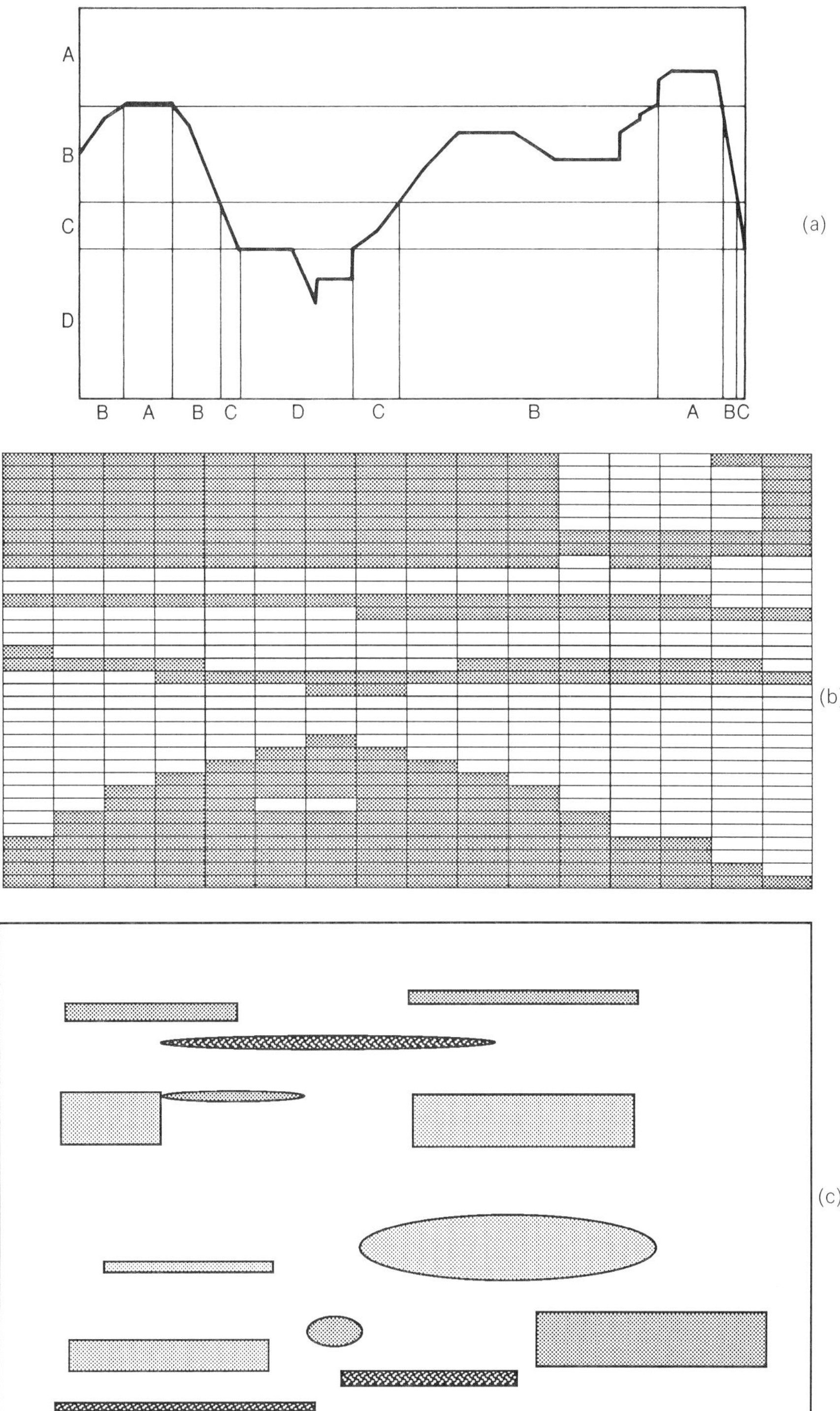
A
B
C
D
B A B C D C B A BC
(a)
(b)
(c)

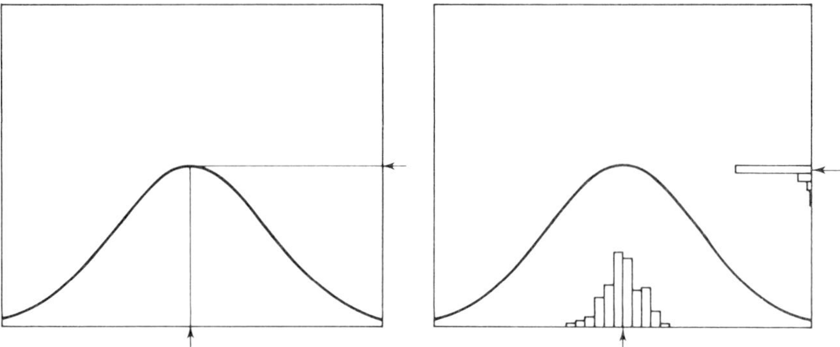

Figure 12.2 The effect of a nonlinear function on a distribution: (a) the function with mean input (arrow) mapped to mean output; (b) a histogram of 100 random inputs mapped to a histogram of outputs with average at the arrow.

limited practical experiments (Omre, 1989) suggest that the production curve is a highly nonlinear function of the values of porosity and (especially) permeability input to the numerical simulation package. The nonlinearity arises from the tendency for fluid flow to take the 'path of least resistance' and from instabilities in the fronts between fluid phases which can grow to form fingering patterns in a heterogeneous medium.

Naïve simulation as illustrated in Figure 12.2(b) can be spectacularly inefficient when the events of interest are rare. There is a wide range of methods available to design simulation experiments more efficiently for specific tasks (see, e.g., Ripley, 1987). Of these, the potentially most useful in dealing with rare events is *importance sampling*. Rather than choose inputs from the natural distribution with density f (here given by the stochastic model of rock geometry), we use a distribution with density g chosen to make interesting events more likely. The results are then down-weighted by the ratio f/g. For example, if we wish to estimate the probability that some function $Z = \phi(X)$ of the experiment lies in a set A, under naïve simulation we would compute the fraction of the simulation runs in which Z lies in A. With importance sampling, we sum $f(X)/g(X)$ over runs with Z in A and divide by the number of runs.

Fluid flow in heterogeneous reservoirs appears to be an ideal domain for the use of importance sampling. The observations of the outputs are very expensive, yet it is easy and relatively cheap to simulate heterogeneous reservoirs from the model. If we can predict which synthetic reservoirs will give rise to extreme flow properties, it will obviously be better to use these for the (numerical) flow experiments. Our suggestion is to generate, say, 1000 synthetic reservoirs, compute simple geometric measurements on these (approximating tracer flow) and select a few reservoirs for the flow experiments.

One nice point about importance sampling is that it very rarely does any harm. Even if the choice of the new density g turns out to be completely inappropriate, the effect is usually only to slightly reduce the efficiency of the experiment. Only if g actually makes rare events even rarer will harm ensue.

12.3 Classes of models

There are a number of basic decisions to be made in choosing a model for the fabric of a reservoir. The numerical flow simulations necessarily choose a fairly coarse grid and thereby discretize the problem in all three dimensions. This is also a valid approach for modeling rock types, although it will be possible to use a substantially smaller size of grid block if necessary, since the computations per block will be *much* faster. However, as the processes of deposition occurred in continuous space, it is also attractive to think of the reservoir containing 'objects' which are beds of one rock type or intrusions such as shales. Between these two extremes are layered models, which are discretized in one direction only.

In all these schemes we will assume that the geometry of the reservoir has been corrected as necessary, so that the vertical axis represents the time sequence of deposition, and the layers are represented as primarily horizontal. This is clearly important for layer-based models, but may be unimportant for some others. Despite the descriptive language used, our models (unlike, say, Bridges and Leeder, 1979) do *not* attempt to model the actual processes of reservoir formation but rather the structures observed when formation is completed.

(1) *Three-dimensional block models.* The only computationally feasible way to specify such models is to use *Markov random fields* (MRFs). The rock type at each block has a distribution which depends only on the rock types at surrounding blocks, its *neighbors*. 'Neighbor' can be defined in several different ways, but usually means *either* the 6 horizontal and vertical neighbors *or* the 26 horizontal, vertical and diagonal neighbors. (The influence of the neighbors can be weighted according to their proximity.) Theory allows much more general neighborhood patterns, but computational speed dictates keeping neighborhoods small. One restriction in MRFs is that the idea of a neighbor must be reflexive (I am your neighbor if you are mine), and that interactions between neighbors are symmetric. Thus it is impossible to allow rock type A to lie above type B and prohibit B from lying above type A.

(2) *Objects in three-dimensions.* This is the most mechanistic description of a reservoir which we shall consider, in which objects are laid down representing rock beds. This can be done in three ways:

(a) objects as intrusions on a single-type background;
(b) objects packing three-dimensional space;
(c) rock types discretizing a continuously-varying random field (as in Figure 12.1a).

Both (a) and (b) can be achieved by the point-process and random-set models of stochastic geometry, which are discussed in detail in Section 12.5. Option (c) typically uses a correlated Gaussian random field with a simple correlogram (or variogram to the *geostatistiques* school). It suffers the (to our mind fatal) flaw of forcing a rigid ordering of rock types. This approach is advocated by Matheron *et al.* (1987), and at the time of writing by the Fontainebleau school.

(3) *Layers.* Once again, there are several possibilities, all based on laying down a reservoir in stochastically dependent layers. For example, one could model a fluvial reservoir by laying down river courses with dependent positions from layer to layer, or link shale intrusion with 'earlier' (i.e. lower) layers. Another possibility is to have a two-dimensional MRF of a rock type with the variable being thickness in number of layers. (Clearly zero must be treated specially.) Yet another way to describe a fluvial reservoir would be to have layers normal to the (supposed) direction of flow at time of deposition.

These classes need not be mutually exclusive. By way of illustration, one could use a three-dimensional MRF block model for sandstones of different permeabilities, overlaid by a two-dimensional thickness MRF model of shale beds and an object model of small-scale shale intrusions. Since the end product will be a model for porosities and permeabilities on the coarse block scale of the flow simulation, the different degrees of discreteness of the components will be immaterial. The ease of interpretability of these components might, however, differ widely.

12.3.1 Permeability models

Thus far we have only discussed rock types. There is also a need to assign porosity and permeability values to each type and hence to each simulation block. The most satisfactory approach appears to be to use a continuous-valued random process on each rock facies to produce an assignment of porosity or permeability throughout the reservoir. A simple correlated Gaussian random field should suffice. The field could be chosen independently for each 'object' (even within grid-based models) or could itself have spatial dependence between objects of the same rock type, for example by having

$$Z(x) = Z(x, I_v(x)) \qquad Z(x, i) \text{ independent}$$
$$I_v(x) \text{ indicator for type } v.$$

The choice between these models seems to depend on the type of reservoir being modeled; there is no 'right answer' but there are good answers. Most of the rest of this paper is devoted to more technical considerations of the two main classes of component models, MRFs and point processes of objects. Nevertheless, it should be constantly borne in mind that these are only parts of a larger structure.

12.4 Markov random fields

Markov random fields are defined in an abstract setting. The abstract version is discussed later in this section; we start with simple examples. Think of a rectangular lattice of pixels, each of which is colored from a palette of k colors. (Readers with a background in statistical physics may like to think of 'spins' on a lattice rather than colored pixels.) Let X_s denote the color at pixel s. Then a Markov random field (MRF) specifies the joint probability distribution of the patchwork of colors, by giving the local conditional probabilities

$$P(X_s = x_s \mid X_t = x_t, t \neq s). \tag{12.1}$$

(This does indeed determine the whole distribution $P(X_s$ for all s).) For a *Markov* random field the probabilities (12.1) depend only on the values of the random variables at neighboring pixels, that is (12.1) is equal to

$$P(X_s = x_s \mid X_t = x_t, t \text{ nhbr of } s). \tag{12.2}$$

A simple example is to let this depend only on the number of neighboring pixels of the same color, for example by

$$P(X_s = c \mid X_t, t \neq s) \propto \exp \beta \#\{\text{nhbrs of } s \text{ with color } c\}. \tag{12.3}$$

In statistical physics this is known as the Potts model, after Potts (1952). Specializing yet again to two colors and the square lattice with horizontal and vertical neighbors, we have the celebrated Ising model of statistical physics.

It is instructive to write down the joint distribution corresponding to (12.3). It is

$$P(X_1 = x_1, \ldots, X_n = x_n) \propto \exp \beta \#\{\text{nhbr pairs of the same color}\}. \tag{12.4}$$

The formal structure needed to define these processes is much less rich than a lattice of pixels. All we need is a notion of *neighbor*, and this is defined by a *graph* G with sites s; two sites are joined by an edge of the graph if and only if they are neighbors. Thus we can define a Markov random field on a graph through the equality of (12.1) and (12.2). The idea of a *clique* is important for MRFs. A clique is a set of sites on the graph each pair from which is connected. A maximal clique is a clique which cannot be enlarged. Less formally, all pairs of sites in a clique are neighbors, and no site not in a maximal clique is a neighbor of *all* the clique members. On the square lattice cliques contain one or two sites, consisting of single sites and neighbor pairs. On the lattice with horizontal, vertical and diagonal neighbors, maximal cliques are squares.

The main theorem of MRFs, sometimes called the Hammersley–Clifford theorem, is that

$$P(X_1 = x_1, \ldots, X_n = x_n) \propto \prod_{\text{cliques}} \phi(X_{s_i}, i = 1, \ldots, j) \tag{12.5}$$

for symmetric positive functions ϕ on the cliques $\{s_1, \ldots, s_j\}$. (A mild technical condition is needed to exclude certain patterns of zero probabilities; the theorem certainly holds if all densities are strictly positive. It is clear that the product can be restricted to maximal cliques.) This result is mainly of theoretical interest in showing how to define MRFs. Almost all examples of the use of (12.5) use only the cliques of size one or two, that is

$$P(X_1 = x_1, \ldots, X_n = x_n) \propto \prod_{\text{sites}} b(x_s) \prod_{s \text{ nhbr of } t} h(x_s, x_t). \tag{12.6}$$

The first product is not strictly needed, but it will be convenient to view h as controlling the (pairwise) interaction between sites, and b as controlling the marginal distribution of X_s. We have as the local specification

$$P(X_s = x_s \mid X_t = x_t, t \neq s) \propto b(x_s) \prod_{t \text{ nhbr } s} h(x_s, x_t). \tag{12.7}$$

On the square or cubic lattice (with only horizontal and vertical neighbors in two and three dimensions respectively), maximal cliques are of size two, so (12.6) is completely general.

Markov random fields have been studied in statistical physics, as model for gases for very large systems ($>10^{10}$ sites) and in statistics for small systems ($n \leqslant 100$) as models for fertility in agricultural field trials and in geographical networks. Interest in moderate n (10^3–10^6) occurs in both image analysis and in our field. Unfortunately this range of n tends to embody the worst features of the two more extreme cases. A change of notation in (12.6) and (12.7) recovers the Strauss model (Strauss, 1977),

$$P(X_1 = x_1, \ldots, X_n = x_n) \propto \exp\left[\sum_s \alpha_{x_s} + \sum_{s \text{ nhbr of } t} \beta_{x_s, x_t}\right] \tag{12.8}$$

and

$$P(X_s = c \mid X_t, t \neq s) \propto \exp\left[\alpha_c + \sum_{t \text{ nhbr } s} \beta_{c, X_t}\right]. \tag{12.9}$$

If all colors c are treated equally, we recover the Potts model (12.3, 12.4).

The well-known difficulty of the Ising model shows just how little theoretical progress has been achieved for these models. There are perhaps simpler models (Bartlett, 1975; Pickard, 1977, 1980) but these are not genuinely spatial. An elegant treatment of what *is* known about Ising models is given by Pickard (1987).

Parts of the study of phase transition in statistical physics are relevant to our purposes. There is a critical value of β in (12.3), β_0, such that for $\beta < \beta_0$ the process has only short-range correlations whereas for $\beta > \beta_0$ there are long-range correlations. In statistical physics β is the inverse temperature, so $1/\beta_0$ is known as the *critical temperature*. Formally,

$$\operatorname{corr}(X_s, X_t) \to \rho_\infty \neq 0 \qquad \text{as} \quad \operatorname{dist}(t, s) \to \infty$$

for $\beta > \beta_0$, where the size of the graph has to grow without bound. Realizations for supercritical β consist of large patches of one color, with occasional isolated pixels of another color and long, fairly smooth boundaries between patches. Similar properties are believed to hold for different neighborhood systems, higher dimensions (but not one-dimensional) and for the Potts–Strauss model. These properties can be used to provide a check on simulations from these models.

We may also generalize the Potts–Strauss model (12.6). For a lattice with horizontal and vertical neighbors, the cliques consist of all pairs of neighboring sites, and (12.4) is the most general form of a Markov random field. (With diagonal neighbors as well, the cliques are squares.) However, (12.4) will provide too many parameters to be useful for even moderate numbers of rock types (values of X_s).

The normalizing constant missing in (12.3), (12.4) is one of the difficulties of this subject. It is denoted by $1/Z$, where Z is known as the *partition function*. In realistic models it cannot be related easily to parameters in $b()$ or $h()$.

Background information on MRFs appears in Kindermann and Snell (1980); an introduction is given by Isham (1981).

Two simulations of three-dimensional MRFs are shown in Figure 12.3. These are conditional simulations of the model with 26 neighbors (equally weighted) on the cubic lattice. To avoid 'phase transition' behavior, a few (about 10) voxels have been conditioned on each of the four colors.

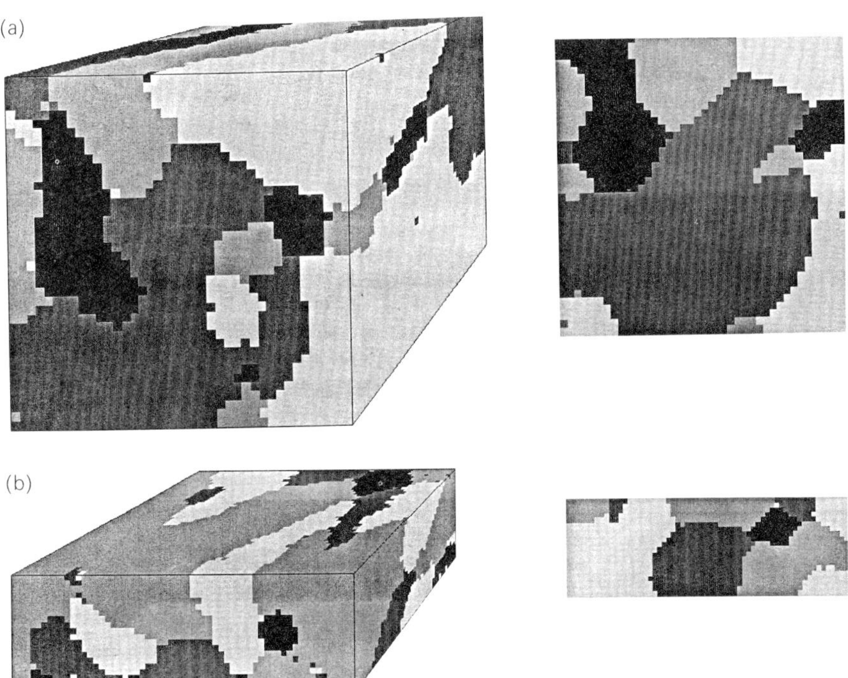

Figure 12.3 Conditional simulations of three-dimensional MRFs. Twenty-six neighbors on the cubic lattice were used: (a) 40^3; (b) $60 \times 20 \times 40$. The right panel is a section from the cuboid at the depth given by the perspective. The conditioning is explained in the text.

12.5 Stochastic geometry

Stochastic geometry was a term coined in the 1960s for the study of randomly placed,interacting, geometrical objects. An early compendium is Harding and Kendall (1974), and a recent introduction is Stoyan *et al.* (1987). The idea is to describe each object by a point in a (high-dimensional) space X. For example, a field of ellipses could be described by their centers, semi-minor and semi-major axes and orientations, and a field of lines by orientation and distance for the origin. The random field is then defined by a *point process* on X.

A point process is a random ensemble of points from X, chosen so that only finitely many points occur in each bounded subset of X. (In general, we can define a bounded subset of X as one in which all objects intersect a bounded subset of the original space.) Some authors refer to *marked* point processes, in which each point from X is marked by a point from the mark space K. They have in mind, for example, a pattern of trees marked by their diameters. However, this is just a special case of the general definition, as a marked point process can be viewed as a point process on $X \times K$.

We will need very little of the general theory of point processes, which can be very demanding mathematically. By confining attention to a bounded region (our reservoir) we need only deal with a finite (but variable) number n of points. The simplest model for a point process is the *Poisson process*, with the following properties:

(1) the number of points $N(A)$ in any (bounded) subset A has a Poisson distribution;
(2) the numbers of points in disjoint subsets A_i are independent;
(3) conditional on $N(A) = n$, the locations of the n points in A are independent.

From properties (1) and (3) we can construct a realization within any bounded set. We have to specify a measure Λ such that $\Lambda(A)$ is the mean number of points within A. To construct the Poisson process on A, we first draw n from a Poisson distribution with mean $\Lambda(A)$. Then we select independently each of n points in A with probability distribution $\Lambda/\Lambda(A)$. This describes a Poisson process with a non-uniform intensity of points. To obtain a stationary Poisson process, Λ must be proportional to the content (area, volume, ...) of A, with constant of proportionality the *intensity* λ. In more general spaces of objects there is usually a unique invariant measure to take the place of the content.

Gibbs point processes are the analogs of Markov random fields for point patterns. They are defined by specifying for each realization $\mathbf{x} = \{x_1, \ldots, x_n\}$ how much more likely it is than under a Poisson process. (Since each realization has zero probability, considerable care is needed in formalizing this notion.) Once again, only pairwise interaction processes are usually considered. These have a density (with respect to the Poisson process) of the form

$$p(\mathbf{x}) = a \prod_i b(x_i) \prod_{i<j} h(x_i, x_j), \tag{12.10}$$

where $b(x)$ affects the local density of points near x, and $h(\,)$ is the interaction function. Often $h(\,)$ will only depend on the distance between the two points. Of course, a is a normalizing constant, determined (in principle) by $b(\,)$ and $h(\,)$.

Much of the basic theory for these processes can be found in Ripley (1976a, 1976b, 1977) and is summarized in Ripley (1981, and 1988b, Chapter 4).

Two simple examples of geometric processes are shown in Figure 12.4. In (a), 100 squares are laid down at random without interaction but solely contained within the bounding square. In (b), 50 squares are randomly packed, that is with the interaction function that they are not allowed to overlap. Both these geometrical processes are of the *germ-grain* type, in which each object is centered on a point (the germ). In fact, we chose the lower left point of the square as the germ in simulating the pictures, and this is uniformly distributed on $(0, 0{\cdot}9)^2$ since the squares are of side 0·1 within a unit square. Viewed as a marked point process, the objects (the grains) are a trivial process here since they are all identical. We could have chosen random sizes or shapes or both. In many cases the germ characteristics are chosen independently from grain to grain.

Gibbs point processes have a local notion analogous to (12.7). The *conditional intensity* $\lambda(\xi; \mathbf{x})$ (of a point at ξ given the other points are the vector $\mathbf{x}$) is defined by

$$\lambda(\xi; \mathbf{x}) = \lim_{A \downarrow \{\xi\}} P(\text{pt in } A \mid \mathbf{x} \text{ on } A^c)/(\text{content of } A)$$

where the limit is over sets containing ξ and shrinking to just that point. For process (12.10) we have

$$\lambda(\xi; \mathbf{x}) = b(\xi) \prod_i h(\xi, x_i). \tag{12.11}$$

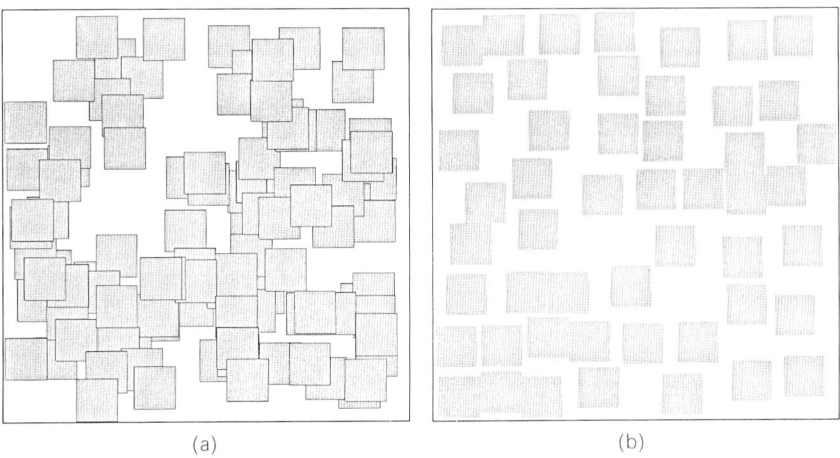

Figure 12.4 Simulations of geometrical processes in a square of unit side: (a) 100 squares of side 0·1 laid down at random; (b) 50 such squares in a random packing.

We can interpret $b(\xi)$ as the underlying intensity of points near ξ, modified by the product which expresses the influence of existing points.

Perhaps the simplest Gibbs point process is the Strauss process defined by

$$\begin{aligned} h(r) &= c \quad \text{for} \quad 0 < r < R \\ &= 1 \quad \text{for} \quad r > R \end{aligned}$$

and which has

$$p(\mathbf{x}) \propto c^{\text{number of } R\text{-close pairs}}. \tag{12.12}$$

Here R is the interaction distance, and c a constant less than 1 measuring the strength of the interaction. For $c = 0$ points closer than R apart are prohibited, so we have a model for randomly packed non-overlapping disks of diameter R.

A useful generalization of the Strauss process is Penttinen's (1984) 'multi-scale' process. This chooses h as a general step function, so

$$\begin{aligned} h(r) &= c_1 \quad \text{for} \quad 0 < r < R_1 \\ &= c_2 \quad \text{for} \quad R_1 < r < R_2 \\ &\vdots \\ &= c_m \quad \text{for} \quad R_{m-1} < r < R_m \\ &= 1 \quad \text{for} \quad r > R_m \end{aligned} \tag{12.13}$$

Once again, the unknown normalizing function proves to be a difficulty in parameter estimation. Note that it does not appear in (12.11).

12.6 Iterative simulation schemes

All the stochastic models considered in the previous sections will be used with parameters which induce quite strong dependence between the random variables, for medium-sized systems. This implies that any attempt to produce simulations by direct rejection sampling will be spectacularly inefficient. For both MRFs and Gibbsian point processes, realizations of the base process (respectively, independent random variables at each site, and a Poisson process) produce realizations with almost vanishing probabilities under the desired model!

The obvious way out is sequential simulation, generating the random variables or random points one at a time. We become stuck at the very first step, as the marginal distribution of the first random variable is unknown. Fortunately there is a clever trick dating from Metropolis *et al.* (1952) which allows a sequential approach, provided it is repeated, theoretically *ad infinitum*. We will state that procedure in complete generality and then specialize it to MRFs and point processes.

12.6.1 Metropolis' method

Metropolis' idea now appears to be rather simple. Suppose we are interested in sampling from a probability distribution π on a finite set $\{1, \ldots, N\}$. (In our examples, the finite set is the complete set of patterns of rock types throughout the whole reservoir. Thus N will be very large, perhaps 6^{100^3}.) We construct a Markov chain with π as its unique equilibrium distribution. (A Markov chain is a stochastic process which moves in discrete time steps, with the future behavior depending on the past only through its current state. It is defined by an initial distribution Π_0 and a transition matrix P giving $p_{ij} = P(X_{t+1} = j \mid X_t = i)$.) If Π_n denotes the distribution of the Markov chain after n steps, it is a theorem that, for aperiodic irreducible chains, Π_n converges to π. Thus sampling the chain for 'long enough' gives a sample from an approximation to π. More formally,

Metropolis' algorithm Choose a *symmetric* irreducible transition matrix Q. When at state i choose a new state from the probability distribution given by row i of Q and move to that state j with probability $\min(1, \pi_j/\pi_i)$, otherwise remain at i.

Here *irreducible* means that we can move from any state to any other state in a finite number of steps with non-zero probability. If a higher probability state is selected, it is used. However, if the proposed move is to a state with lower probability, the move is accepted with probability less than one.

Metropolis's algorithm defines a new transition matrix P by

$$P_{ij} = \min(1, \pi_j/\pi_i)q_{ij}$$

for $j \neq i$ and so

$$p_{ii} = q_{ii} + \sum_{j \neq i} \max(0, 1 - \pi_j/\pi_i)q_{ij},$$

from which we find that

$$\pi_i p_{ij} = \pi_j p_{ji} \qquad \text{for all} \quad i \neq j, \tag{12.14}$$

the so-called detailed balance conditions. These suffice (Kelly, 1979, Section 1.2) for π to be an equilibrium distribution of P. The conditions on Q suffice to make P irreducible and aperiodic unless π is constant (Ripley, 1987, pp. 113–4), so Π_n then converges to π as asserted.

Other formulas for accepting the jump from i to j also work; Barker (1965) suggested jumping with probability $\pi_j/(\pi_i + \pi_j)$. Note that Barker's method jumps less often than Metropolis' and so might be thought to converge more slowly. Many such remarks are part of the folklore of iterative simulation but have not been proved rigorously. Some are false (Kirkland, 1989). Barker's method was defined earlier by Flinn and McManus (1961).

Hastings (1970) generalized the method to allow nonsymmetric transition matrices Q. Let

$$a_{ij} = \pi_i q_{ij}/\pi_j q_{ji}$$

and

$$\alpha_{ij} = \frac{s_{ij}}{1 + a_{ij}} \qquad \text{for} \quad j \neq i,$$

where $\{s_{ij}\}$ is an arbitrary nonnegative symmetric matrix chosen so that $\alpha_{ij} \leqslant 1$. Then if we select a move from i to j with probability q_{ij} and accept the move with probability α_{ij}, the resulting transition matrix satisfies the detailed balance equations (12.14).

There are also continuous-space methods closely related to Metropolis'. Some stochastic processes arise as the equilibrium distributions of diffusions and birth-and-death processes, and both of these methods have been used. In particular, the most popular method to simulate spatial point processes (Section 6.3) is of this type.

The main questions in the implementation of Metropolis' method are

- the choice of Q;
- how long does the process need to be run to be acceptably close to equilibrium?
- efficient implementation.

These are discussed in the following sections.

12.6.2 Applications to MRFs

It will be helpful to have a specific example in mind in discussing the family of methods based on Metropolis' principle. The Potts and Ising models are known to exhibit long-range behavior for $\beta > \beta_0$ in two or more dimensions; this is the basis of their interest. Unfortunately, very little of their distribution is understood analytically, so simulation is an essential tool in gaining understanding about them.

Metropolis' name would be associated with a specific method for the Potts process. The transition matrix Q is defined by choosing a site s with uniform probability and then selecting a value of X_s uniformly amongst the k possibilities (or perhaps amongst the possibilities other than the current state). It is easy to see that this has the required properties. Conditional on the choice of s and for $a_0 = X_s$, the probabilities assigned to $X_s = a$ are proportional to $\min(h(a), h(a_0))$ for $a \neq a_0$, where

$$h(a) = P(X_s = a \mid X_t,\ t \text{ a neighbor of } s).$$

Thus if the process is currently in the conditionally most probable state, the conditional choice for a is from a distribution proportional to the conditional distribution except that the probability of $X_s = a_0$ is increased.

The 'heat bath' or Gibbs sampler (Geman and Geman, 1984) selects a site at random and chooses X_s from the current conditional distribution

$$P(X_s \mid X_t,\ t \text{ a neighbor of } s).$$

This can be justified directly, as the transition matrix P is easily shown to be irreducible and to leave π unchanged. (It can also be shown that the detailed balance equations (12.14) hold.) In contrast to the raw form of

Metropolis' method, the Gibbs sampler will usually move to a conditionally high probability value of X_s. In the special case $k = 2$, the Gibbs sampler is exactly equivalent to Barker's method with Q choosing a site at random and flipping its state to the other possibility (Ripley, 1987, p. 114).

The general Gibbs sampler fits into Hastings' more general framework. Let $p_s = P(\text{pixel } s \text{ is chosen})$. Then if j differs from i at pixel s, let $q_{ij} = p_s P(X_s = j_s \mid X_t = i_t, t \neq s)$. Then

$$a_{ij} = \frac{P(X_s = i_s \mid X_t = i_t, t \neq s) q_{ij}}{P(X_s = j_s \mid X_t = i_t, t \neq s) q_{ji}} = 1$$

and we can take $s_{ij} \equiv 2$.

Yet another idea is Flinn's (1974) 'spin exchange' method, which has achieved an undeserved popularity through Cross and Jain (1983). Two sites are selected at random and the interchange of their values considered, using the Metropolis principle. If the sites s and t which are selected are not themselves neighbors,

$$\frac{\pi_j}{\pi_i} = \frac{P(X_s = \text{old } x_t \mid \text{rest}) P(X_t = \text{old } x_s \mid \text{rest})}{P(X_s = \text{old } x_s \mid \text{rest}) P(X_t = \text{old } x_t \mid \text{rest})}$$

and this can be modified when s and t *are* neighbors (Ripley, 1987, p. 116). The problem is that Q is *not* irreducible and so the equilibrium value is not unique. A little thought shows that the irreducible classes under Q are given by the marginal distribution of $\{X_s\}$, that is by fixing the number of sites with $X_s = c$ for each value of c. This conditioning is occasionally useful, but is not valid in Cross and Jain's applications. (Although usually attributed to Flinn, this method was mentioned by Hammersley and Handscomb (1964, pp. 122–3).)

In practice these methods are performed by making a systematic sweep over the sites, and this can be shown to be sufficient for convergence to the desired π by the theory of nonhomogeneous Markov chains. (This is *provided* irreducibility can be proved to hold. This is true for the Gibbs' sampler, and for the original Metropolis' method for more than two colors. With two colors there are some simple counter-examples.) Simple experiments demonstrate that convergence is rather faster with a systematic sweep than with random selection. Further, with a systematic sweep the method can be implemented more efficiently. Each time we return to a site on the graph we make a decision as to whether or not to change its value (and also what the new value might be). In equilibrium this decision is for most sites not to change, since the conditional distribution will be heavily concentrated on just one value. The probability of change remains unchanged until a neighboring random variable changes value, and so if the neighbors do not change there will be a geometrically distributed time to the next change. We can simulate that geometrically distributed time very easily, and so set a clock as to when next to visit (and change) that site. When a site does change value, the clocks of its neighbors are marked as invalid. This clock method (Ripley, 1988b, p. 99) speeds simulations up dramatically for large β (by a factor of 10–100).

(a) Speed of convergence

The major difficulty with all these methods is that they alter only one site at a time. Thus the simulation will make global changes only slowly. Remember (Section 12.4) that realizations of the Ising model for large β are dominated by blocks of predominantly black or white. It will take a site-by-site method an *extremely* long time to flip the color of the part of the process being simulated. Similarly, boundaries between two colors can persist for a very long time. Thus the process can appear to have converged when it has merely become trapped for a while in a 'quasi-equilibrium' state. This is illustrated in Figure 12.5.

12.6.3 Refined methods for specific MRFs

An alternative is to alter more than one site at once. On a square two-dimensional lattice we have considered updating a 2×2 or 3×3 block of sites simultaneously, but so far this appears unpromising, since much more work is needed per sweep, and the rate of convergence appears to be essentially the same (Kirkland, 1989). We could also simulate a row at a time. It is well known that complete rows of a Markov random field on a lattice form a Markov chain. This method is advocated by Qian and Titterington (1989), but it is quite clear that they have not run the process to convergence, and our limited experience suggests that it is no faster than site-by-site methods.

(a) Cluster methods

Swendsen and Wang (1987) introduced a completely different Q matrix and update procedure for the symmetric Potts model (only). Given the current state of the system, set up a random subgraph of the graph G by deleting edges joining pairs of sites with different values, and independently deleting edges joining sites of the same value with probability $1 - \exp(-\beta)$. Find the connected subsets of this subgraph, and assign each subset (and X_s for all s in that subset) a value chosen independently from the uniform distribution on $\{1, \ldots, k\}$. This completes the description of the P matrix, or indeed of the Q matrix since the coloring is over equally probable states.

One step of this process is illustrated in Figure 12.6. A formal algorithm in pseudocode is as follows:

Repeat forever:

(1) Search the entire graph. Whenever two neighboring sites have the same color, join them by a bond with probability $p = 1 - \exp(-\beta)$.
(2) Find the clusters of bonds.
(3) For each cluster, independently choose a color for all sites randomly from the set of colors (or, from the colors other than the current color for the cluster).

A formal proof that this converges to the desired process is given by Kirkland (1989). This amounts to demonstrating that the P matrix does satisfy the detail balance equations (12.14). The idea of the method came from a

(a)

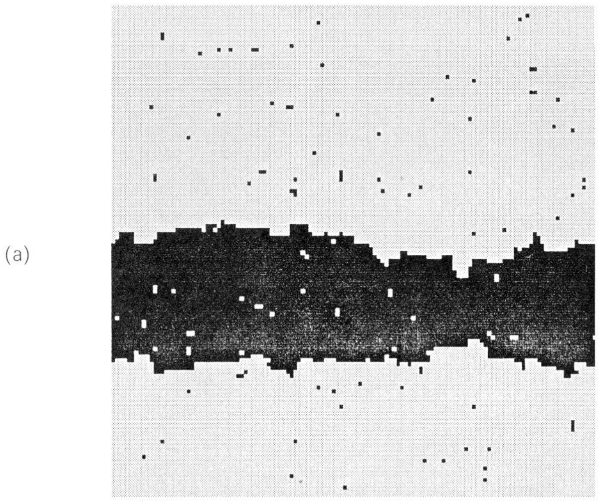

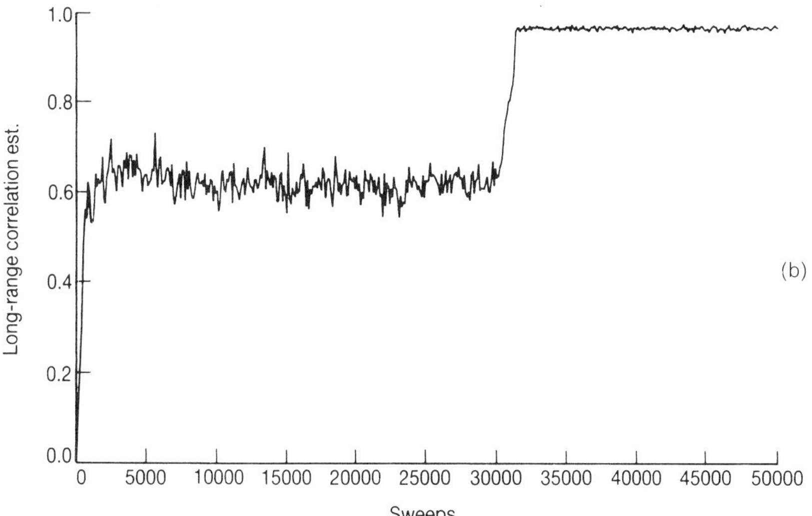

(b)

Figure 12.5 Quasi-equilibrium behavior in Metropolis' method: (a) shows the realization of an Ising model with $\beta = 2{\cdot}6$ after 20 000 sweeps. The local characteristics have converged, but as (b) shows, the long-range correlations are in quasi-equilibrium at the wrong value until about 32 000 sweeps, when the image became mainly white with a few isolated black values.

representation of the symmetric Potts model as a percolation process by Kasteleyn and Fortuin (1969).

A very closely related algorithm is given by Wolff (1989). (See also Gould and Tobochnik, 1989.) Instead of finding all clusters on the lattice, this chooses a site at random and constructs the cluster to which it belongs. A formal algorithm is as follows:

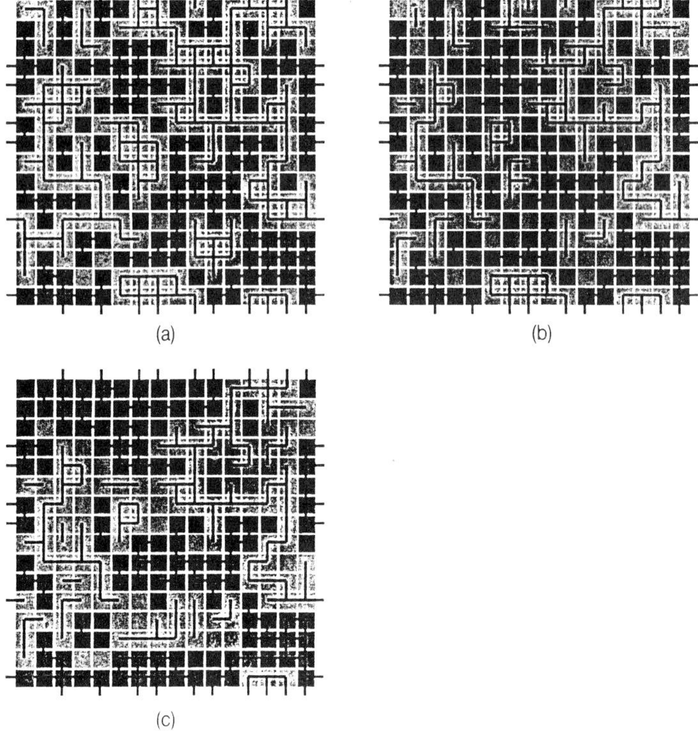

Figure 12.6 One step of the Swendsen–Wang simulation method for the Ising model: (a) all neighbor pixels of the same color are joined; (b) bonds are retained independently with probability p; (c) the clusters are independently colored.

Repeat forever:

(1) Choose a site at random. Set up an array of 'perimeter sites' consisting of its neighbors.
(2) Repeat
    Remove the first perimeter site and replace by the last. With probability p make this a site of the cluster.
    If a site was added to the cluster, append all its neighbors to the set of perimeter sites.
  Until no perimeter sites remain.
(3) Choose a color for all sites in the cluster at random from the set of colors (or from the rest of the set of colors).

This has the advantage over Swendsen–Wang of a more 'local' calculation, but appears to lack the advantages of convergence speed.

The Swendsen–Wang method has been extended to more general models ('with an external field' in the terminology of statistical physics) (Edwards and Sokal, 1988; Martinelli *et al.*, 1990). Consider the specific Strauss model (a special case of a nonstationary generalization of (12.8)):

$$P(X_1 = x_1, \ldots, X_n = x_n) \propto \exp\left[\sum_s \alpha_{s,x_s} + \sum_{s \text{ nhbr of } t} \beta_{st} 1(x_s = x_t)\right],$$

where 1( ) is the indicator function. This model allows a preference for different colors but has a symmetric interaction between colors. Step (3) of the Swendsen–Wang procedure is replaced by

(3) For each cluster C, independently choose a color for all sites randomly from the set of colors with probability proportional to $\exp(\Sigma_{s \in \text{cluster}} \alpha_{s,c})$ for color c.

And in step (1) p is replaced by $p_{st} = 1 - \exp(-\beta_{st})$ for the bond from s to t.

(b) Multilevel methods

The problem with site-by-site methods is that they take too long to reach equilibrium, and to make large changes in the realization. This can be overcome by using a separate and possibly approximate method to find a suitable pattern for the sites. There are two related but distinct ideas here.

Gidas (1989) has a 'multi-scale' method, in which an MRF on a lattice is approximated by an MRF on a coarser lattice, say one in which sites are combined into squares of fours. This process can be repeated until a small lattice is obtained. For this small lattice iterative methods will converge rapidly, and exact methods may even be possible. Thus a simulation on the final scale can be performed by successively refining the realization, replacing each square by four of the same color and running a few sweeps of the (site-by-site) iterative simulation process. The disadvantage of Gidas' approach is that the CPU time needed to generate the approximations is very large. Thus his method is most appropriate when many simulations are needed from the same (stationary) process.

The alternative is to produce an MRF on a graph consisting of a set of linked lattices. Each lattice is an aggregation of that immediately below it. Usually four 'daughter' squares are linked to each parent 'square'. Then we have a graph on which the desired MRF is run at each level, but there is a link parameter γ for the parent–daughter links. This defines an MRF which can be simulated by site-by-site methods. When a square at a high level in the hierarchy changes color it biases its 'descendants' and so tends to cause large changes at the lowest level. This process does *not* simulate the correct MRF on the lowest lattice, but it can be run for a short while, reducing the parameter γ to zero. How to choose and reduce γ seems to be a matter for trial and error. For the Ising model (Kirkland, 1989) it proved to be very effective.

Kandel *et al.* (1988) independently use multi-level methods, but applied to the Swendsen–Wang approach. Further approaches are given by Goodman and Sokal (1989).

(c) Discussion

A direct comparison of these methods is difficult. They will all have different rates of convergence and take different times to complete a number of steps. For example, we can compare the clock method and the Swendsen–Wang method on the number of 'sweeps' across the graph. On that basis the clock method is *much* faster per sweep, especially for large β, but appears to converge more slowly.

Where they are applicable, cluster methods appear to converge very rapidly and so are cheap. However, their applicability is very limited, and methods to find a good starting point followed by a few sweeps of site-by-site methods are at least as efficient and much more general. These appear to be the first choice, given our current rather limited knowledge.

12.6.4 Point processes

Let us consider multidimensional point processes on a bounded set $D \subset \mathcal{R}^d$. A Poisson point process then chooses a number n of points from a Poisson distribution and distributes them uniformly and independently on D. Gibbs point processes are defined by their density p (Section 12.5). One relatively simple example is the class of pairwise interaction processes with, for n points $\mathbf{x} = \{x_1, \ldots, x_n\}$,

$$p(\mathbf{x}) = ab^n \prod_{i<j} h(d(x_i, x_j)), \tag{12.15}$$

where $h(t)$ measures the interaction at distance t and is zero for very small t, one for large t. This specializes (12.10) to be position-independent, and therefore as stationary as is possible within a bounded set.

These point processes can be simulated via a spatial birth-and-death process. A traditional birth-and-death process models a population with the birth rate and death rate dependent on the number of individuals in the population. A spatial birth-and-death process allows the birth and death rates to depend on the spatial configuration of the points as well; intuitively this allows higher death and/or lower birth rates in regions of overcrowding.

One version of the birth-and-death process is to allow deaths to occur at unit rate for each point, and a birth rate at ξ of

$$\lambda(\xi; \mathbf{x}) = b \prod_{i} h(d(\xi, x_i)), \tag{12.16}$$

which defines a Markov chain on the (continuous) state space D. It is easy to see that the continuous version of the detailed balance conditions (12.14) holds, and this suffices to show that the Gibbs point process is the unique equilibrium distribution. The Markov process is easily simulated by viewing the process only at the sequence of time points at which jumps occur. (This is known as using the jump chain.) The jump is either a death or a birth with odds $1 : \Lambda(\mathbf{x})$, where

$$\Lambda(\mathbf{x}) = b \int_D \prod_i h(d(\xi, x_i))\, \mathrm{d}\xi$$

and if a birth occurs, its distribution in D has density proportional to the integrand. Note that we must sample the process at a fixed time t rather than after a fixed number of jumps which would introduce a bias.

If we want a simulation conditional on n, the process alternately deletes a point at random and adds a point ξ with p.d.f. proportional to $\lambda(\xi)$. There are two possibilities; either a point is added to make $n+1$ and then one of these is deleted, or a point of the n is deleted and a new point added. As the first of these alternatives frequently deletes the point it has just added, the second is usually preferred. It is coded in FORTRAN in Ripley (1979). Some care is needed for the conditional version, as irreducibility can fail. For example, in packing disks within a square ($h(r)=0$ for $r<R$) it will not be possible to move between any two configurations if n is fixed at a large enough value.

12.6.5 When is 'long enough'?

Simulators frequently ignore the methods of their own discipline in designing and analysing simulation experiments (Ripley, 1988a). Here is a case in which we can make good use of theory to analyse iterative simulation methods. Since Π_n comes from a Markov chain, we know from the theory surrounding the Perron–Frobenius theorem (Seneta, 1973) that for any norm $\|\ \|$,

$$\|\Pi_n - \pi\| = O(r^n)$$

for a constant $0<r<1$, the second-largest modulus of an eigenvalue of P. Thus if we can measure a suitable characteristic of Π_n, θ say, we can estimate r, the rate of convergence and hence how long is 'long enough'. More specifically, we will fit

$$\hat{\theta}_n = \theta_\infty + \alpha r^n + \varepsilon_n,$$

where $\{\varepsilon_n\}$ is a correlated sequence of errors. We can usually model the error sequence by a stationary time-series model and so estimate r by maximum likelihood.

It is important to choose a characteristic θ which brings out the right aspect of Π_n, as some will depend very weakly on the largest eigenvalues. This is dramatically demonstrated for the Potts process. We considered both pseudo-likelihood (Besag, 1975) and asymptotic maximum likelihood (Pickard, 1987) estimators of β and the long-range correlation (Bartlett, 1975, Pickard, 1987). These are defined in more detail in Section 12.7. The maximum likelihood estimator equates the observed nearest-neighbor correlation to its expectation, whose asymptotic value (as the square lattice expands in both directions) is known as a function of β. Similarly, the limit of the correlation of sites very far apart is known, and is non-zero for $\beta > \beta_c$. By definition, both the MLE and the long-range correlation are (approximately) unbiased and so provide a good check for convergence to known values for the simulated process.

Some examples of this process are shown in Figure 12.7.

Peskun (1973) compared the speed of convergence for methods with a fixed matrix Q. Let f be an arbitrary function, and suppose we wish to estimate $I = E_\pi f(X)$. As usual we use the Monte-Carlo estimate $I_N = \Sigma f(X_t)/$

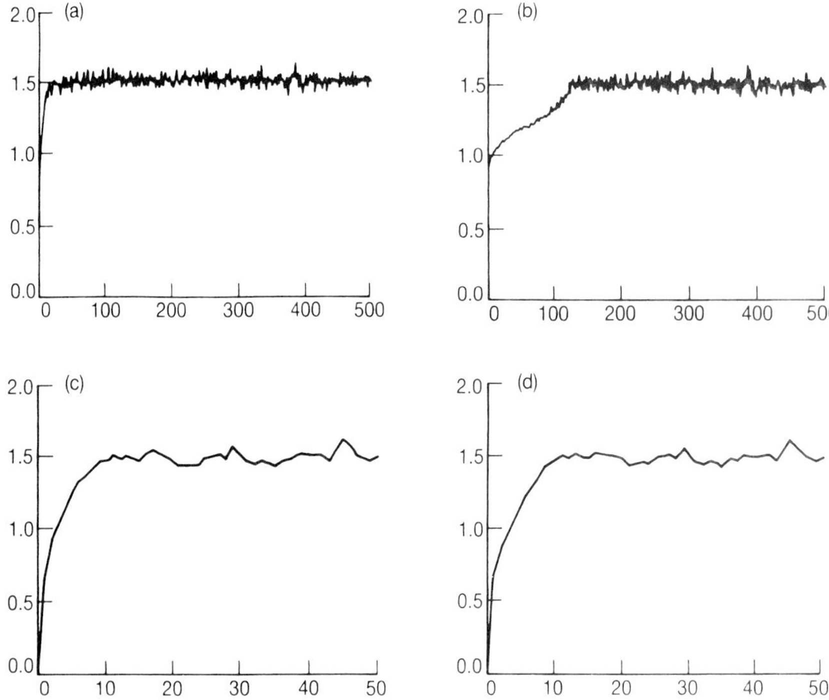

Figure 12.7 Estimation of convergence of iterative methods for the Ising model with true $\beta = 1{\cdot}5$: (a), (b) pseudo-likelihood and asymptotic maximum likelihood estimates of β for 500 sweeps of Metropolis' method; (c), (d) ditto for 50 sweeps of the Swendsen–Wang method.

N. This has bias of $O(N^{-1})$ and variance of the same order. Peskun showed that the variance is smallest for Metropolis' method, for fixed Q.

These results do not determine the rate of convergence of Π_n to π since Metropolis' method for I_N is best essentially because it makes most jumps. Some recent results by Frigessi *et al.* (1990) show that for the Ising model in two dimensions, Metropolis is best at large β and worst at small β; at small β Gibbs' is better than Metropolis' but even better methods can be found.

12.6.6 Conditioning

Some types of conditioning are easy. If the rock type of a particular block in an MRF model is known (for example, at a well), the conditional distribution at that block is fixed and the color will never alter. Effectively we never visit that block in the iterative scheme. Other types of conditioning are not at all easy! There are, however, some general strategies set out below. These (and their description here) are 'for experts only'.

Conditioning a simulation on some event amounts to throwing away all realizations for which that event is false, and rescaling the probabilities of the remainder so they sum (or integrate) to one. The iterative simulation process in equilibrium at each time step produces a sample from the process being simulated. Throwing away realizations amounts to waiting until a realization is generated which meets the sample condition. However, if the event is rare, we may have to be very patient indeed.

A way to reduce the wait is to modify the process on the set of realizations for which the event is false so as to make them less likely. This is a form of *importance sampling* as discussed in Section 12.2. Suppose we have a function $D(\mathbf{x})$ which is zero when the desired event is true, and positive when it is false. Then we could simulate from the process given by

$$p_\lambda(\mathbf{x}) \propto p(\mathbf{x})e^{-\lambda D(\mathbf{x})} \tag{12.17}$$

and wait until the event happens. For large λ it should be much more likely, and so the wait may be less. This process is precisely that employed in *simulated annealing* (Ripley, 1987, Section 7.2).

The difficulty with (12.17) is that it may be much more difficult to use for iterative simulation. In particular, p_λ may not be an MRF defined on either the same graph or any simple related graph. However, we *can* easily use (12.17) for MRFs to include preference information from seismic studies and for including ordering constraints on the rock types. For stochastic geometry models it will be effective to condition on the rock type at wells and the likely ordering of rock strata.

The idea of using (12.17) for simple constraints is in Green (1986). It has also been used much more generally by Farmer (1989) with a minimally specified $p(\mathbf{x})$ but many constraints on the *statistical* nature of the realization. Of course, it is usually much faster to build such details into $p(\mathbf{x})$ and tailor the simulation method accordingly.

(a) Relevant observables

The information available for conditioning could come either from direct measurement (rock types down wells from core samples) or from other measurements on the reservoir (well logs, seismic data). The first is relatively easy to incorporate, the second less so. A major difficulty is that the certainty of seismic data, for example, is difficult to ascertain. It is important not to condition precisely on imprecise data, as this would lead to over-estimation of the precision of the final results of the numerical flow experiments.

12.7 Parameter estimation

This section is intended only to give an overview to a technically rather difficult and little developed area. It *is* important in that parameters in the models have to be chosen somehow, and estimating them from data on the patterns of real rock structures (e.g. outcrops) is clearly desirable.

12.7.1 MRFs

Parameter estimation in all types of MRFs have been fraught with difficulties. Gaussian MRFs, such as might be used for permeability models, caused enough difficulties in the 1970s. These are defined by giving a mean μ_s for each site, and the conditional mean of X_s conditional on all other X_t as

$$E[X_s \mid X_t, t \neq s] = \mu_s + \sum_{t \neq s} B_{st}(X_t - \mu_t) \tag{12.18}$$

and conditional variance κ. This is an MRF if $B_{st} \neq 0$ only when s and t are neighbors.

When these models were first proposed, people shied away from maximum likelihood estimation of the parameters in B and in the mean vector. The computational difficulty comes from the normalizing constant which we have conveniently ignored up to now. For a conditional autoregression the joint probability density of $(X_1, \ldots, X_N)$ is (from its specification as a multivariate normal distribution)

$$\frac{|I - B|^{1/2}}{(2\pi\kappa)^{N/2}} \exp\left[-\frac{1}{2\kappa}(\mathbf{X} - \mu)^{\mathrm{T}}(I - B)(\mathbf{X} - \mu) \right],$$

so minus twice the log likelihood is given by

$$N \ln 2\pi\kappa - \ln |I - B| + (\mathbf{X} - \mu)^{\mathrm{T}}(I - B)(\mathbf{X} - \mu)/\kappa. \tag{12.19}$$

This is easily maximized over κ to give

$$\hat{\kappa} = N^{-1}(\mathbf{X} - \mu)^{\mathrm{T}}(I - B)(\mathbf{X} - \mu),$$

so parameters in B are chosen to minimize

$$N \ln \hat{\kappa} - \ln |I - B|. \tag{12.20}$$

The perceived difficulty is the determinant $|I - B|$, but for realistically sized problems it is quite straightforward to minimize (12.20) numerically.

However, we should ask whether maximum likelihood is desirable *per se*. The classical justifications of maximum likelihood in statistics are asymptotic, for an infinite sequence of identically distributed *independent* observations. This is not a natural asymptotic regime in spatial statistics, although Mardia and Marshall (1984) have proved that for a related type of asymptotics some classical results hold. In general, we do not even know if (12.19) is a well-behaved likelihood function. Ripley (1988b, Chapter 2) demonstrates that it can fail to be concave, but that with a known mean vector and $B = \rho W$ is *is* unimodal. (Much of the statistical theory of likelihood functions depends on concavity.) Thus although maximum likelihood estimation is almost always possible computationally, we must not assume that it is necessarily statistically desirable. The literature on this point is often misleading or plain wrong.

An alternative, *pseudo-likelihood* estimation, was introduced by Besag (1975). The pseudo-likelihood, PL, is defined as the product of the conditional densities $P(X_s \mid X_t, t \neq s)$, and is treated like a likelihood. There seems no simple explanation as to why this is a sensible idea, but it has proved to work well in practice. Besag (1975) sketched a proof that the pseudo-likelihood estimator would be consistent (converge to the true value) as the problem size is increased, and an elegant general proof is given by Geman and Graffigne (1987). For a conditional autoregression we have

$$\ln PL = -\frac{N}{2}\ln(2\pi\kappa) - \frac{1}{2\kappa}\sum_s [(I-B)(\mathbf{X}-\mu)]_s^2,$$

so pseudo-likelihood estimation amounts to least-squares fitting to the 'residuals'

$$\eta_s = X_s - \mu_s - \sum_{t\neq s} B_{st}(X_t - \mu_t).$$

Be warned that (12.18) cannot be treated like an ordinary regression, and that the residuals (η_s) will themselves be spatially autocorrelated.

Now consider the Potts–Strauss process (12.4). Then the log pseudo-likelihood is given by

$$\ln PL = \sum_s \left[\beta\#\{\text{nhbrs of } s \text{ of same color}\} - \ln\left(\sum_c \exp \beta\#\{\text{nhbrs of } s \text{ of color } c\}\right)\right], \tag{12.21}$$

which is a fairly simple function of β and can certainly be maximized numerically.

The maximum likelihood estimator of β for (12.4) depends on the unknown normalizing constant, and so is not known exactly. However, we know for general statistical theory that the MLE of β is chosen by equating the neighboring correlation to its theoretical value as a function of β. For the Ising model, Onsager (1944) gave a formula for the asymptotic value of the theoretical correlation as the lattice becomes infinitely large in both directions, and this can be used to give an approximate MLE.

The MLE method depends on a theoretical result which is only known for the 2-color symmetric 4-neighbor case in two dimensions. Thus it is of little general use. The PL estimator can, however, be easily generalized. Against this, it proves to be a much less sensitive statistic to use than the methods of Section 12.6.5, since it depends only on 'local' properties of the pattern used for estimation.

12.7.2 Point processes

Parameter estimation in pairwise-interaction point processes proved to be difficult for a decade, and an obstacle to their wider use. One can of course use trial-and-error methods, matching some aspect of the simulated patterns to the data. This was illustrated for some data on the pattern of Spanish market towns in Ripley (1977), using my K-function to measure fit, and there are more extensive comparisons in Ripley (1988b, Figure 4.2). However, it is both non-intuitive and rather computer-intensive.

Even the simplest cases have shown difficulties. The maximum likelihood estimate of R in a hard-core process is $d_{\min}$, the smallest distance between a pair of points. Since pairs closer than R cannot occur, we know $d_{\min} > R$ and hence the estimator is biased. Ripley and Silverman (1978) showed how to correct this estimator for bias. Even if R is known for the Strauss process (defined at (12.12)), the maximum likelihood estimator of c is not straightforward to derive. The density is

$$p(\mathbf{x}_1, \ldots, \mathbf{x}_n) = a(b, c) b^n c^{y(R)} \tag{12.22}$$

and it is the unknown normalizing constant $a(b, c)$ which causes the trouble. To ease the notation, let us write $Y(R)$ for the random variable measuring the number of pairs of points within D closer than R. Then after some calculation we find that the maximum likelihood estimators are given by the solution in b, c to the equations

$$n = E_{b,c} N(D), \qquad y(R) = E_{b,c} Y(R) \tag{12.23}$$

if the number of points is allowed to vary, otherwise b is irrelevant and c solves

$$y(R) = E_c Y(R). \tag{12.24}$$

Here $y(R)$ is the fixed number of observed close pairs, and the right-hand side is the expected number of pairs. It is intuitively obvious that this increases from zero for $c = 0$ up to the value for a Poisson process for $c = 1$. The latter depends on the shape of the domain D, but is known for several common shapes (Ripley, 1988b, pp. 28–9); for small R it is approximately $n(n-1)\pi R^2/2$. It is perfectly possible for us to observe more close pairs than are the average for a Poisson process; in such a case we take $\hat{c} = 1$, although we ought to consider why we are fitting an inhibition process with interaction distance R!

Maximum likelihood estimation depends on our being able to calculate the means in (12.23). They are not known analytically, but can be estimated by simulation (Penttinen, 1984; Ripley, 1988b). An alternative is to make approximations. When interactions are rare, we can derive the approximations (Ripley, 1988b, pp. 56–7)

$$\hat{c} \approx y(R) \times \frac{2a}{n(n-1)\pi R^2} \tag{12.25}$$

for fixed n, and for variable n

$$\hat{b} \approx n/\mu, \qquad \hat{c} \approx y(R) \times \frac{2a}{n^2 \pi R^2},$$

where a denotes the area of D. These assumptions correspond to fitting a straight line in c to $E_c Y(R)$, although they are derived by assuming that any point enters into just one close pair. More sophisticated approximations have been considered, notably by Ogata and Tanemura (1981, 1984), who borrowed approximations from statistical physics for 'non-ideal' gases. Details are given in Ripley (1988b, pp. 59–62). For the Strauss process, they give an approximation to the log-likelihood as

$$y(R) \ln c - \frac{n^2}{2a}(c-1)\pi R^2 - \frac{5{\cdot}79 n^3 (c-1)^3 R^4}{6a^2}$$

corresponding to

$$E_c Y(R) \approx \frac{n^2 \pi R^2}{2a} c[1 + 1{\cdot}84(nR^2/a)(c-1)^2],$$

a cubic in c. (There is an error of a factor of 3 in the constants in Ripley, 1988b.)

(a) Pseudo-likelihood

We have seen that even in the simplest case maximum likelihood estimation causes difficulties, and just as in the regional data case, there is no reason to accord maximum likelihood methods special status from a theoretical viewpoint. This encourages us to consider pseudo-likelihood as it was so successful there. There are technical problems in the conditioning, but these can be overcome either by approximating by a lattice process (Besag, 1977; Besag *et al.*, 1982) or via the theory of conditional intensities. The problem is that we want the probability of a point occurring at $\mathbf{x}$ given the locations of the remaining points. This probability is zero, but we can consider the conditional intensity $\lambda(\mathbf{x}; \mathbf{x}_1, \ldots, \mathbf{x}_n)$. If $\mathbf{x}$ is the location of one of the existing points it is omitted from the conditioning. Then the pseudo-likelihood is the product of the conditional intensity over all points in D, occupied or not, which leads to the log pseudo-likelihood as

$$\sum_1^n \ln \lambda(\mathbf{x}_i, \mathbf{x}_1, \ldots, \mathbf{x}_{i-1}, \mathbf{x}_{i+1}, \ldots, \mathbf{x}_n) - \int_D \lambda(\mathbf{x}; \mathbf{x}_1, \ldots, \mathbf{x}_n)\, \mathrm{d}\mathbf{x}. \tag{12.26}$$

For a pairwise interaction process the conditional intensity is (12.11)

$$b(\mathbf{x}) \prod_{\text{pts other than x}} h(\mathbf{x}, \mathbf{x}_i)$$

so the log pseudo-likelihood is

$$\ln\left[\prod_i b(\mathbf{x}_i) \prod_{i,j} h(\mathbf{x}_i, \mathbf{x}_j)\right] - \int_D b(\mathbf{x}) \prod_i h(\mathbf{x}, \mathbf{x}_i)\, \mathrm{d}\mathbf{x}. \tag{12.27}$$

At first sight the first term is the likelihood (12.15) without the normalizing constant, but the second product is different in that each pair $\{i, j\}$ occurs twice. Nevertheless, pseudo-likelihood methods are very similar to maximum-likelihood ones, except that the normalizing function is replaced by much simpler integrals. For the Strauss process the pseudo-likelihood estimators $\tilde{b}$ and $\tilde{c}$ solve

$$\tilde{b} \int_D \tilde{c}^{t(\mathbf{x})}\, \mathrm{d}\mathbf{x} = n$$

$$\tilde{b} \int_D t(\mathbf{x}) \tilde{c}^{t(\mathbf{x})}\, \mathrm{d}\mathbf{x} = 2y(R) \tag{12.28}$$

for variable n, where $t(\mathbf{x})$ denotes the number of points of the pattern within distance R of the test point $\mathbf{x}$. (Conditional intensities, hence pseudo-likelihood, make no sense if n is fixed. The intensity will be zero or infinite depending on whether there are n or $n-1$ points elsewhere.) The integrals in (12.28) can be estimated by sums over a grid of points within D, and sophisticated methods are available to find $t(\mathbf{x})$ rapidly using auxiliary data structures.

These pseudo-likelihood methods proved to be a special case of a family of moment measures derived by Takacs (1986) and Fiksel (1984). These compare the expected values of observations on the process with similar expectations conditional on a point of the pattern at $\mathbf{x}$. However, whereas the Takacs–Fiksel work depends on arbitrary choices of moments to compare (and is mathematically forbidding), pseudo-likelihood has some theoretical rationale.

(b) Edge effects

All the parameter estimation methods mentioned up to now have been for a process defined only on the domain D. This is frequently inadequate, in that we imagine the underlying process placing the points to occur within a much larger domain D' but which is observable (or has only been recorded) within the window D. In such a case the number of points is necessarily variable. A rigorous treatment of this case is very difficult, as it would involve averaging over the positions of points outside D which might interact with those within D. Rather, we choose to correct the estimators we have seen so far for edge effects. This is particularly easy for the Strauss process, since the only statistic which occurs is $y(R)$, the number of R-close pairs. This occurs in the second-moment K-function which is often used to describe a spatial point pattern, and so much work has been done on correcting for edge effects. Full details of the corrections proposed and of their efficacy are given in Ripley (1988b, Chapter 3).

12.7.3 Choosing an interaction function

How will we know what shape of interaction function to pick? Almost never is there any appropriate theory to suggest a particular functional form, and it is really the shape rather than the mathematical form which we seek. Two ideas have recently been suggested. The conceptually simplest is that of Ripley (1988b, p. 73). Fit a multi-scale process (12.13) to obtain a step-function (histogram-like) estimate of the interaction function h, and choose a suitable functional form to fit parametrically. This is just like plotting a histogram in univariate statistics, and choosing a family of distributions (normal, gamma, ...) from its shape.

Fortunately, estimating the parameters in the multi-scale process by pseudo-likelihood is just as easy as for the Strauss process. The equations are similar to (12.28), with c replaced by c_i and $t_i(\mathbf{x})$ denoting the number of points of the pattern whose distance from $\mathbf{x}$ is between R_{i-1} and R_i. That is, $(\tilde{c}_i)$ solves

$$\tilde{b}\int_D t(\mathbf{x})\prod_j \tilde{c}_j^{t(\mathbf{x})}\,\mathrm{d}\mathbf{x} = \sum_{\text{pts}} t_i(\mathbf{x}_j) = 2[y(R_i) - y(R_{i-1})]$$

for each i together with

$$\tilde{b}\int_D \prod_j \tilde{c}_i^{t(\mathbf{x})}\, \mathrm{d}\mathbf{x} = n,$$

and the ratio is a function of $(c_1, \ldots, c_m)$ which is easily estimated numerically, and so the equations are solved numerically.

Care is needed in interpreting the shape of the function, as the estimates at small distances are very variable, unlike a histogram. For a Poisson process we would have that

$$\tilde{c}_i \approx [y(R_i) - y(R_{i-1})] \times \frac{2a}{n(n-1)\pi[R_i^2 - R_{i-1}^2]}$$

by an extension of (12.25), and the count of pairs $[y(R_i) - y(R_{i-1})]$ is approximately Poisson. Thus $\tilde{c}_i$ has a standard deviation approximately equal to the square root of the second term, and this will be large if the area of the annulus of points between R_{i-1} and R_i away from a fixed point is small. This will inevitably be the case for distances between 0 and R_1.

The other idea borrows from statistical physics a relationship between the interaction function h and the second-moment function K known as the Percus–Yevick formula. Whereas in physics this is used to calculate K from h, Diggle *et al.* (1987) had the idea to reverse the process to obtain a non-parametric estimator of h; full details are in their paper. The Percus–Yevick formula is approximate, and something of a mystery to me, but simulations have shown it to be reasonably accurate. It relates the interaction function h to the pair-correlation function g. This is defined by

$$g(t) = (2\pi t)^{-1}\frac{\mathrm{d}K(t)}{\mathrm{d}t}$$

where $K(t)$ is my reduced second-moment function (Ripley, 1977, 1981). Then the formula is

$$h(t) \approx g(t)/[g(t) - c(t)]$$

where $c(\cdot)$ is the solution to

$$c(t) = g(t) - 1 - \lambda\int_0^{2\pi}\int_0^{\infty}[g(s) - 1]c(\sqrt{[t^2 + s^2 - 2ts\cos\theta]})s\,\mathrm{d}s\,\mathrm{d}\theta.$$

Details of how $g(t)$ is estimated and how the equations are solved are given in the paper. The procedures are *not* straightforward, but in their authors' hands give good results for simulations of patterns of a few hundred points.

12.8 Examples

Some simple examples are given above in Figures 12.3 and 12.4. In this section we illustrate the procedures with some more realistic examples; these are still simple compared to what is computationally possible.

12.8.1 MRF for the height of a layer

This takes up an idea mentioned in Section 12.3, with an 'object' being a bed of one rock type of varying thickness. One- and two-dimensional examples are shown in Figures 12.8 and 12.9. The thickness of the layer takes one of the values $\{0, \ldots, nlayer\}$, and zero is treated specially. The joint probability distribution for the heights $\{Z_s\}$ is

$$P(\{Z_s\}) \propto \exp \sum_{\text{nhbr pairs}} [-\beta(Z_s - Z_t)^2 + \gamma I(Z_s = Z_t = 0)] \tag{12.29}$$

where $I()$ denotes the indicator function. From (12.29) we can deduce the conditional distribution given the neighboring heights. If r neighbors are nonzero and have heights summing to t, we find

$$P(Z_s = z \mid Z_t, t \neq s) \propto \begin{cases} \exp \gamma(\#\text{nhbrs} - r) & z = 0 \\ \exp - \beta(\#\text{nhbrs}\, z^2 - 2zt), & z > 0. \end{cases} \tag{12.30}$$

Positive values of β encourage a locally smooth surface. The value of γ alters only the effect of a zero height: positive values encourage larger regions of zero, negative values discourage zeros.

These models suffer some of the problems of k-class MRFs. In particular a strong local dependence (to give local smoothness) gives rise to too strong a global dependence (so the rock bed turns out to be *very* flat).

For efficient simulation by the Gibbs sampler, (12.30) is used to precompute probabilities for each combination of conditional variables. As these only depend on r and t, there are only

$$\sum_{r=0}^{\#\text{nhbrs}} (r \cdot nlayer) = \frac{\#\text{nhbrs}\,(\#\text{nhbrs} + 1)}{2}\, nlayer$$

sets of conditional distributions. These are discrete distributions, so an efficient general-purpose method such as the alias method (Ripley, 1987,

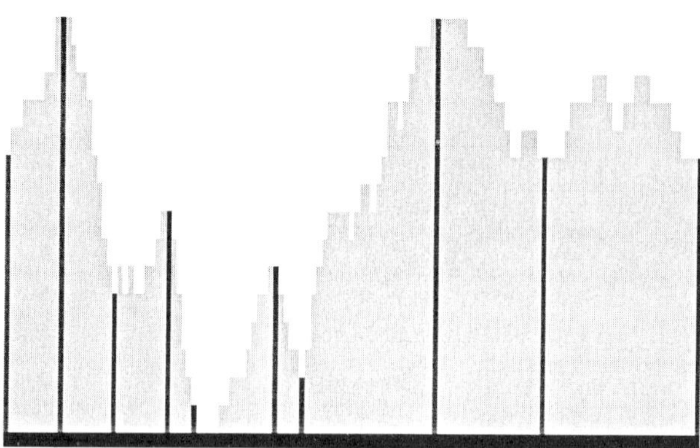

Figure 12.8 A one-dimensional example of a layer-height MRF (gray) with conditioning heights (black).

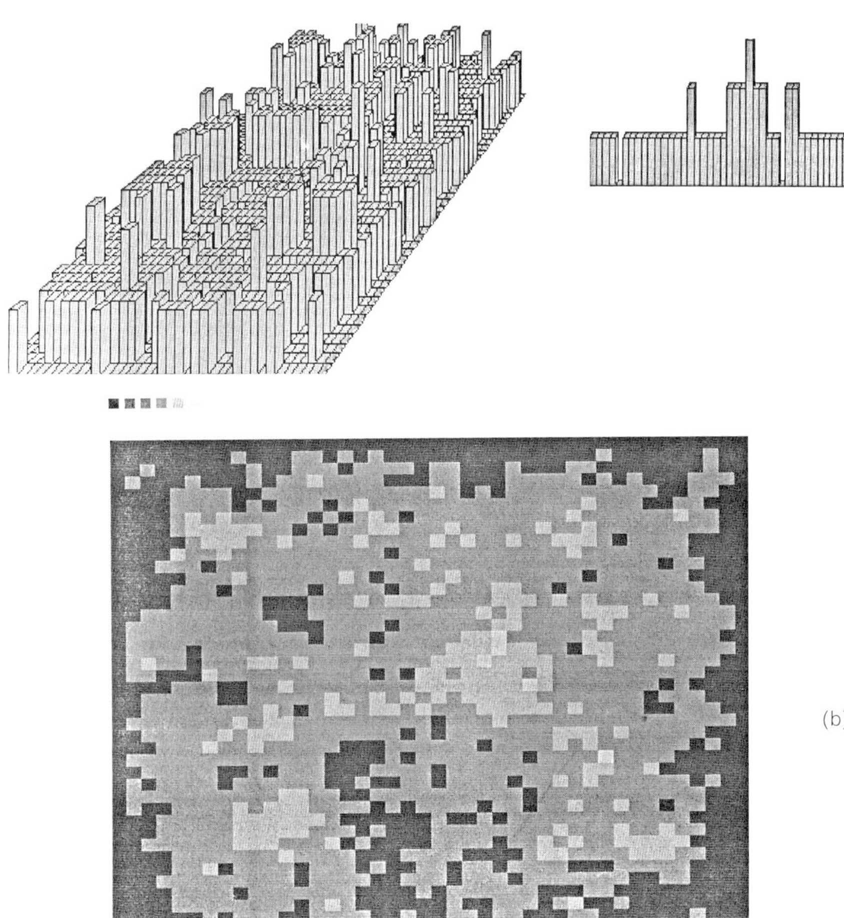

Figure 12.9 A two-dimensional layer-height MRF. (a) A perspective view and a slice. The boundary is zero (black); no other conditioning was used. (b) 5 heights shown as gray levels.

Section 3.3) should be used. The program visits each (horizontal) pixel in turn, calculates r and t, looks up the appropriate alias method tables for those values and returns a new height of the pixel.

In our experience the simulation process stabilizes after a handful of sweeps *provided* there are enough conditioning events. For a one-dimensional example such as Figure 12.8 with $\beta = 1$ and $\gamma = 1$ (and clearly two neighbors per pixel), the simulations settle down to (statistical) equilibrium after less than ten sweeps. On the other hand, the two-dimensional simulation of Figure 12.9 (with $\beta = 0{\cdot}3$ and $\gamma = 0$) is constrained only by having a zero boundary, and needs hundreds of sweeps to approach equilibrium.

12.8.2 Rectangular and cuboid objects

The remaining examples are of point processes of objects as described in Section 12.5. In two dimensions the objects are rectangles, in three dimensions cuboids. In both cases they are randomly distributed *without* overlaps. (Their rectangular nature makes it easier to check for overlaps.) The reservoir is taken as a unit square. In two dimensions the objects are $W \times H$ rectangles with $H < W$. Their sizes (and shapes) are independent from object to object. In Figure 12.10, W has a gamma(2) distribution with mean 0·20, and $H = UW$, where $U \sim U(0, 1)$. In three dimensions the cuboids are $W \times H \times D$, where W is gamma(2) with mean 100, $H = UW$, $D = VW$ with U, V independent uniform on (0, 1).

The simulations were performed by the spatial birth-and-death process of Section 12.6.4, with death rate μ and birth rate $\lambda(\xi; \mathbf{x})$, 1 if ξ fits without overlap, 0 otherwise. For the examples shown $\mu \approx 10^{-4}$. (Note that it is only the ratio of the rates that determines the object process.) If overlaps were allowed, the expected number of objects would be $1/\mu$. The effect of forbidding overlaps is to reduce this very considerably, since births become much rarer as soon as the reservoir is well-packed with objects.

Figure 12.11 gives some views of a three-dimensional example. The objects are contained within a $216 \times 128 \times 216$ cuboid with their longest dimension horizontal and parallel to the front face. The cube is shown both cut at an arbitrarily chosen vertical slice (Figure 12.11(a)) and as a nearly transparent medium containing opaque objects (Figure 12.11(b)). Three-dimensional simulations are best viewed on a graphics workstation. These plots were

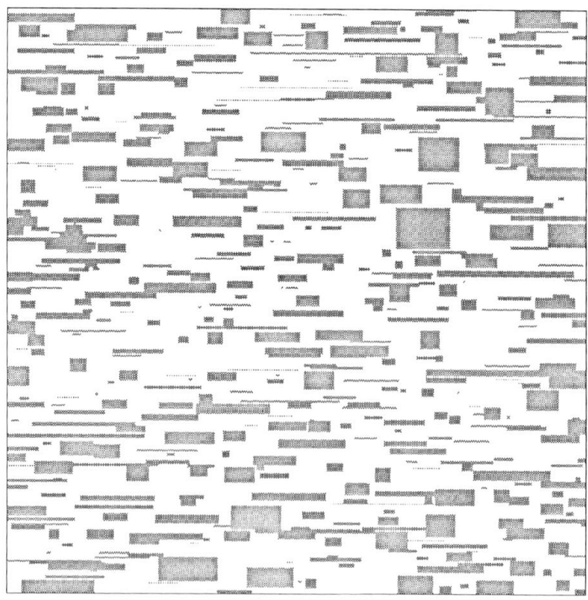

Figure 12.10 A two-dimensional object simulation.

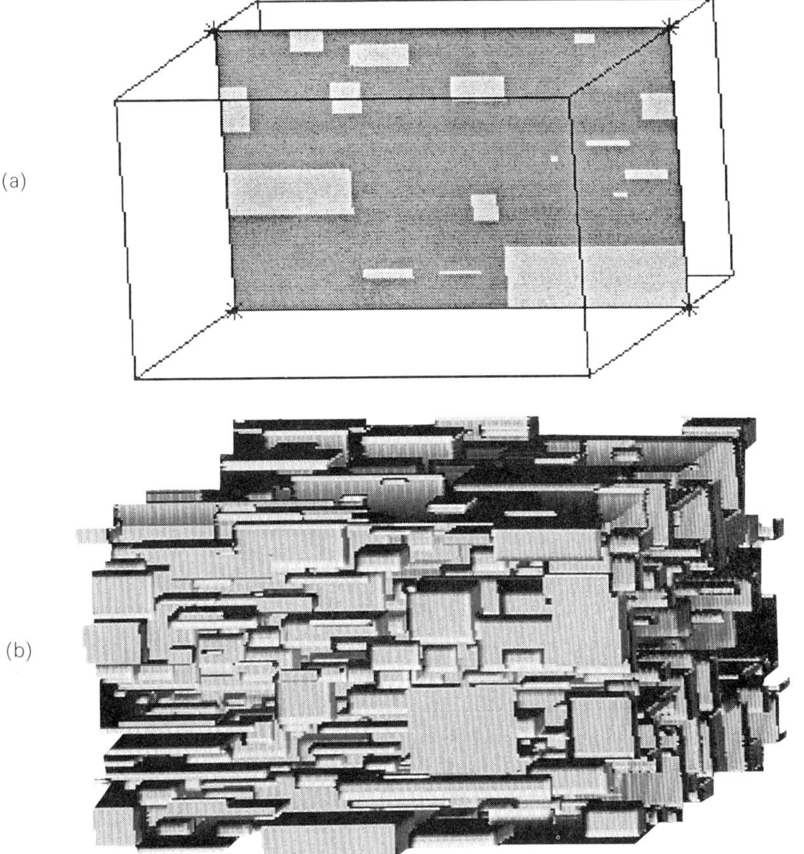

Figure 12.11 A three-dimensional simulation of non-overlapping cuboid objects: (a) sliced through at a vertical plane; (b) as objects in a nearly transparent medium.

done on a Sun 4/370 with TAAC-1 image accelerator which allows real-time interactive slicing and changes of viewpoint.

The two classes of examples shown here are but elements of the overall strategy. They can be combined, for example to have layers of relatively impermeable rock and shale intrusions in a sandstone bed.

Acknowledgements

The computational work described here was only possible by Sun workstations provided by the SERC under its 'Complex Stochastic Systems' and 'Computational Science' initiatives. Figures 12.5 and 12.7 were computed by Mark Kirkland, Figures 12.6, 12.8–11 by Ruth Ripley.

Comments on the petroleum engineering background by Drs Stephen Begg and Peter King of BP Research, Sunbury and Dr Henning Omre of

Norwegian Computing Center, Oslo, have been much appreciated. This work was partly financed by BP Research Centre, Sunbury, and is published with their permission.

This work was done and the paper prepared whilst the author was at the University of Strathclyde.

References

Barker, A.A. (1965). Monte Carlo calculations of radial distribution functions for a proton-electron plasma. *Aust. J. Phys.*, **18**, 119–33.

Bartlett, M.S. (1975). *The Statistical Analysis of Spatial Pattern*, Chapman and Hall, London.

Besag, J. (1975). Statistical analysis of non-lattice data. *The Statistician*, **24**, 179–95.

Besag, J. (1977). Some methods of statistical analysis of spatial data. *Bull. Int. Statist. Inst.*, **47** (2), 77–92.

Besag, J., Milne, R., and Zachary, S. (1982). Point process limits of lattice processes. *J. Appl. Prob.*, **19**, 210–16.

Bridge, J.S. and Leeder, M.R. (1979). A simulation model of alluvial stratigraphy. *Sedimentology*, **26**, 617–44.

Clementsen, R., Hurst, H., Knarad, R., and Omre, H. (1989). *A Computer Program for the Evaluation of Fluvial Reservoirs*, Norsk Regnesentral, Oslo note SAND/06/1989.

Cross, G.R. and Jain, A.K. (1983). Markov random field texture models. *IEEE Trans. PAMI*, **5**, 25–39.

Diggle, P.J., Gates, D.J., and Stibbard, A. (1987). A nonparametric estimator for pairwise-interaction point processes. *Biometrika*, **74**, 763–70.

Edwards, R.G. and Sokal, A.D. (1988). Generalization of the Fortuin–Kasteleyn–Swendsen–Wang representation and Monte Carlo algorithm. *Phys. Rev. D*, **38**, 2009–12.

Farmer, C.L. (1988). The generation of stochastic fields of reservoir parameters with specified geostatistical descriptions. In *Mathematics in Oil Production*, Edwards, S. and King, P.R. (eds), Clarendon Press, Oxford, pp. 235–52.

Farmer, C. (1989). *Numerical Rocks: the Mathematical Generation of Reservoir Geology*. Preprint, Winfrith Petroleum Technology.

Fiksel, T. (1984). Estimation of parameterized pair potentials of marked and non-marked Gibbsian point processes. *Elektron. Inform. Kybernet.*, **20**, 270–8.

Flinn, P.A. (1974). Monte Carlo calculation of phase separation in a 2-dimensional Ising system. *J. Statist. Phys.*, **10**, 89–97.

Flinn, P.A. and McManus, G.M. (1961). Monte-Carlo calculation of the order–disorder transformation in a body-centred cubic lattice. *Phys. Rev.*, **10**, 89–97.

Frigessi, A., Hwang, C.-R., Sheu, S.-J., and di Stefano, P. (1990). *Convergence Rates of the Gibbs Sampler, the Metropolis Algorithm and Other Single Site Updating Dynamics*. Preprint, Istituto per le Applicazioni del Calcolo 'Mauro Picone', CNR, Rome.

Geman, S. and Geman, D. (1984). Stochastic relaxation, Gibbs distributions and the Bayesian restoration of images. *IEEE Trans. PAMI*, **6**, 721–41.

Geman, S. and Graffigne, C. (1987). Markov random fields and their applications to computer vision. *Proc. Inter. Cong. Math. 1986*, Gleason, A.M. (ed.), Amer. Math. Soc., Providence, RI. 1496–1517.

Gidas, B. (1989). A renormalization group approach to image-processing problems. *IEEE Trans. PAMI*, **11**, 164–80.

Goodman, J. and Sokal, A.D. (1989). Multigrid Monte Carlo method. Conceptual foundations. *Phys. Rev. D*, **40**, 2035–71.
Gould, H. and Tobochnik, J. (1989). Overestimating critical slowing down. *Computers in Physics*, July/August, 82–6.
Green, P.J. (1986). Contribution to the discussion of Prof. Besag's paper. *J. Roy. Statist. Soc.* B, **48**, 284–5.
Haldorsen, H.H. and MacDonald, G.J. (1987). *Stochastic Modelling of Underground Reservoir Facies* (SMURF). Soc. Pet. Eng.-16751, Dallas.
Haldorsen, H.H., Brand, P.J., and Macdonald, C.J. (1988). Review of the stochastic nature of reservoirs. In *Mathematics in Oil Production*. Edwards, S. and King, P.R. (eds), Clarendon Press, Oxford, pp. 109–209.
Hammersley, J.M. and Handscomb, D.C. (1964). *Monte Carlo Methods*, Methuen, London.
Harding, E.F. and Kendall, D.G. (1974). *Stochastic Geometry*, Wiley, Chichester.
Hastings, W.K. (1970). Monte-Carlo sampling methods using Markov chains and their applications. *Biometrika*, **57**, 97–109.
Isham, V. (1981). An introduction to spatial point processes and Markov random fields. *Inst. Statist. Rev.*, **49**, 21–43.
Kandel, D., Domany, E., Ron, D., Brandt, A., and Loth, E. (1988). Simulations without critical slowing down. *Phys. Rev. Lett.*, **60**, 1591–4.
Kasteleyn, P.W. and Fortuin, C.M. (1969). Phase transition in lattice systems with random local properties. *J. Phys. Soc. Japan Suppl.*, **26**, 11–14.
Kelly, F.P. (1979). *Reversibility and Stochastic Networks*, Wiley, Chichester.
Kinderman, R. and Snell, J.L. (1980). *Markov Random Fields and their Applications*, Amer. Math. Soc., Providence, RI.
Kirkland, M.D. (1989). *Simulation Methods for Markov Random Fields*, University of Strathclyde, Ph.D. thesis.
Martinelli, F., Oliveri, E., and Scoppola, E. (1990). *On the Swendsen Wang Dynamics I: Exponential Convergence to Equilibrium*, CARR reports in mathematical physics 5/90, Università 'La Sapienza', Rome.
Matheron, G., Beucher, H., de Fouquet, C., Galli, A., Guerillot, D., and Ravenne, C. (1987). *Conditional Simulation of the Geometry of Fluvio-deltaic Reservoirs*. Soc. Pet. Eng.-16753, Dallas.
Metropolis, N., Rosenbluth, A.W., Rosenbluth, M.N., Teller, A.H., and Teller, E. (1953). Equations of state calculations by fast computing machines. *J. Chem. Phys.*, **21**, 1087–92.
Omre, H. (1989). Presentation at *Statistics, Earth and Space Sciences*, Bernoulli Society, Leuven, Belgium.
Onsager, L. (1944). Crystal statistics I. A two-dimensional model with an order–disorder transition. *Phys. Rev.*, **65**, 117–49.
Penttinen, A. (1984). Modelling interaction in spatial point patterns: parameter estimation by the maximum likelihood method. *Jyväsklyä Studies in Computer Science, Economics and Statistics*, **7**.
Peskun, P.H. (1973). Optimal Monte-Carlo sampling using Markov chains. *Biometrika*, **60**, 607–12.
Pickard, D.K. (1977). A curious binary lattice process. *J. Appl. Prob.*, **14**, 717–31.
Pickard, D.K. (1980). Unilateral Markov fields. *Adv. Appl. Prob.*, **12**, 655–71.
Pickard, D.K. (1987). Inference for discrete Markov random fields: the simplest nontrivial case. *J. Amer. Statist. Assoc.*, **82**, 90–6.
Potts, R.B. (1952). Some generalized order–disorder transformations. *Proc. Camb. Phil. Soc.*, **48**, 106–9.
Qian, W. and Titterington, D.M. (1989). On the use of Gibbs Markov chain models

in the analysis of images based on second-order pairwise interactive distributions. *J. Appl. Statist.*, **16**, 267–81.

Ripley, B.D. (1976a). Locally finite random sets: Foundations for point process theory. *Ann. Probab.*, **4**, 983–94.

Ripley, B.D. (1976b). The foundations of stochastic geometry. *Ann. Prob.*, **4**, 995–8.

Ripley, B.D. (1977). Modelling spatial patterns. *J. Roy. Statist. Soc.*, B, **39**, 172–212.

Ripley, B.D. (1979). Algorithm AS137. Simulating spatial patterns: Dependent samples from a multivariate density. *Appl. Statist.*, **28**, 109–12.

Ripley, B.D. (1981). *Spatial Statistics*, Wiley, New York.

Ripley, B.D. (1987). *Stochastic Simulation*, Wiley, New York.

Ripley, B.D. (1988a). Uses and abuses of statistical simulation. *Math. Prog.*, **42**, 53–68.

Ripley, B.D. (1988b). *Statistical Inference for Spatial Processes*, Cambridge University Press, Cambridge.

Ripley, B.D. and Kirkland, M.D. (1990). Iterative simulation methods. *J. Comp. Appl. Math.*, **31**, 165–72.

Ripley, B.D. and Silverman, B.W. (1978). Quick tests for spatial interaction. *Biometrika*, **65**, 641–2.

Seneta, E. (1973). *Non-Negative Matrices: An Introduction to Theory and Applications*, Allen and Unwin, London.

Stoyan, D., Kendall, W.S., and Mecke, J. (1987). *Stochastic Geometry and its Applications*, Akademic/Wiley, Berlin/Chichester.

Strauss, D.J. (1977). Clustering on coloured lattices. *J. Appl. Probab.*, **14**, 135–43.

Swendsen, R.H. and Wang, J.-S. (1987). Nonuniversal critical dynamics in Monte Carlo simulations. *Phys. Rev. Lett.*, **58**, 86–8.

Takacs, R. (1986). Estimator for the pair potential of a Gibbsian point process. *Statistics*, **17**, 429–33.

Wolff, U. (1989). Collective Monte Carlo updating for spin systems. *Phys. Rev. Lett.*, **62**, 361–4.

Chapter 13

Analysis of geochemical data sampled on a regional scale

K. Conradsen, A. A. Nielsen and K. Windfeld

When working with the analysis of geochemical data on a regional scale one will often encounter calibration problems arising from the chemical analysis of samples that might be collected over several years in geographical units. A procedure for removal of such geographical unit patterns is described. Stream sediment samples are collected at 33 992 sites and analyzed for the content of 26 elements. Semivariograms, the geostatistical analogue of spatial autocorrelation functions, are estimated, modeled and applied in interpolating to a regular grid. The interpolation method used is kriging, in which the mean square prediction error is minimized taking the spatial correlation into account. Heavy mineral samples are collected at 3094 sites and analyzed for the content of several economically interesting minerals. Based on the nonparametric method, CART (classification and regression trees) probability maps for the occurrence of gold and cassiterite are constructed using the kriged stream sediment data as predictors. This method is very useful for geologists when studying geological structures on a regional scale.

13.1 Geostatistics

The basis of geostatistics is the idea of considering the observed values of a geochemical, a geophysical or another natural variable at a given set of positions as a realization of a stochastic process in space. For each position $\mathbf{x}$ in a domain $\mathscr{D}$ there exists a measurable quality $z(\mathbf{x})$, a so-called *regionalized variable*. $\mathscr{D}$ is typically a subset of $\mathscr{R}^2$ or of $\mathscr{R}^3$. z is considered a particular realization of a *random variable* $Z(\mathbf{x})$. The set of random variables $\{Z(\mathbf{x}) \mid \mathbf{x} \in \mathscr{D}\}$ constitutes a random function. $Z(\mathbf{x})$ has mean value $E\{Z(\mathbf{x})\} = \mu(\mathbf{x})$ and covariance $\operatorname{cov}\{Z(\mathbf{x}), Z(\mathbf{x}+\mathbf{h})\} = C(\mathbf{x}, \mathbf{h})$. If the covariance is translation invariant over $\mathscr{D}$, i.e. $C(\mathbf{x}, \mathbf{h}) = C(\mathbf{h})$, Z is said to be second-order stationary. Often $Z(\mathbf{x})$ is assumed to follow a normal or a lognormal distribution. This statistical view on natural phenomena was introduced by Georges Matheron in 1962–63 and is described in great detail in David (1977) and in Journel and Huijbregts (1978). An introductory textbook is Clark (1979). David (1988) looks back on ten years of application of geostatistics.

The classical application of geostatistics has been the calculation of ore reserves. Here it is applied to describe the spatial distribution of geochemical elements over large areas (in the order of tens of kilometers by tens of kilometers). The spatial autocorrelation structure is described by means of semivariograms and if such autocorrelation is present it is utilized in interpolation performed by kriging, a best linear unbiased estimator (BLUE).

Point measurements of geochemical, geophysical or other natural variables or measurements taken over areas or volumes, also known as *supports*, are in principle continuous phenomena in space. If 'dense' sampling is performed the continuous nature of the variable in question will be reflected in the covariation of neighboring samples. If taken further apart from each other there will be little or no covariation between samples. Whether samples are 'dense' depends on the variable in question and sample sizes. Also, the autocorrelation revealed will depend on the scale at which one is operating. Different autocorrelation structures can be present simultaneously at different scales (mineralizations at the size of a few meters vs. regional variations at the size of tens of kilometers); this is referred to as *nested structures*.

The above remarks on autocorrelation applies to cross-correlations also if more than one variable is studied at a time.

13.1.1 The semivariogram

Consider two scalar values $z(\mathbf{x})$ and $z(\mathbf{x}+\mathbf{h})$ measured at two points in space $\mathbf{x}$ and $\mathbf{x}+\mathbf{h}$ separated by $\mathbf{h}$. z is considered a particular realization of a random variable Z. The variability is described by the autocovariance function

$$C(\mathbf{x},\mathbf{h}) = E\{(Z(\mathbf{x})-\mu)(Z(\mathbf{x}+\mathbf{h})-\mu)\}.$$

The *variogram* is defined as

$$2\gamma(\mathbf{x},\mathbf{h}) = E\{[Z(\mathbf{x})-Z(\mathbf{x}+\mathbf{h})]^2\}.$$

In general the variogram will depend on the location in space $\mathbf{x}$ and on the displacement vector $\mathbf{h}$. Note that the variogram represents a more general concept than that of the covariance function since the increment process $Z(\mathbf{x})-Z(\mathbf{x}+\mathbf{h})$ may have well-defined properties which the basic process $Z(\mathbf{x})$ does not possess. The *intrinsic hypothesis* in geostatistics is that the variogram is independent of the location in space and that it depends on the displacement vector only:

$$2\gamma(\mathbf{x},\mathbf{h}) = 2\gamma(\mathbf{h}).$$

Second-order stationarity of $Z(\mathbf{x})$ implies the intrinsic hypothesis. $\gamma(\mathbf{h})$ is called the *semivariogram*. The autocovariance function and the semivariogram are related by

$$\gamma(\mathbf{h}) = C(\mathbf{o}) - C(\mathbf{h}).$$

Note that $C(\mathbf{o}) = \sigma^2$.

An estimator for the semivariogram is the mean of the squared differences between any two measurements $z(\mathbf{x}_i)$ and $z(\mathbf{x}_i+\mathbf{h})$:

$$\hat{\gamma}(\mathbf{h}) = \frac{1}{2N(\mathbf{h})}\sum_{i=1}^{N(\mathbf{h})}[z(\mathbf{x}_i)-z(\mathbf{x}_i+\mathbf{h})]^2,$$

where $N(\mathbf{h})$ is the number of point pairs separated by $\mathbf{h}$. $\hat{\gamma}$ is called the *experimental semivariogram.* Pooling in both magnitude and argument of $\mathbf{h}$ is often performed. Pooling in the magnitude of $\mathbf{h}$ – i.e. the distance between samples – is done to obtain a sufficiently high $N(\mathbf{h})$ to ensure a small estimation variance ($N(\mathbf{h})$ is proportional to the estimation variance). Pooling in the argument of $\mathbf{h}$ – i.e. the direction – is done to check for anisotropy.

There is an extensive literature on the problems one encounters when estimating experimental semivariograms on real world data, cf. e.g. Cressie (1985).

In order to be able to define characteristic quantities for the semivariogram (and in order to apply the semivariogram in kriging, see below) a model is often assumed. An often used semivariogram model is the spherical model with nugget effect. A reason for this is the easy interpretability of the parameters. Assuming isotropy and setting $|\mathbf{h}| = h$ the form of this model is

$$\gamma^*(h) = \begin{cases} 0 & \text{if } h = 0 \\ C_0 + C_1\left[\dfrac{3}{2}\dfrac{h}{R} - \dfrac{1}{2}\left(\dfrac{h}{R}\right)^3\right] & \text{if } 0 < h < R \\ C_0 + C_1 & \text{if } h \geqslant R, \end{cases}$$

where C_0 is the *nugget* effect and R is the *range of influence.* $C_0/(C_0 + C_1)$ is the relative nugget effect and $C_0 + C_1$ is the *sill* ($= \sigma^2$). The nugget effect is a discontinuity in the autocorrelation function at $h = 0$ due to both measurement errors and to micro-variability, the structure of which is not available at the scale of study. This variability thus turns up as noise. The range of influence is the distance at which covariation between measurements stops; measurements taken further apart are uncorrelated. The spherical semivariogram model with nugget effect can easily be extended to e.g. a double spherical model with nugget effect to allow for nested structures:

$$\gamma^*(h) = \begin{cases} 0 & \text{if } h = 0 \\ C_0 + C_1\left[\dfrac{3}{2}\dfrac{h}{R_1} - \dfrac{1}{2}\left(\dfrac{h}{R_1}\right)^3\right] + C_2\left[\dfrac{3}{2}\dfrac{h}{R_2} - \dfrac{1}{2}\left(\dfrac{h}{R_2}\right)^3\right] & \text{if } 0 < h < R_1 \\ C_0 + C_1 + C_2\left[\dfrac{3}{2}\dfrac{h}{R_2} - \dfrac{1}{2}\left(\dfrac{h}{R_2}\right)^3\right] & \text{if } R_1 < h < R_2 \\ C_0 + C_1 + C_2 & \text{if } h \geqslant R_2, \end{cases}$$

where C_0 is the nugget effect and R_2 is the range of influence. $C_0/(C_0 + C_1 + C_2)$ is the relative nugget effect and $C_0 + C_1 + C_2$ is the sill. Other models for the semivariogram such as linear, bilinear and exponential models are often used also.

The parameters in the above semivariogram models γ^* can be estimated from the experimental semivariograms $\hat{\gamma}$ by means of iterative, nonlinear least squares methods. Different weights of the estimated values in the experimental semivariogram $\hat{\gamma}$ may be considered. A weighting with the number of point pairs included in the estimation for each lag distance seems natural. Also, if one is interested in a good model for small lags a weighting with the inverse lag distance applies.

Another important concept in geostatistics is *regularization*, i.e. the averaging of a random function over a domain $\mathscr{D}$. Let $\mathbf{x} \in \mathscr{D}$. The regularized value $Z_{\mathscr{D}}$ of Z is

$$Z_{\mathscr{D}} = \frac{1}{|\mathscr{D}|} \int_{\mathscr{D}} Z(\mathbf{x})\, \mathrm{d}\mathbf{x},$$

where $|\mathscr{D}|$ is the area (or volume) of $\mathscr{D}$. Similarly for the moment functions, e.g.

$$\bar{\gamma}(\mathscr{A}, \mathscr{B}) = \frac{1}{|\mathscr{A}||\mathscr{B}|} \int_{\mathscr{A}} \int_{\mathscr{B}} \gamma(\mathbf{x} - \mathbf{y})\, \mathrm{d}\mathbf{y}\, \mathrm{d}\mathbf{x}.$$

Thus $\bar{\gamma}(\mathscr{A}, \mathscr{B})$ is the regularized semivariogram when one end of the displacement vector $\mathbf{h} = \mathbf{x} - \mathbf{y}$ varies in $\mathscr{A}$ and the other end of $\mathbf{h}$ varies in $\mathscr{B}$. This integral can be either solved analytically for certain semivariogram models and supports of simple geometry or solved numerically.

What is said above about autocovariance functions and variograms can easily be extended to covariance functions and cross-variograms if more variables are studied simultaneously.

13.1.2 Kriging

Suppose that the random variable $Z(\mathbf{x})$ is sampled on a number of supports (could be points) $\mathscr{D}_1, \ldots, \mathscr{D}_n$ giving the following scalar measurements $z(\mathbf{x}_1), \ldots, z(\mathbf{x}_n)$. We now want to estimate $Z_{\mathscr{D}}$ on a support $\mathscr{D}$ where Z is not sampled (or Z is sampled on a part of $\mathscr{D}$ only). We are looking for a linear, unbiased estimator:

$$\hat{Z}_{\mathscr{D}} = \sum_{i=1}^{n} w_i Z_{\mathscr{D}_i}$$

$$E\{\hat{Z}_{\mathscr{D}} - Z_{\mathscr{D}}\} = 0.$$

The unbiaseness condition of the estimator gives

$$\sum_{i=1}^{n} w_i = 1.$$

The estimation variance (or the mean squared error) is

$$\sigma_E^2 = E\{(\hat{Z}_{\mathscr{D}} - Z_{\mathscr{D}})^2\}.$$

The *kriging estimate* is defined by the values of the weights w_i that minimize the estimation variance σ_E^2 subject to the constraint that the sum of the (kriging) weights is unity. This can be done by introducing a Lagrangian multiplier and setting each of the n partial derivatives

$\partial[\sigma_E^2 - 2\lambda(\Sigma_{i=1}^n w_i - 1)]/\partial w_i = 0$ leading to the $(n+1)\times(n+1)$ set of equations

$$\sum_{i=1}^{n} w_i \bar{\gamma}(\mathcal{D}_j, \mathcal{D}_i) + \lambda = \bar{\gamma}(\mathcal{D}_j, \mathcal{D}), \qquad j = 1, \ldots, n$$

$$\sum_{i=1}^{n} w_i = 1$$

with the *kriging variance* (or the minimum mean squared error)

$$\sigma_K^2 = \sum_{i=1}^{n} w_i \bar{\gamma}(\mathcal{D}_i, \mathcal{D}) + \lambda - \bar{\gamma}(\mathcal{D}, \mathcal{D}).$$

Of course this can be expressed in terms of the covariance function also

$$\sum_{i=1}^{n} w_i \bar{C}(\mathcal{D}_j, \mathcal{D}_i) - \lambda = \bar{C}(\mathcal{D}_j, \mathcal{D}), \qquad j = 1, \ldots, n$$

$$\sum_{i=1}^{n} w_i = 1.$$

with the kriging variance

$$\sigma_K^2 = -\sum_{i=1}^{n} w_i \bar{C}(\mathcal{D}_i, \mathcal{D}) + \lambda + \bar{C}(\mathcal{D}, \mathcal{D}).$$

Both kriging systems can be written in matrix form, here expressed by means of the covariance functions

$$\mathbf{Cw} = \mathbf{c}$$

with

$$\mathbf{C} = \begin{bmatrix} \bar{C}(\mathcal{D}_1, \mathcal{D}_1) & \ldots & \bar{C}(\mathcal{D}_1, \mathcal{D}_n) & 1 \\ \vdots & & \vdots & \vdots \\ \bar{C}(\mathcal{D}_n, \mathcal{D}_1) & \ldots & \bar{C}(\mathcal{D}_n, \mathcal{D}_n) & 1 \\ 1 & \ldots & 1 & 0 \end{bmatrix},$$

$$\mathbf{w}' = (w_1, \ldots, w_n, -\lambda)$$

and

$$\mathbf{c}' = (\bar{C}(\mathcal{D}_1, \mathcal{D}), \ldots, \bar{C}(\mathcal{D}_n, \mathcal{D}), 1).$$

The solution to the kriging system expressed in matrix form by means of the covariance function is

$$\mathbf{w} = \mathbf{C}^{-1}\mathbf{c}.$$

If the supports $\mathcal{D}, \mathcal{D}_1, \ldots, \mathcal{D}_n$ can be considered points the kriging performed is referred to as *point kriging*, otherwise it is referred to as *block kriging* or *panel kriging*.

A few remarks on some very important properties of kriging:

- Kriging is an interpolation form that provides us with not only an estimate based on the covariance structure of the variable in question but also an estimation variance.
- The kriging system has a unique solution if and only if the covariance matrix $\{\mathbf{C}_{ij}\}$, $i, j = 1, \ldots, n$ is positive definite; this also ensures a nonnegative kriging variance.
- The kriging estimator is a best linear unbiased estimator (BLUE) and it is also exact, i.e. if the support to be estimated coincides with any of the supports of the data included in the estimation, kriging provides an estimator equal to the known measurement and a zero kriging variance.
- The kriging system and the kriging variance depend only on the covariance function (semivariogram) and on the spatial lay out of the sampled supports and not on the actual data values. If a covariance function is known (or assumed) this has important potential for minimizing the estimation variance in experimental design (i.e. in the planning phase of the spatial layout of the sampling scheme).

In all applications of kriging the problem of assuming stationarity arises. *Universal kriging* is a technique that allows for some forms of nonstationarity part of which is modeled as a trend in the mean value, that is described either as a linear combination of known functions or by means of local Taylor expansions. The type of spatial irregularity we are facing in the application below is not suitable for that type of solution. The nonstationarity is very nonlinear and all attempts at 'fitting' with 'regular' functions and models have failed. Another possible approach when analyzing multivariate observations is *co-kriging*. Here one could take the spatial covariation between different variables into account in the 'adjustment'. Again, the spatial irregularity in the case study below is not suited for that type of solution. Hence, the *ordinary kriging* method described above is applied in the case study. Universal kriging, co-kriging and other advanced types of kriging will not be elaborated on here; good references are Journel and Huijbregts (1978), Myers (1982) and Carr (1985).

13.1.3 Case study: central Spain

This section describes applications of geostatistical methods in the analysis of stream sediment geochemical data from a large area in central Spain. The work reported constitutes part of a project described in Conradsen *et al.* (1990).

(a) Geochemical data

Samples were collected at 33 992 sites in 16 mapsheets (a mapsheet covers approximately 30 km × 20 km). All samples were analyzed for the contents of 27 geochemical elements by inductively coupled plasma emission spectrometry (ICP). The elements are

P, As, Sb, Sn, Pb, B, Zn, Cd, Hg, Cu, Ag, Ni, Co, Fe,
Mn, Cr, Mo, W, V, Nb, Y, Be, Ba, Al, Mg, Ti and Sc.

For proprietary reasons, all contents have been scaled by a constant factor.

Each observation holds information on UTM coordinates (universal transverse Mercator projection), element contents and lithology code.

Sc is a multiplicative correction element added manually and not present in nature in these samples. For several elements there is a detection limit problem. Based on information from geologists the following procedure was decided upon:

- If the content z of an element is <5 p.p.m. replace that value with 3·5 p.p.m. which is the approximate center of mass for a triangle with baseline [0, 5] p.p.m.
- If the content z_i of an element at sample site i is $\geqslant 5$ p.p.m. replace that value with $z_{i,\text{corr}} = z_i \text{Sc}_{\text{mean}} / \text{Sc}_i$.

The value of Sc_{mean} can be chosen as the overall mean or as a mean for individual mapsheets. As no visual difference was observed the simpler method applying the overall mean was preferred.

When inspecting sample site images of the individual elements a conspicuous mapsheet effect is noted, cf. plate. The effect is due to calibration problems in the chemical analysis of the samples. To facilitate a regional study the effect is removed by means of a method explained below.

(b) Semivariograms and kriging

To reveal the spatial structure of the individual elements and to prepare the interpolation by means of kriging experimental semivariograms were calculated. To allow for the mapsheet effect semivariograms were calculated before the removal of this effect but without allowing interaction between mapsheets, i.e. no point pair across any mapsheet border was included in the calculation. Spherical models with nugget effect were estimated by means of a weighted, iterative, nonlinear least-squares estimation procedure. Figure 13.1 shows semivariograms for the natural logarithm of the content of some elements (measured in p.p.m.).

Point kriging to a regular UTM grid with points in a 500 m × 500 m grid was performed for individual elements using the nearest 20 neighbors resulting in 192 lines with 234 samples each. A speed-up feature used is the exploitation of the fact that all elements have the same neighbors, based on the sampling layout only to build a neighborhood table. The application of this speed-up feature and point (vs. block) kriging reduced the CPU time for kriging one element in the study area from approximately 26 hours to approximately 2 hours and 10 minutes on a DEC MicroVAX II. The plate shows kriged Cr and kriging variance for Cr.

In a small part of the study area (one mapsheet) a pilot study on cokriging and kriging of principal components was carried out. This study showed very little visual difference between these more advanced forms of kriging and ordinary kriging.

(c) Mapsheet calibration

One of the major problems in the analysis of stream sediment data in the study area is the very distinct mapsheet pattern observed when displaying

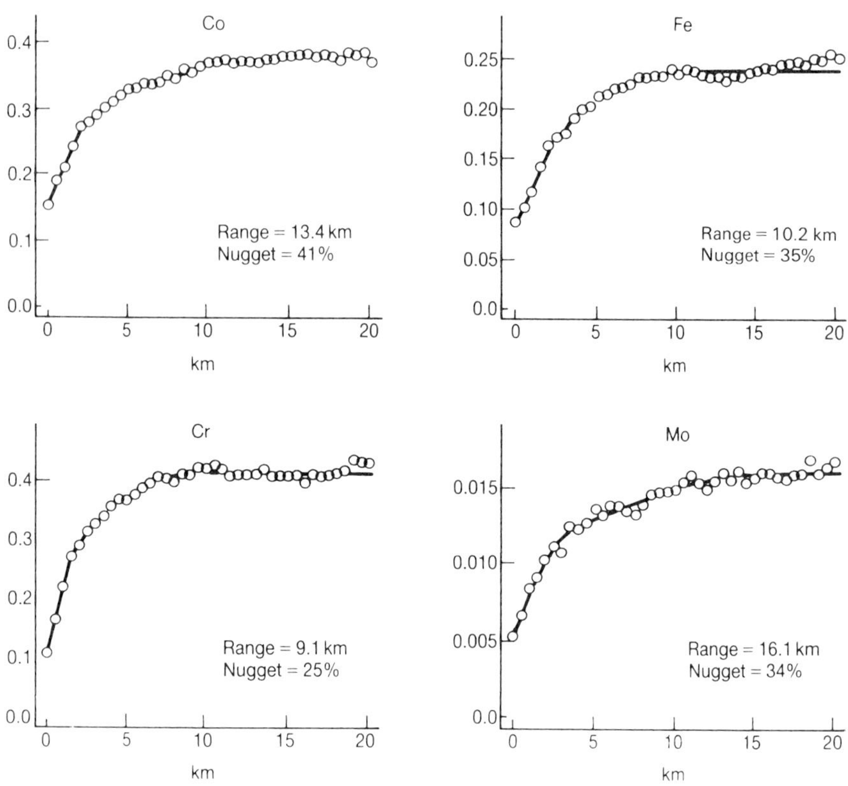

Figure 13.1 Experimental and modeled semivariograms for some elements in central Spain.

the samples as images. This obvious nonstationarity is due to problems in the chemical analysis of the samples and it violates the assumptions underlying many of the statistical techniques used in the analysis of spatial patterns. Therefore a method for removing the mapsheet pattern and thus calibrating the data is called for when the object is to analyze the area as a whole.

The general idea in the methods considered for mapsheet pattern removal is to transform observed values in the mapsheets using monotone transformations in order to preserve structure such that observations in neighboring mapsheets that are close to each other in distance have similar histograms. Linear transforms are simple monotone transformations. It turns out that linear transformations are not sufficient for removing the mapsheet pattern. The phenomenon is more complex.

A method based on local histogram matching was found to have a satisfactory performance, removing the mapsheet pattern while preserving the spatial structure in the images. The method starts out in one mapsheet, transforming the neighboring mapsheets to match at the common borders. Afterwards the neighboring mapsheets of these transformed mapsheets are transformed. The scheme propagates throughout the whole area until all mapsheets conform to one another in the sense that there are no discontinuities along the mapsheet limits in the image.

Consider for example the mapsheet layout in Figure 13.2.

The transformation scheme can start out in any mapsheet. For simplicity let us start in mapsheet 1 (MS1). Then the method works as follows:

Leave the observations in MS1 unchanged. Now, consider the observations in a narrow band on each side of the mapsheet limt shared by MS1 and MS2. To transform the observations in MS2 use the transformation that makes the histogram of the narrow band in MS2 equal to that of the narrow band in MS1.

In determining the transformation a number of empirical quantiles – for example, the 5, 10, 15, . . . per cent quantiles – are determined from each of the two narrow bands. The transformation takes the quantiles in MS2 to the corresponding quantiles in MS1. Between the quantiles we have used linear interpolation. Outside the range specified by the chosen quantiles a linear transformation from [min, min] to the lowest quantile and from the highest quantile to [max, max], respectively, is used; min denotes the overall minimum, max the overall maximum of the element in question. This transformation removes a possible discontinuity along the mapsheet limit between MS1 and MS2. Analogously, MS3 is transformed to remove discontinuities along the mapsheet limit shared by MS1 and MS3.

Now, in transforming MS4, two transformations are determined: one corresponding to MS2 and one corresponding to MS3. The actual transformation applied is a convex combination of these two. For each observation in MS4 the weight on each transformation is inversely proportional to the distance to the corresponding mapsheet limit. Thus, the observation z in MS4 with distance d_2 and d_3 to MS2 and MS3 respectively gets transformed

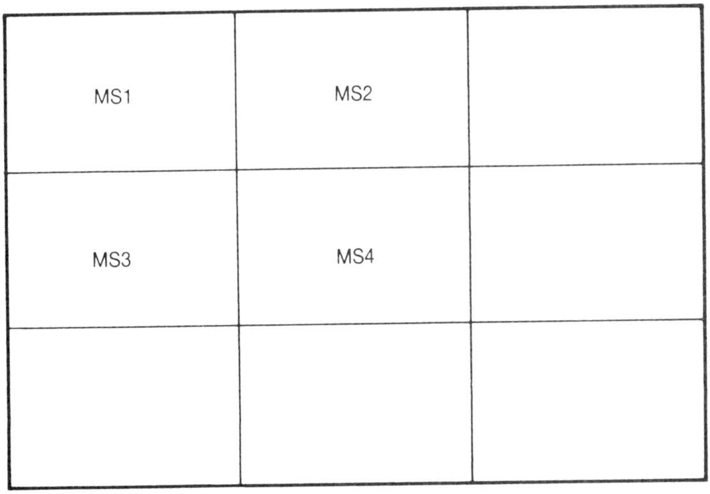

MS1	MS2	
MS3	MS4	

Figure 13.2 Example mapsheet layout.

with the transformation T:

$$T(z) = \frac{d_3}{d_2 + d_3} T_2(z) + \frac{d_2}{d_2 + d_3} T_3(z)$$

where T_2 and T_3 are the two transformations that correspond to MS2 and MS3. It is easy to generalize this idea to transforming a mapsheet to match more than two neighboring mapsheets by using convex combinations of transformations.

The transformation scheme propagates this way until all mapsheets have been transformed.

The mapsheet smoothing is carried out on the logarithm of the Sc-corrected variables. Observations below 5 p.p.m. have been simulated from a triangular distribution before Sc-correction in order to avoid problems with degenerate histograms in the determination of smoothing transformations.

Two examples of the transformation are shown in Figure 13.3. The result of the calibration can be seen in the plate. Note that the very distinct mapsheet pattern present before the calibration is removed. At the same time the overall spatial structure is preserved.

13.2 Prediction of heavy minerals

CART (Classification And Regression Trees) is a new and often powerful alternative to classical parametric methods in classification and regression. The methodology was developed in the 1970s and early 1980s and it is described in Breiman *et al.* (1984). In this section we will briefly describe what CART does, focusing only on the classification part.

In classification, one has measurements on an object and then one uses some sort of *decision rule* to decide to which class the object belongs. In our case we have measurements of concentrations of some geochemical variables

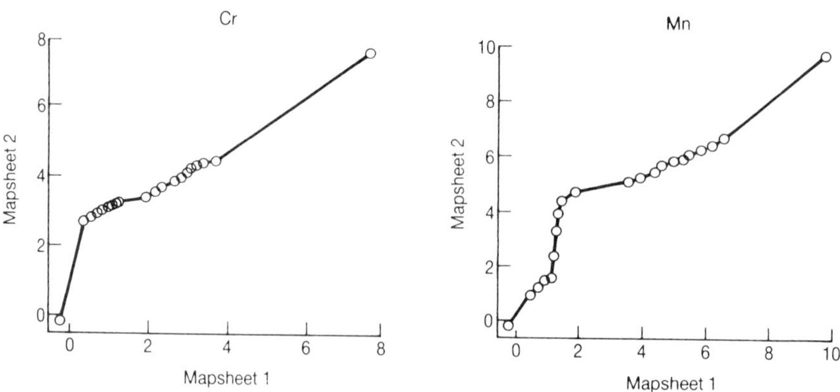

Figure 13.3 Examples of mapsheet transformations.

and we want to predict whether the concentration of gold, for example, is detectable or not. To construct a decision rule we need some data for which we know to what classes the cases belong. This is called the *training set* or the *learning set*. The decision rule that CART produces is in the form of a *binary decision tree*. An example is given in Figure 13.4.

CART grows trees by operating on the training set of the data. A *node* in the tree corresponds to a subset of the data. To begin with, we have a node containing all the cases in the learning set (training set).

In the tree-growing process, we need a strategy for determining the *splits* in the tree. Suppose we only have two classes: 0 and 1. A split of a node is good if it does a good job of separating class 0 and class 1 in the two descending nodes. One rule is to maximize decrease in node *impurity* for each split. For this we need a measure $i(t)$ of node impurity for a node t. The *Gini index* has been found useful. It has the form

$$\begin{aligned} i(t) &= \sum_{j \neq k} p(k \mid t) p(j \mid t) \\ &= \left(\sum_{j \neq k} p(k \mid t) \right)^2 - \sum_k p^2(k \mid t) \\ &= 1 - \sum_k p^2(k \mid t). \end{aligned}$$

This function has the desired properties of symmetry, concavity and non-negativeness required, and it is computationally attractive. This is not insig-

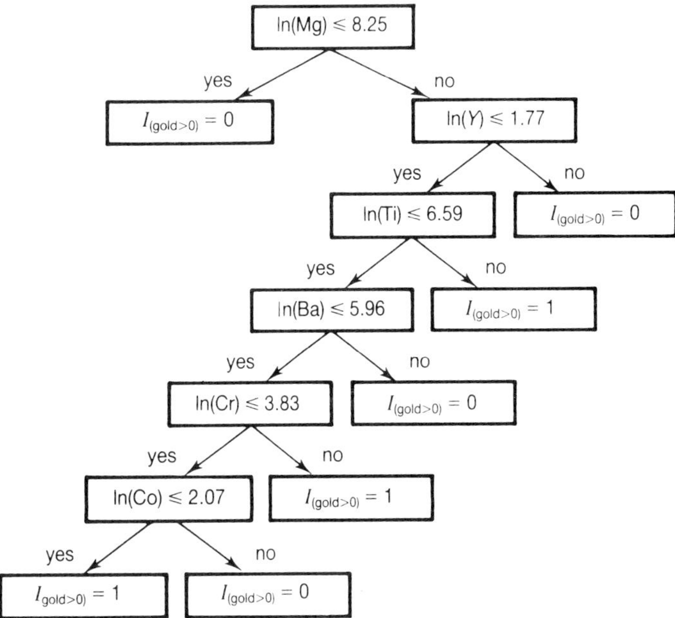

Figure 13.4 CART tree for gold.

nificant since the function is evaluated many times in the process of growing the tree.

CART examines all the coordinate splits of the form $x < c$, where $c \in [\min(x), \max(x)]$ and x is a measured variable. By this we mean: all cases satisfying $x < c$ go left and the rest go right.

For each x there will exist a value of c that yields the best split in terms of separating the classes. The split chosen by CART is the best of the best individual variable splits, resulting in the best overall split on a single variable. This procedure is repeated in a recursive manner for the descendant nodes, until a very large tree is constructed with pure terminal nodes or very few cases in these. This means that if we run all cases down the tree, use the *plurality rule* for predicting the class (i.e. the rule that assigns to a terminal node the class for which the proportion of learning set cases is largest), all cases will be predicted correctly, i.e. the resubstitution estimated misclassification rate is 0! Of course this is too optimistic. We have modeled the 'noise' as well as the structure in the training data. Hence, we need a more 'honest' estimate of the true error rate. This can be accomplished by setting aside some cases from the learning set in a *test* set not used in the actual tree-growing process. This test set is used to estimate the true error rate by running it down the tree and recording the proportion of misclassified cases. The concept of setting aside learning set cases for model selection and error estimation is generally denoted *cross-validation*, a common technique in the context of non-parametric computer-intensive models.

The large initial tree grown is too large so we need to prune it in a reasonable manner. A *cost-complexity* measure is introduced of the form

$$\text{total cost} = \text{misclassification cost} + \alpha \times \text{complexity of tree.}$$

For each value of α there exists a unique subtree of the initial tree, minimizing the total cost. By increasing α starting at 0 we obtain a sequence of smaller and smaller subtrees. The final tree selected in the sequence is the one that minimizes the error rate as estimated by use of the test set.

CART has features for handling missing data, linear combinations splits, variable importance ranking, class priors and misclassifications costs as well; for further details cf. Breiman *et al.* (1984).

13.2.1 Case study: Central Spain

The CART methodology has been used to construct tree-structured predictors for heavy minerals of economic interest in central Spain. The motivation for constructing models for prediction of heavy mineral data using the stream sediment data is that we want to estimate structural maps of heavy minerals in the area. Only the structural map for cassiterite is described below.

Conventional methods for interpolation such as kriging are not feasible because of the extreme skewness of the distribution of heavy minerals concentration at the sample points and also because of sparsity of heavy minerals samples. Essentially an *indicator* variable for the occurrence of a heavy mineral is recorded. The stream sediment data are sampled more densely so that if we can build a model for predicting heavy mineral occurrences based on stream sediment variables, we can get an interpolated image using the kriged images of the stream sediments as predictors.

Stream sediment samples were collected at 33 992 sites in the study area. At some of these sites heavy mineral samples were collected also. Two different laboratories were used for the chemical analysis of heavy minerals. For calibration reasons only the samples analyzed by one laboratory were used in constructing the models. There are 3094 samples from this laboratory, i.e. there are 3094 observations of stream sediment and heavy mineral data at the same locations.

The CART methodology was chosen because it is a new and powerful technique used with success in other research areas. It handles different types of data in an elegant and unified way and interactions between predictors are automatically taken into account. Furthermore results have shown a good agreement between the structural images obtained using the CART trees and geological information from the area.

(a) CART trees for cassiterite

The distribution of cassiterite concentration is extremely skewed so an indicator $I_{(\text{cass} > 0{\cdot}05)}$ for cassiterite concentration greater than 0·05 g/10 l rather than the original concentrations was used in the analysis. The model should then predict whether the cassiterite concentration is detectable or not based on geochemical variables concentrations. Using the CART methodology, the tree shown in Figure 13.5 was selected.

In the construction process of the tree, 1/3 of the observations (994 cases, the test set) were used in the cross-validation of the model and for choosing

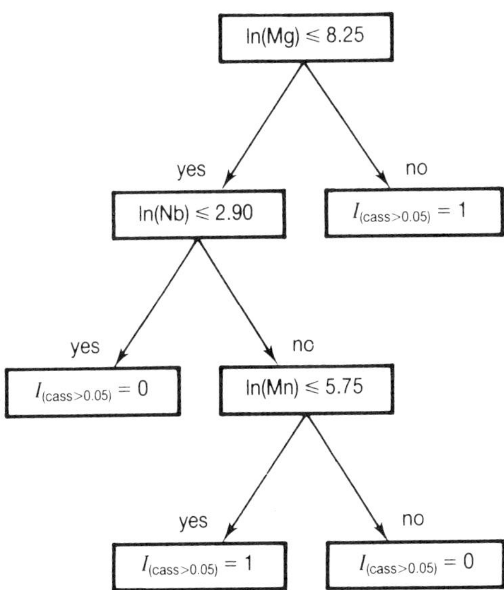

Figure 13.5 CART tree for cassiterite.

a right-sized tree. Running these cases down the tree yielded the classification matrix shown in Table 13.1.

As a standard feature CART makes a variable importance ranking based on a measure of predictive value of the individual predictors (here the stream sediment variates). The variable importance list for predicting cassiterite is given in Table 13.2.

Now, having constructed the tree we are able to run new cases down the tree and predict cassiterite occurrence. Also associated with each terminal node in the tree is an estimate of the probability of cassiterite occurrence, given that a case ends up at that particular node. These estimates can be obtained from running the test set down the tree.

To construct a structural image of the probability of cassiterite occurrence we run each pixel down the tree using the kriged images of stream sediment variables. The image for cassiterite is shown in the plate. Also shown in the plate is a structural image of the probability of gold occurrence. Different gray tones in these images correspond to different terminal nodes in the classification trees.

Table 13.1 Test set classification matrix for cassiterite.

	Predicted class		
	0	1	Total
True 0	768 (79%)	204 (21%)	972 (100%)
Class 1	6 (27%)	16 (73%)	22 (100%)
Total	774 (78%)	220 (22%)	994 (100%)

Table 13.2 CART variable importance ranking for predicting cassiterite.

Variable	Relative importance
Ti	100
Mg	52
Nb	46
Al	36
Zn	35
Cu	35
Ni	33
Cr	31
V	31
Ba	28
Co	28
Y	26
Fe	26
As	25
Mn	25
V	24
Sb	21
Pb	21
Sn	12
Be	12

13.3 Conclusion

A method for handling nonstationarity problems due to calibration problems for spatial data has been presented. The method uses a propagating weighted transformation scheme. It is simple and works well in practice.

Also we have shown an application of a combination of classical geostatistical interpolation and CART, a modern tree-structured classification method. We have found an interesting connection between the tree-structure of the classifier and the geological structures in the images. This can be seen in the structural image produced by running the pixels of the interpolated images down the classification tree.

One should be careful in interpreting the classification trees. Spatial correlation can be responsible for the correlation between the heavy minerals response and the stream sediment sample values. This means that a classification tree cannot be extrapolated to another area outside the images analyzed.

Acknowledgements

The work reported is part of a larger project funded by the Commission of the European Communities under Contract No. MA1M-0015-DK(B) and Minas de Almadén y Arrayanes, S. A. Also, the MOBS programme under the Danish Technical Research Council contributed under grant number 5.26.09.07.

The fruitful cooperation with other members of the IMSOR Image Group, the Minas de Almadén y Arrayanes, S. A., Geology Department and employees of the Commission is acknowledged. Especially, we would like to thank Enrique Ortega, MAYASA, for his enthusiasm and helpful comments.

References

Breiman, L., Friedman, J.H., Olshen, R.A., and Stone, C.J. (1984). *Classification And Regression Trees*, Wadsworth.

Carr, J.R. (1985). Co-kriging – a computer program. *Computers and Geosciences*, **11** (2).

Clark, I. (1979). *Practical Geostatistics*, Elsevier Applied Science, London.

Conradsen, K., Ersbøll, B.K., Nielsen, A.A., Pedersen, J., Stern, M. and Windfeld, K. (1990). *Development and Testing of New Techniques for Mineral Exploration Based on Remote Sensing, Image Processing Methods and Multivariate Analysis*. IMSOR, Technical University of Denmark.

Cressie, N. (1985). Fitting Variogram Models by Weighted Least Squares. *J. Int. Assoc. Math. Geol.*, **17** (5).

David, M. (1977). *Geostatistical Ore Reserve Estimation*, Developments in Geomathematics 2, Elsevier, Amsterdam.

David, M. (1988). *Handbook of Applied Advanced Geostatistical Ore Reserve Estimation*, Developments in Geomathematics 6, Elsevier, Amsterdam.

Journel, A.G. and Huijbregts, Ch.J. (1978). *Mining Geostatistics*. Academic Press, London.

Myers, D.E. (1982). Matrix Formulation of Co-kriging. *J. Int. Assoc. Math. Geol.*, **14** (3).

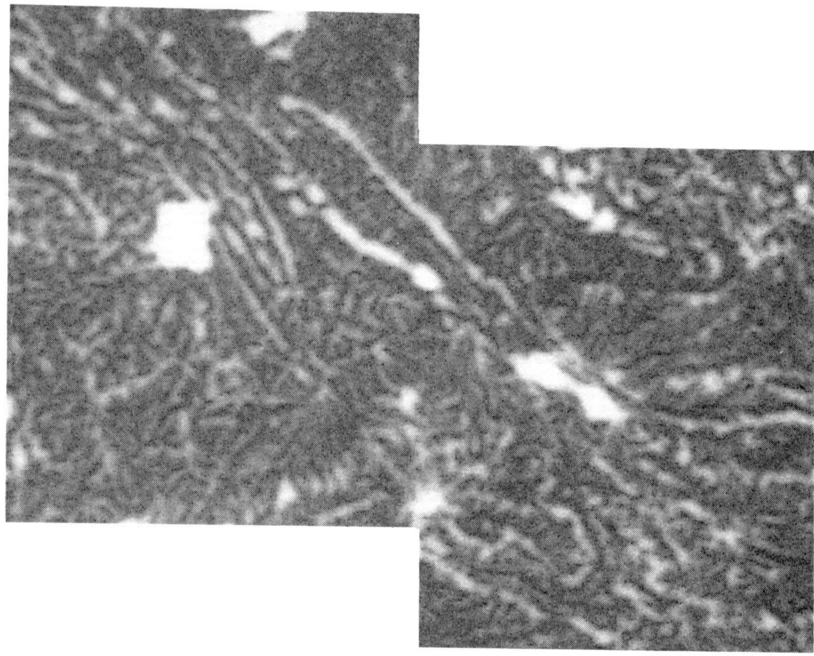

Figure 13.6 Cr before mapsheet pattern removal.

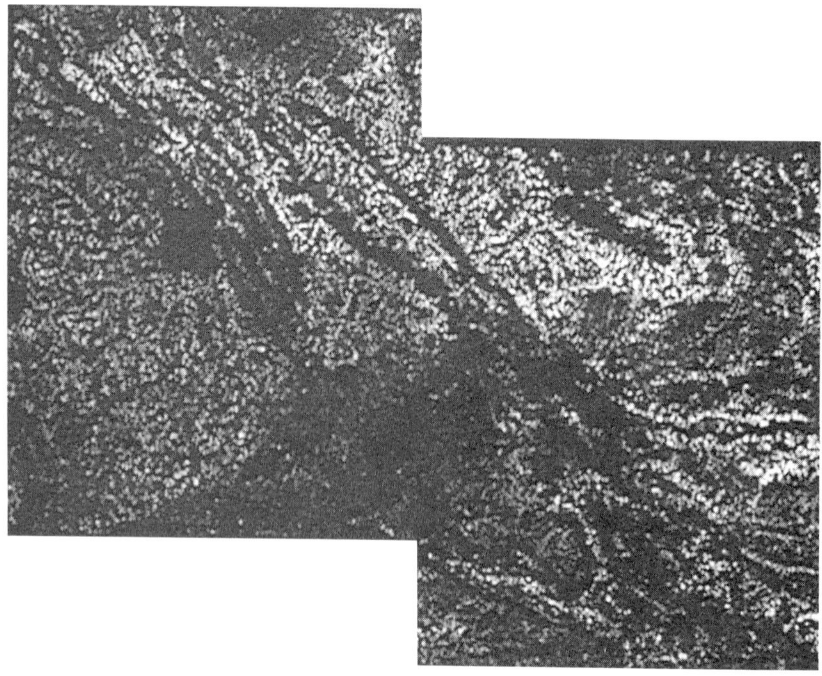

Figure 13.7 Cr after mapsheet pattern removal.

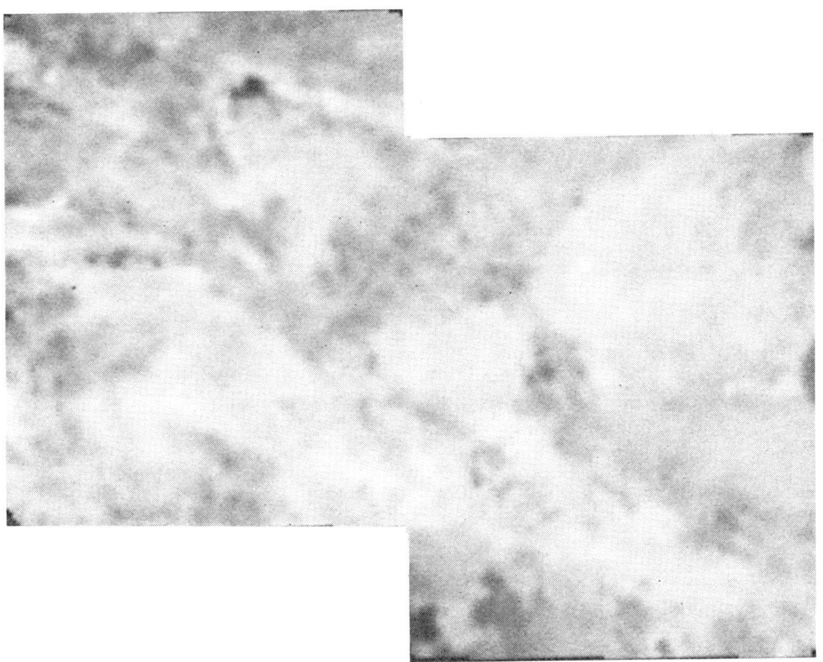

Figure 13.8 Kriged Cr.

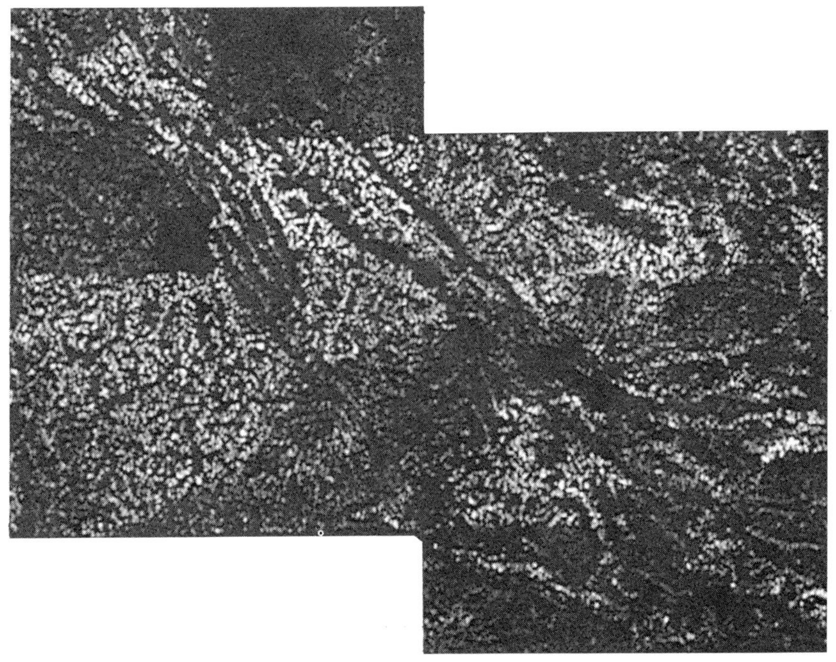

Figure 13.9 Kriging variance for Cr.

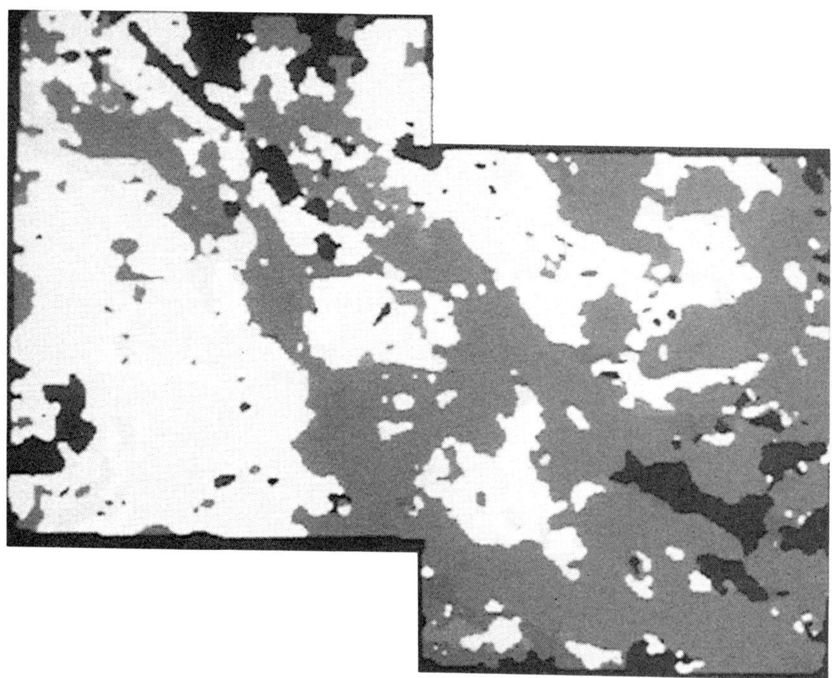

Figure 13.10 CART probability map for gold.

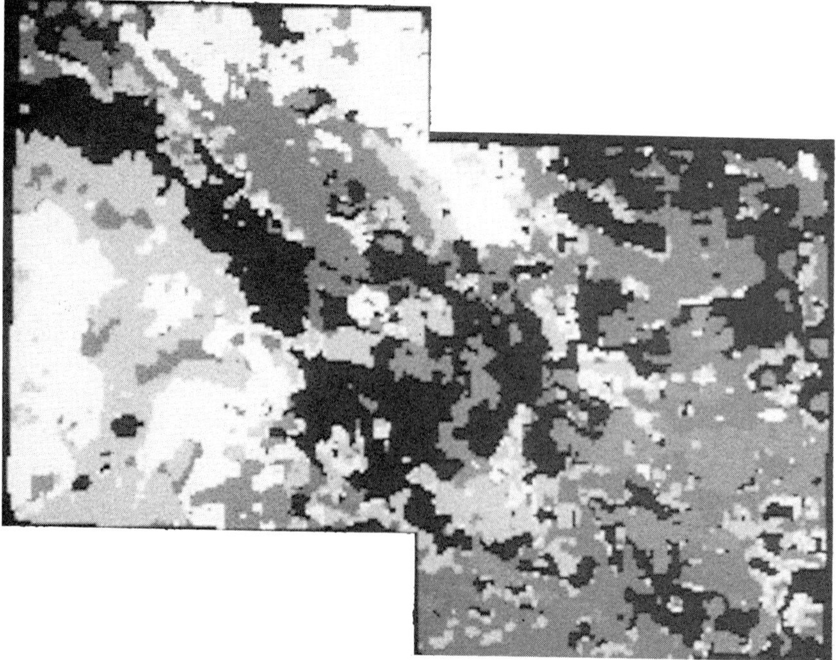

Figure 13.11 CART probability map for cassiterite.

Index